NMR

Basic Principles and Progress

26

Editors: P. Diehl E. Fluck H. Günther
R. Kosfeld J. Seelig

In-Vivo Magnetic Resonance Spectroscopy I: Probeheads and Radiofrequency Pulses Spectrum Analysis

Guest-Editor: M. Rudin

With contributions by
R. de Beer, E. B. Cady, M. Garwood,
J. Link, Ch. T. W. Moonen, P. G. Morris,
S. Müller, D. van Ormondt, M. D. Schnall,
K. Ugurbil, P. C. M. van Zijl

With 123 Figures

Springer-Verlag
Berlin Heidelberg GmbH

Library of Congress Cataloging-in-Publication Data.
In vivo magnetic resonance spectroscopy I : probeheads and radiofrequency pulses spectrum analysis / guest-editor, M. Rudin ; with contributions by R. de Beer ... [et al.]. p. cm. -- (NMR ; 26)
ISBN 978-3-642-45699-2 ISBN 978-3-642-45697-8 (eBook)
DOI 10.1007/978-3-642-45697-8
1. Nuclear magnetic resonance spectroscopy. I. Rudin, M. (Markus), 1953- . II. Beer, R. de. (Ronald) III. Series: NMR (Series) ; v. 26.
QC490.N2 vol. 26 [QP519.9.N83] 538'.362 s--dc20 [543'.0877] 91-30593 CIP

Softcover reprint of the hardcover 1st edition 1992

Typesetting: Thomson Press (India) Ltd, New Delhi

51/3020-543210 - Printed on acid-free paper

Foreword

Since the first report on NMR in living cells in 1974, in vivo magnetic resonance spectroscopy (MRS) has evolved into a potent technique for studying metabolic processes in an undisturbed biological system. In principle, MRS should provide well-resolved spectra of a selected region of interest allowing the identification and quantitation of as many metabolites as possible. However, a number of methodological problems have to be solved in order to generate useful MRS data. For example, the spectra should contain only little or no contribution from signals from outside the region of interest. Furthermore, proper quantification of metabolite concentrations is often difficult since the large number of compounds contributing to the spectra produces overlapping resonances. Similarly, the resolution of the spectra is often impaired by variations in tissue magnetic susceptibilities, leading to large linewidths. Many metabolites occur at low concentrations and a poor signal-to-noise ratio is another major obstacle in proper quantification. The use of radiofrequency (RF) probes with an inhomogeneous RF field profile, such as the surface coil, causes problems when defined RF excitation has to be used, e.g. for proper signal inversion.

During the last 10 to 15 years, a number of solutions to the above mentioned problems have been developed and a proper review of the methodological achievements in in vivo MRS has long been overdue. Rather than writing a textbook, we decided to invite the acclaimed experts to review their respective speciality. We did not intend to discuss in vivo MRS from its basics, but rather refer to textbooks and review articles which have covered these aspects. Basic principles e.g. in NMR probe design or data processing are considered as 'prior knowledge', with which the reader should be familiar to some extent.

The book is divided into three volumes. Volume I contains sections on MR probe construction, design of RF pulses, and spectrum analysis. Volume II is devoted to localization techniques, i.e. methods for selecting a region of interest, and techniques for spectral editing. Volume III describes potential applications of in vivo MRS. Due to the rapid growth of the field during the last few years it would have been an almost 'infinite' task to provide a comprehensive coverage of all ongoing activities. We thus have asked the respective authors to focus on methodological and conceptional aspects in order to illustrate the potential of MRS more than the breadth of application.

We would like to thank all contributors for their efficient and pleasant collaboration and for keeping within the short deadlines imposed on them by the editors. We are also indebted to Springer-Verlag, in particular Dr. R. Stumpe and Ms. P. Frank, for their assistance in general and the editorial supervision.

Basel, September 1991

M. Rudin
J. Seelig

Guest-Editor

Professor M. Rudin
Preclinical Research 386/206
Sandoz Pharma Ltd.
Ch-4002 Basel

Table of Contents

RF Probeheads

RF Pulses

Spectrum Analysis

Table of Contents of Volume 27
In Vivo Magnetic Resonance Spectroscopy II:
Localization and Spectral Editing

Table of Contents of Volume 28
In Vivo Magnetic Resonance Spectroscopy III:
In Vivo MR Spectroscopy

RF Probeheads

During the NMR experiment, radio- frequency (RF) pulses have to be transmitted to the sample while the weak NMR signals emanating from the excited nuclei have to be detected with a sensitive receiver device. The quality of the transmitting and receiving RF probes, therefore, determines to a large extent the performance (sensitivity) of the spectrometer. The NMR receiver is discussed in great detail in [1].

Classical NMR probes are the solenoid and the saddle coil. These probes, however, are of limited use for many in vivo applications. As soon as the length of the lead becomes comparable to the RF wavelength, phase problems arise and alternative solutions have to be found in order to guarantee a homogeneous RF field distribution in the probe structure. This situation is analogous to spectroscopy involving microwaves, such as electron spin resonance (ESR), and many of the concepts developed for the design of microwave probes can be transferred to in vivo MR. For example the probe design described by Alderman and Grant [2] is a modified slotted tube resonator [3], or the loop-gap resonator has been originally designed for pulsed ESR experiments [4]. The design of RF probes with homogeneous RF field distribution, as needed for imaging and many localization techniques, is reviewed in the following by Link.

Many in vivo investigations are of the multinuclear type, i.e. the MR signals of, for instance, hydrogen and phosphorus are monitored simultaneously. This requires RF probes, which are simultaneously tuned to multiple frequencies. Such probes, however, always sacrifice sensitivity compared to the single coil approach. Therefore, a crucial step in probe construction is to minimize this loss in sensitivity. The design of multi-frequency NMR probes is discussed by Schnall. The main focus will be on the construction of multiple tuned surface coils.

1. Hoult DI (1977) Prog NMR Spectrosc 12: 41
2. Alderman DW, Grant DM (1979) J Magn Reson 36: 447
3. Schneider HJ, Dullenkopf P (1977) Rev Sci Instrum 48: 68
4. Froncisz W, Hyde JW (1982) J Magn Reson 47: 515

The Design of Resonator Probes with Homogeneous Radiofrequency Fields

Jürgen Link

Spectrospin AG, Industriestr. 26, CH-8117 Fällanden

Table of Contents

The following contribution is meant to provide an overview of the wealth of existing NMR resonator designs, and an introduction into practical probe design.

Two sections are devoted to theoretical aspects of NMR probe design: "Temporal field properties" deals with the electromagnetic interaction of resonator probes with spins in a magnetic field. Probes capable of generating linearly and circularly polarized fields are considered. "Spatial field properties" addresses the manifold geometries of resonators developed to date, classifying them roughly with respect to transversal and longitudinal field properties, and in a more refined mesh according to their homogeneity and operation frequency range.

The following two sections focus on practical aspects of probe construction: In "Design Practicalities", purpose and methodical approaches to probe shielding and coupling are considered. In "Performance Tests", a number of commonly used tests to determine resonator performance during the construction process and after completion, are discussed.

NMR Basic Principles and Progress, Vol. 26

Conventions

Vectors ($\bar{\boldsymbol{M}}$) are represented by bold characters using the overscore sign. In some contexts cartesian coordinates are attributed as $\bar{\boldsymbol{r}} = (x, y, z)$ or $\bar{\boldsymbol{r}} = \begin{pmatrix} x \\ y \\ z \end{pmatrix}$. The "circonflex" sign (*e.g.* $\hat{x}$) is used throughout this text to denote unit vectors, i.e. vectors of length 1. In slight generalization. $\hat{\boldsymbol{B}}_1$ denotes the magnetic field $\bar{\boldsymbol{B}}_1$ produced by unit current in a coil.

1 Introduction

NMR experiments are designed to selectively perturb an ensemble of nuclear spins that have been aligned to a strong static magnetic field B_0 using a second radiofrequency (rf) field B_1, and to monitor the specific electromagnetic message sent out by these spins upon their return to equilibrium. The rf probe has a dual role in this game: it delivers the perturbation energy to the spin system, and it receives the rather weak response signal from the spin system under test. Desirable properties of this rf "interface" include good power efficiency for the excitation, and high sensitivity for signal reception, with an homogeneous distribution of the field and reception sensitivity across the whole sample space.

The setup has been compared to a radiobroadcasting system, involving a high-power transmitting antenna and a receiving antenna sensitive to very low rf signals. Many well-known, old antenna design principles found in classical radio electronics handbooks such as Terman [1] are therefore directly applicable to the design of NMR probes. Nevertheless, there are marked differences between a radiobroadcasting antenna system and an NMR probe setup that prohibit the unrestricted use of the term "antenna" for NMR probes [2]. The comparison shall be elaborated somewhat further, as some confusion in the literature stems from improper use of this analogy.

In the radio domain transmit antennas are designed to "radiate" their energy as far as possible, and reception antennas are designed to pick up signal from distant sources. In both cases the field of interest is far from the antenna body and can therefore be described by simplified Maxwell equations disregarding current and charge effects (so-called far-field approximation). The field can be characterized in terms of travelling waves, where the spatial and temporal, as well as the electric and magnetic behavior, are inseparable.

NMR probes, in contrast, are designed to confine their field and energy density to a well-defined volume; any distant field distribution contributes to power and sensitivity losses, but not to the desired spin interaction. Sensitivity to the region outside the probe results in a low "filling factor" (defined by the fraction of the sensitive volume occupied by the sample), in susceptibility to external noise, and in reduction of overall sensitivity. As a consequence, NMR probes are characterized much more by the electric and magnetic fields caused by the current and charge distribution on the conductor structure than by free-field effects (near-field approximation). The full Maxwell equations including charge and current distribution have to be applied to describe the fields. NMR probes produce standing waves, with regions of high electric field more-or-less separated from regions of high magnetic field.

Thus, one will generally build radio antennas as open structures with large dimensions of quarter to half wavelengths, but NMR probes as closed structures with smallest dimensions compatible with the target object. To illustrate the different behavior, consider a straight wire carrying current J (Fig. 1a), representing an elementary radio antenna. The magnetic field created by this

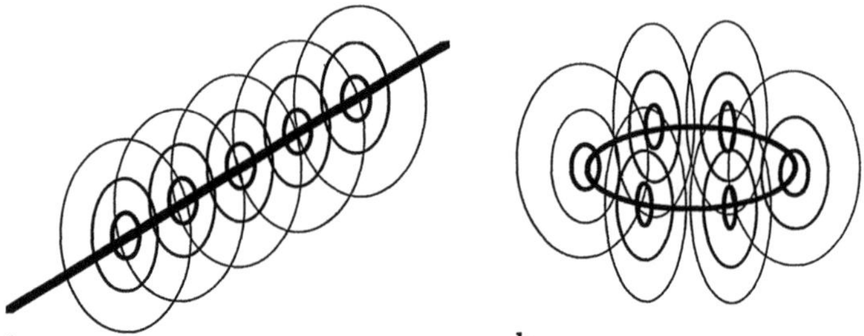

Fig. 1 a, b. The magnetic fields created by **a**) a straight wire, representing of prototype radio antenna, and **b**) a circular wire, representing a prototype NMR probe

current is arranged circularly around the wire and decays reciprocally with distance. Bending this wire to a loop results in a prototype NMR probe (Fig. 1b): inside the loop the field is reinforced by contribution of all parts of the wire, while outside the loop it is "diluted". Within the loop plane the magnetic field is fairly constant; outside it decays much more rapidly than the field of the straight wire.

Both in radio-technology and in NMR the transmission antenna is principally suitable to be used as a reception antenna as well (e.g. citizen band antennas). In both cases the major difference is that the transmission antenna has to be able to handle powers in the range from 1 W to several kW, while the reception antenna will be happy to receive as little as μW to mW powers. In the radio domain this power drop is simply due to geometric "dilution". The solid angle at which the reception antenna appears from the transmit antenna, and thus the power partition, drops very rapidly with distance. In NMR there is ideally no energy radiation into space; but energy is stored and concentrated in the confined volume. The power continuously delivered to the NMR probe during a pulse does not, however, lead to infinitely strong fields in the sensitive volume, as resistive coil and sample losses increase with field energy density. Equilibrium is reached when the coil and sample losses match the power delivered per cycle. In both directions, to and from the spin system, energy transport is taxed very heavily by energy dissipation in the antenna structure and in the sample. Thus the energy available to perturb the desired spins is unfortunately only a small fraction of the total energy delivered to the system. Correspondingly, only a minor part of the total spin energy released upon relaxation is available to the receiver unit.

In many biomedical applications the transmission and reception functions are accomplished by two different probes: a large resonator (body coil) is used to excite the whole sample space evenly. A smaller reception coil which focusses its sensitive volume to the region of interest is applied to achieve higher sensitivity at the cost of a more inhomogeneous sensitivity profile. This approach introduces distinct problems resulting from the mutual coupling of transmission and reception antenna, as has been discussed in Ref. [3]. But apart from the electronic decoupling circuitry to be introduced into the probe structures, this approach does not alter the design goals of resonators and does not lead to design differences between transmission and reception probes. In the following, therefore, no distinction will be made between resonators used for transmission, for reception, or both.

2 NMR Probe Design: Temporal Field Properties

To understand design goals for resonator probes, it is helpful to derive required properties from first principles. Let us use the more descriptive classical picture, employing an ensemble of spins polarized in a magnetic field $\bar{\boldsymbol{B}}_0 = (0, 0, B_0)$ to a macroscopic magnetic moment $\bar{\boldsymbol{M}} = (0, 0, M_z)$. Let us further assume that all spins resonate at the Larmor frequency ω, such that off-resonance effects may be neglected. Desired is a device to perturb this magnetization in a controlled manner at minimal power loss. In most experimental situations the desired motion is a deflection of $\bar{\boldsymbol{M}}$ from the z axis towards the xy plane (as only spins oriented in the xy plane can produce a NMR signal, see later in text). This is achieved by employing a magnetic field $\bar{\boldsymbol{B}}_1$ to create a torque:

$$\bar{\boldsymbol{D}} = -\bar{\boldsymbol{M}} \times \bar{\boldsymbol{B}}_1. \tag{1}$$

$\bar{\boldsymbol{M}}$ is precessing with the Larmor frequency ω at an angle ϑ (initially $\vartheta = 0$) around the z axis

$$\bar{\boldsymbol{M}} = M_0 \begin{pmatrix} \sin\vartheta \cos\omega t \\ \sin\vartheta - \sin\omega t \\ \cos\vartheta \end{pmatrix} \tag{2}$$

and must be rotated towards the equator plane along a meridian (Fig. 2). As only $\bar{\boldsymbol{B}}_1$ components perpendicular to $\bar{\boldsymbol{M}}$ contribute to the desired torque, it is most efficient to locate the $\bar{\boldsymbol{B}}_1$ field in the equatorial plane orthogonal to the projection of $\bar{\boldsymbol{M}}$ onto this plane, thus ensuring continuous orthogonality of $\bar{\boldsymbol{M}}$ and $\bar{\boldsymbol{B}}_1$ independent from the deflection angle ϑ:

$$\bar{\boldsymbol{B}}_1 = B_1 \begin{pmatrix} \sin\omega t \\ \cos\omega t \\ 0 \end{pmatrix}. \tag{3}$$

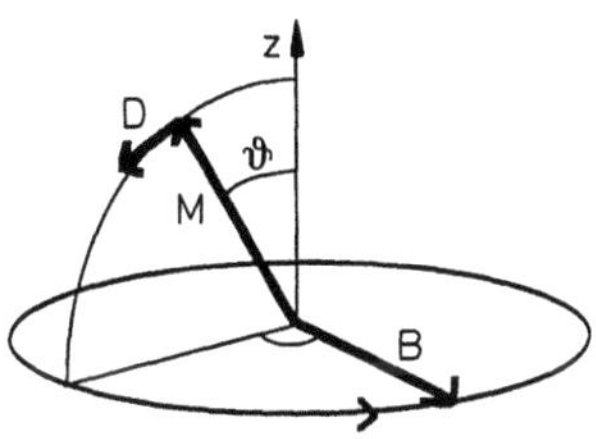

Fig. 2. A magnetic field $\bar{\boldsymbol{B}}$ rotating in the xy plane with the Larmor frequency ω produces a torque $\bar{\boldsymbol{D}}$ and effects a nutation of coherent magnetization $\bar{\boldsymbol{M}}$ from the z axis towards the "equator"

Technically, creating such a rotating field is not trivial, but, as already suggested by the cartesian vector representation, it can be synthesized by two orthogonal fields, each oscillating linearly with a 90° phase difference:

$$\bar{\boldsymbol{B}}_{1x} = B_1 \begin{pmatrix} \sin \omega t \\ 0 \\ 0 \end{pmatrix} \quad \text{and} \quad \bar{\boldsymbol{B}}_{1y} = B_1 \begin{pmatrix} 0 \\ \cos \omega t \\ 0 \end{pmatrix}. \tag{4}$$

A linearly polarized field can be easily produced by a resonant circuit comprising a coil of inductance L to create the desired magnetic field, and a capacitance C to resonate the circuit at the desired Larmor frequency

$$\omega = 1/\sqrt{LC}. \tag{5}$$

In fact, it is not mandatory to synthesize the rotating field by superposition of two linear fields. A single, linearly oscillating field creates the necessary torque as well, but for half of the transmission cycle only (in a crude, but numerically correct approximation). To achieve the same nutation of 90° during a time τ by a single linearly polarized field, the magnitude of B_1 and therefore the coil current has to be doubled. This requires twice the voltage and thus 4 times the power necessary in each of a set of two orthogonal linearly polarized coils. Thus, by employing two orthogonal coils, 1/2 transmission power can be saved. This point is of increasing importance at higher fields and larger dimensions, and may be decisive in whole body applications where power deposition approaches physiological or legal limits.

Similar considerations apply for the reception process. According to the principle of reciprocity [4], the EMF ξ induced in an antenna by magnetic moment M from a voxel dV at a specific position $\bar{\boldsymbol{r}}$ is determined by the field $\hat{\boldsymbol{B}}_1$ produced by unit current in the coil, at the respective position.

$$d\xi(dV) = -\delta/\delta t(\hat{\boldsymbol{B}}_1 \cdot \bar{\boldsymbol{M}}) dV \tag{6}$$

Assuming that a reception coil in an arbitrary spatial direction produces the field $\hat{\boldsymbol{B}}_1(\bar{\boldsymbol{r}}) = (B_{1x}, B_{1y}, B_{1z})$ when driven by unit current, the induced emf results to:

$$\begin{aligned} \delta\xi(dV) &= -d/dt \left(\begin{pmatrix} B_{1x} \\ B_{1y} \\ B_{1z} \end{pmatrix} \begin{pmatrix} \sin \vartheta \cos \omega t \\ \sin \vartheta(-\sin \omega t) \\ \cos \vartheta \end{pmatrix} M_0 \, dV \right) \\ &= M_0 \omega \sin \vartheta (B_{1x} \sin \omega t + B_{1y} \cos \omega t) \, dV. \end{aligned} \tag{7}$$

Note that the M_z component is not time dependent and therefore does not contribute to ξ. A coil with its axis oriented along the main field in z direction is therefore not capable of receiving any signal from the precessing spins. For reception, again, it is desirable to orient the coil with its field direction orthogonal to the z axis. A coil in x direction picks up the signal voltage.

$$d\xi = M_0 \omega \sin \vartheta \quad \hat{B}_1 \sin \omega t \, dV, \tag{8}$$

An orthogonal coil in y direction receives the complementary signal:

$$d\xi = M_0 \omega \sin \vartheta \quad \hat{B}_1 \cos \omega t \, dV. \tag{9}$$

Introducing a 90° phase shifter into the second channel and adding the signals

$$d\xi = d\xi_1 + i d\xi_2 = 2 M_0 \omega \sin \vartheta \quad B_1 \sin \omega t \, dV \tag{10}$$

doubles the phase-correlated signal of both channels, while the associated noise voltage, being uncorrelated, only increases by a factor of $\sqrt{2}$. Employment of circularly polarized probes therefore yields an increase in signal-to-noise ratio of $\sqrt{2}$.

This time behavior underlies all known probe designs, and differences can be found only in one respect: while most probes support linear field polarization, and an orthogonal pair is necessary to create or detect circularly polarized fields, one resonator design is capable of intrinsic circular drive and reception due to its distinct symmetry in the transverse plane. This feature will be discussed in detail in the following section.

3 NMR Probe Design: Spatial Field Properties

There exists a wide variety of geometric resonator structures which differ strongly in their field homogeneity and sensitivity. The ideal structure with respect to both criteria is known from electromagnetic theory to be a sphere supporting surface currents parallel to the equator plane with the current density modulated as $J(\vartheta) \sim \sin \vartheta$ (Fig. 3). This design has recently been cast in a real resonator version [5]. It is, however, of very limited applicability as it requires a spherical sample and allows no access to the sample within the probe. More realistic conditions are offered by resonators mounted on cylindrical probe bodies. The wealth of known resonator designs adapted to cylindrical geometries can be grouped into two categories, according to the field orientation with respect to the cylindrical probe body.

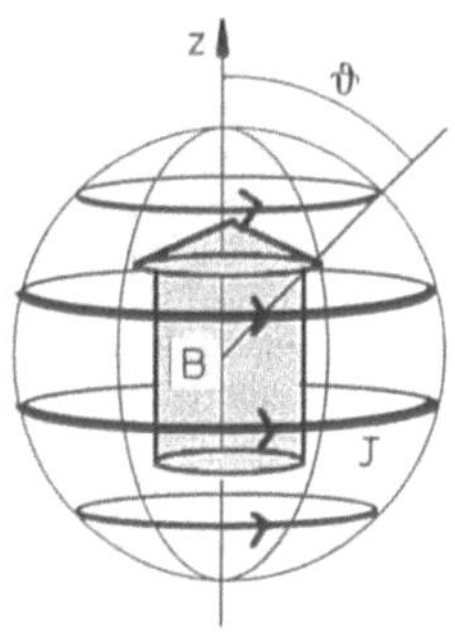

Fig. 3. A sinusoidally modulated current density J running tangentially on a spherical surface in planes parallel to the equator plane creates a homogeneous magnetic field oriented along the normal axis inside the enclosed volume. Displayed is an approximation to the continuous current distribution by four current filaments

3.1 Resonator Design for Longitudinal Field

A homogeneous current running tangentially on the surface of an infinitely long cylinder (Fig. 4a) creates a perfectly homogeneous magnetic field parallel to its axis inside the cylinder volume [6]. Achieving such a tangential current distribution might appear difficult but in fact is very trivial: winding a wire carrying current J onto a cylinder surface creates a nearly homogeneous current distribution practically orthogonal to the cylinder axis (Fig. 4b). Assuming its length is substantially longer than its radius, this so-called solenoid concentrates virtually all its field energy at a very homogeneous field distribution in its interior and is the "ideal" coil design with respect to sensitivity and field homogeneity. Practical consideration such as optimum wire thickness and

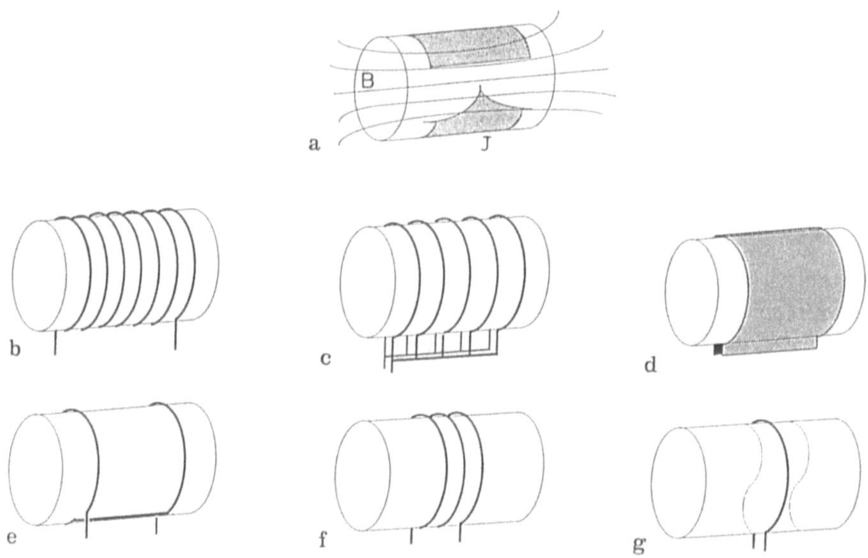

Fig. 4a–g. The family of cylindrical resonators with longitudinal magnetic field. **a**) The ideal solution: A homogeneous current running tangentially on the surface of a cylinder creates a homogeneous axial magnetic field inside the cylinder. **b**) The solenoid: The ideal current distribution of Fig. 4. is approximated by a multiturn wire loop on a cylinder surface. Winding a "solenoid" in parallel loops (**c**) shortens the ratio conductor length vs. wavelength for each conductor element and renders the coil applicable at higher frequencies. **d**) The Loop-Gap Resonator: A continuous tangential current distribution is achieved by a conducting cylindrical surface, with a longitudinal gap formed to a capacity. This design operates with practicable capacitance values only at very high frequencies. **e**) The Helmholtz coil: A "two-turn solenoid", conveniently reshaped, leaves radial access to the sample space. Field homogeneity is optimized, if the loop diameter equals the loop distance. Reduction of the number of turns of a solenoid leads to multiturn (**f**) and single turn (**g**) surface coils. These coils can be freely positioned in the magnet's clear bore as long as their axis is orthogonal to the main field axis. They are best used in combination with large odd-shaped samples, when only a small region of interest is to be examined

winding spacing are discussed in Refs. [7, 8]. To support the desired oscillating time behavior a capacitor is connected parallel to the solenoid. A variable capacitor allows fine tuning of the resonance frequency in the desired frequency range.

There are two reasons why such a solenoid can be inappropriate despite its outstanding electromagnetic performance: a) Self-resonance, and b) Geometry conflicts.

a) Self-resonance. If the total conductor length l of the solenoid approaches the wavelength $\lambda = 2\pi c/\omega$, the current on the conductor cannot be assumed constant over the whole conductor length any more. The current in the wire and the fields around it form standing wave patterns:

$$J(x, t) = J_0 \cos(2\pi x/\lambda) \cos \omega t \tag{11}$$

$$B(x, t) = B_0 \cos(2\pi x/\lambda) \cos \omega t. \tag{12}$$

In the extreme case $l = \lambda/2$, currents and fields at both ends of the solenoid (at $x = 0$ and $x = l$) are opposite in phase and tend to cancel (Fig. 5). While for small probes at low fields there is no risk of approaching this so-called self-resonance condition, both increasing the field strength (as in modern analytical instruments) or increasing the probe dimensions (as for whole-body resonators) leads to unfavorable l/λ ratio. This effect can be countered by decreasing the number of turns, however at the cost of decreased homogeneity. To reduce l/λ without performance degradation one can wind multiple turns in parallel, not in series, thus decreasing the length of individual conductors and the phase shifts (Fig. 4c). This general approach is described in Ref. [9]. Rigorous application of this parallel-concept has led to the development of the split-ring resonator [10] and loop-gap-resonator [11]. In this design a homogeneous current distribution is achieved by a cylindrical sheet-metal surface which is resonated by an integrated capacitor provided by the gap (enhanced by dielectric materials) (Fig. 4d). Due to its low inductance this design operates with resonable capacitance values only at high frequencies.

b) Geometry conflicts. Superconductive magnets are generally built as solenoids as well, offering a cylindrical clear bore with axial main field. The B_1

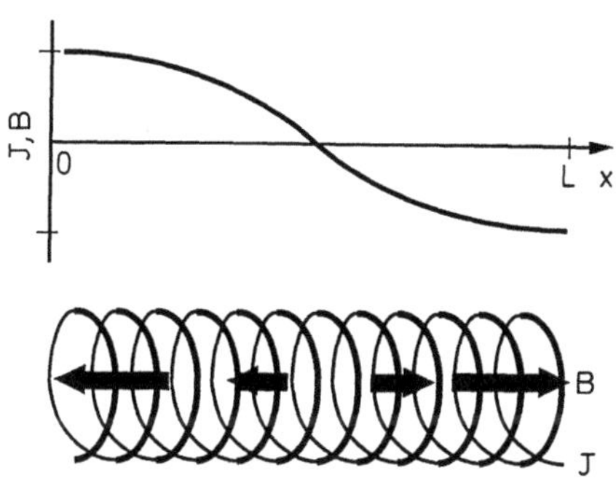

Fig. 5. Distribution of current J and magnetic field B in a solenoid whose total conductor length equals 1/2 of the wavelength. The magnetic field in the coil center is cancelled

field to be created by the probe has to be in the spin precession plane and therefore must be orthogonal to the main field. Choosing a cylindrical solenoid for the probe results in conflicting geometries: the sample cannot be accessed upon introduction into the magnet, and clear bore space is certainly not used economically. Reasonable free access is granted if the solenoid is reduced to two turns and conveniently reshaped (Fig. 4e), corresponding to another universal and long-known coil design: the Helmholtz coil. Its geometry is optimized to yield maximum homogeneity if the distance between the two coil planes equals the coil diameter [12]. While the Helmholtz coil provides lower sensitivity and homogeneity than a solenoid of equal dimensions, it allows samples to extend beyond its sensitive volume and is thus reasonably well suited to examine living objects.

If the sample extends far beyond the desired region of interest in several dimensions (as is often the case in medical applications), the solenoid can be compressed, simultaneously reducing the number of turns, to form a so-called surface coil (Fig. 4f, g). While the homogeneity in the axial direction of such a surface coil is very poor, its dimension can be chosen according to the region of interest and thus it has very good filling factor and sensitivity.

Recently, arrays of surface coils have been developed to increase planar field homogeneity, as compared to a single loop of the same area. These include the petal resonator [13], the switched array [14], and the phased array [15]. Their geometries are sketched in Fig. 6. The switched array allows switching of the desired size and position of the active probe partition to optimize coil

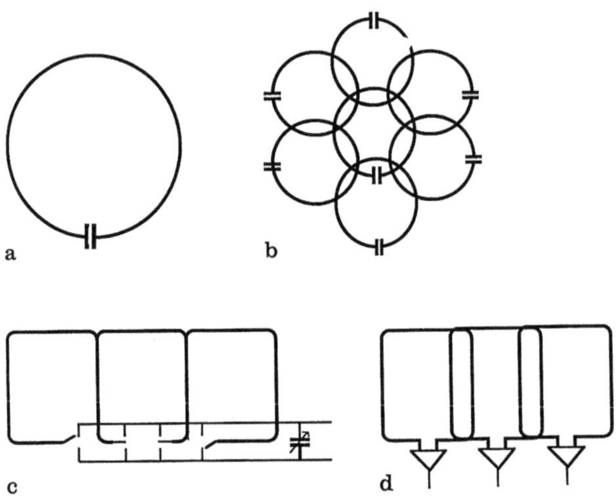

Fig. 6 a–d. Single large surface coils (**a**) can be replaced by arrays of smaller surface coils to improve the parallel-plane field homogeneity (petal resonator, **b**). The switched-array coil (**c**) allows selection of the active coil region and thus adaptation of the sensitive volume to the region-of-interest. The phased-array coil (**d**) employs an array of geometrically uncoupled loops, each connected to its own preamplifier unit, to achieve the SNR of a single coil on a much larger field-of-view sensitivity with respect to the desired region of interest. The phased array concept includes separate reception circuitry for each coil element. This allows to apply narrow-banded filters adapted to the field-of-view of each coil element, reducing the noise contribution of each coil and substantially increasing the overall signal-to-noise ratio. A general feature of the array designs is the flexibility to adapt coil geometry more closely to the region of interest than classical designs. The array concept can be expected to significantly gain in importance in the near future.

3.2 Resonator Design for Transversal Field

Any probe design in the longitudinal field group that allows access to the sensitive volume of the probe along the axis of the main magnet has to sacrifice field homogeneity because the theoretical concept of homogeneous tangential current distribution is compromised. In *in vivo* and in medical applications this access is of crucial importance, however. Fortunately there exists a design supporting homogeneous magnetic fields transversal to a conducting cylinder, allowing the probe cylinder to be co-axial to the main field:

A current on a cylindrical surface running parallel to the cylinder axis in z direction with its intensity modulated sinusoidally with the angle φ in the transversal plane (Fig. 7a) such that:

$$\bar{\boldsymbol{J}} = J_0 \cos \varphi \hat{\boldsymbol{z}} \tag{13}$$

generates a homogeneous transversal magnetic field [16]:

$$\bar{\boldsymbol{B}}_1 = (0, B_1, 0). \tag{14}$$

In a more general description, a current distribution:

$$\bar{\boldsymbol{J}} = J_0 \cos (\varphi - \alpha) \hat{\boldsymbol{z}} \tag{15}$$

can be applied, where α describes an offset angle of the current distribution. By proper choice of α the magnetic field direction can then be chosen anywhere in the xy plane:

$$\bar{\boldsymbol{B}}_1 = (B_1 \sin \alpha, B_1 \cos \alpha, 0). \tag{16}$$

Realisation of such a current distribution is less trivial, than just winding a single wire on top of the surface. The simplest approximation to this ideal distribution consists of a pair of wires running along the cylinder at the positions of highest current density (Fig. 7b). This design is very similar to the previously discussed surface coil. Connecting the wires at the far end by a capacitor instead of an inductor symmetrizes the coil with respect to electric fields, and shortens the ratio l/λ for individual conductor segments, as is desirable for high-frequency operation.

A better approximation of the desired current distribution is achieved by placing 4 wires carrying equal current at angles of 60° and $-60°$ and reversed

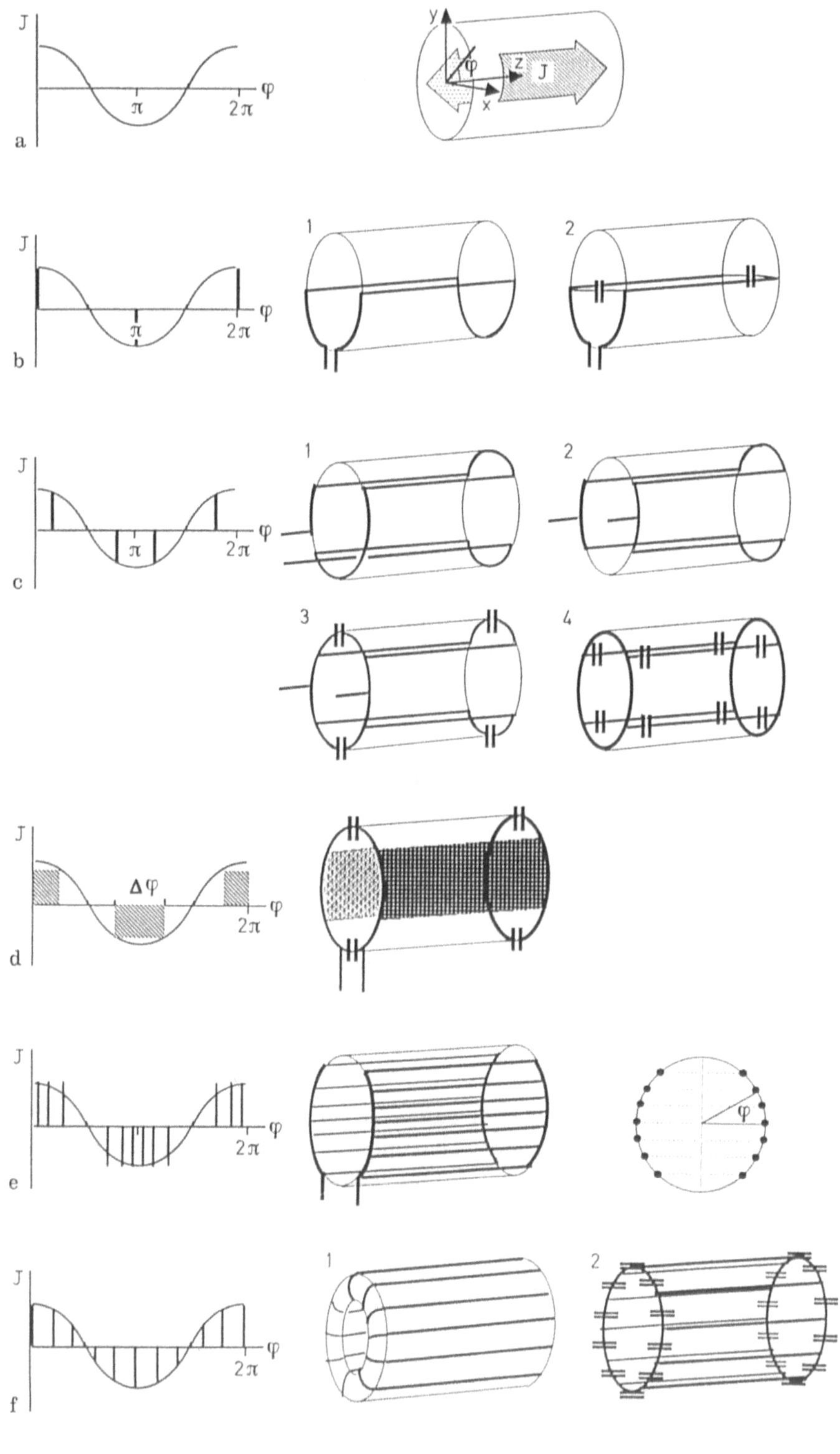
J
φ
π
2π
a
y
z
x
J
b
1
2
c
3
4
d
Δφ
e
f

current at 120° and −120°, respectively (Fig. 7c). This design mimics the ideal current distribution at 6 positions, including two virtual wires carrying 0 current at 0° and 180° [17]. The assembly can be wired into serial loops or into two parallel loops, depending on the operation frequency. It has been termed saddle resonator due to its characteristic shape (including the connecting arcs that do not contribute to the field in the desired manner).

At higher frequencies or larger dimensions the connecting arcs can be replaced by capacitors to reduce the length l/λ of individual conductors and to create symmetric electric field distributions so that electric probe-sample interaction is minimized. Designs are in use where capacitors do not connect current elements directly, but connect individual elements to common ground planes. All elements must then be driven individually by a complex coupling circuit providing currents in the desired phase behavior. In this layout the four elements can be visualized as individual Hertz antenna resonators arranged in a phased array to create constructive interference of their magnetic fields inside the enclosed volume.

A highly successful design, known under the names "Alderman–Grant Resonator" [18] and "Slotted Tube Resonator" (STR) [19, 20], approaches the ideal current distribution by distributing the field-producing current into a rectangular density profile. This is accomplished by using a pair of conducting sheets which subtend a substantial angle $\Delta\varphi$ on the cylinder surface (Fig. 7d). This design results in a very usable field homogeneity over much of the sensitive volume. Capacitances have been recommended to be distributed on all ends of the structure, although at low-frequency operation direct inductive connection

Fig. 7 a–f. The family of cylindrical resonators with transversal magnetic field. **a**) The theoretical solution: A homogeneous magnetic field in the transversal planes of a cylinder is generated by a longitudinal current distribution modulated sinusoidally with the angle φ in the transversal plane. **b**) The "Half-Saddle coil": The sinusoidal current distribution is approximated at two radial positions to achieve a transversal field of not very pronounced homogeneity. This layout is the transversal field analog to the surface coil design. Employing symmetric capacitances (2), the ratio of conductor length vs. wavelength can be shortened to achieve higher resonance frequencies. **c**) The Saddle Coil: The desired sinusoidal current distribution is approximated at six radial positions at an angle increment of 60°. To increase the resonance frequency the serial version (1) can be rewired into a parallel version (2), and the connection leads can be replaced by symmetrically arranged capacitors (3). The high-frequency version (4) consists of four independent symmetrized resonance elements which have to be driven independently by a suitable coupling network. **d**) The Slotted-Tube or Alderman–Grant resonator: The sinusoidal current distribution is approximated by a suitable rectangular current profile using longitudinal conducting sheets. Interconnections are provided by symmetrically placed capacitances. **e**) Arranging a number of parallel longitudinal conductors such that the number of current filaments in a transversal segment of angle $\Delta\varphi$ is proportional to cos φ results in an arbitrarily good approximation to the desired sinusoidal current distribution. The proper layout is achieved if the projections of the filaments on the vertical diameter of the cylinder divide the diameter in sections of equal length. **f**) The "Birdcage" and its low-frequency predecessor. A wire that is exactly one wavelength long, wound on a toroid, produces the desired sinusoidal current distribution on the inner cylinder (1). A cylindrically arranged ladder network of coupled self-resonating elements provides sinusoidally modulated currents in its longitudinal elements (2) would be possible as well. (Note the similarity to the loop-gap resonator, Fig. 8). Historically the Alderman–Grant resonator was the first design that strictly symmetrized the electric fields and used two ring structures known as "guard rings" to shield electric fields at the resonator ends from the sensitive volume. With respect to magnetic field-producing conductive elements, this design behaves as a single current loop and is thus very easily tuned by a single variable capacitor across the connection leads, without breaking the transversal symmetry. Practical descriptions are provided in Refs. [21, 22].

There still exists some controversy regarding the ideal angle $\Delta\varphi$ subtended by the arcs [23]. This can be attributed to the undefined current behavior at the boundaries of the conductors (the assumption of a rectangular current profile neglects the current enhancement at the conductor boundaries due to the skin effect), and to the influence of a shielding cylinder on the field distribution. It is generally agreed, however, that an angle $\Delta\varphi = 80°$ is close to the optimum.

A further approach to model the sinusoidal current distribution has been described in Ref. [16]. A number of longitudinal conductors is arranged along a cylinder surface at varied distances with narrow spacing at $\varphi = 90°$ and 270° and wide spacing at $\varphi = 0°$ and 180°. The construction algorithm is outlined in Fig. 7e. Placing wires along the cylinder surface such that their projections onto a diameter appear equidistant results in the desired current distribution ΔJ within a subtended angle $\Delta\varphi$:

$$\Delta J/\Delta\varphi(\varphi) \sim \cos\varphi. \tag{17}$$

The approximation can be arbitrarily improved by increasing the number of filaments. The individual elements are combined to single feed points at both ends such that again tuning and matching by a simple two-capacitor network does not influence transversal current distribution and homogeneity. This design can be viewed as a high-frequency version of a multiturn distributed saddle coil [24], with conductors placed in parallel, not in series.

A quite clever design [25] that is useful only at lower frequencies and therefore has not found widespread application uses the fact that a standing wave on a wire of length $l = \lambda$ produces a current distribution $J(x) = J_0 \cos 2\pi x/\lambda$ along the wire (see Eq. (11)). If this wire is wound on a cylindrical toroid (Fig. 7e), the inner conductors produce the desired current distribution while the outer conductors effectively shield the probe.

A new dimension was added to the field of probe design by an ingenious approach based on phase delay networks [17]. If fed with an alternating input

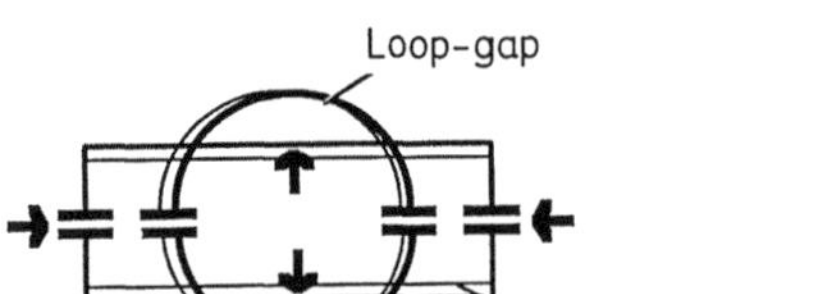

Fig. 8. The Alderman–Grant resonator and the Loop-Gap resonator are closely related through a simple geometrical transformation. Note that such an analogy applies to most of the discussed longitudinal and transversal field resonators

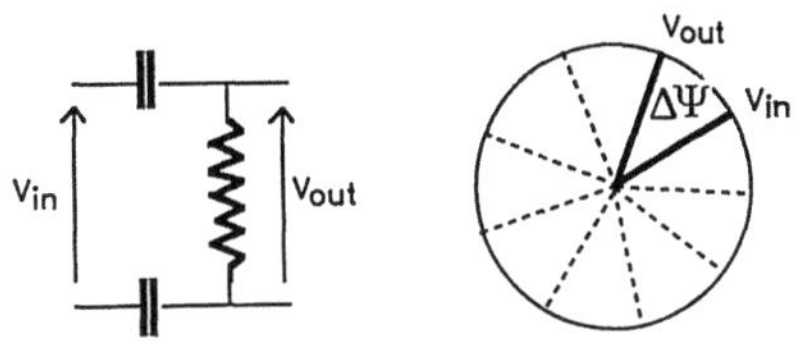

Fig. 9. A phase-delay LC element produces a frequency dependent phase shift $\Delta\Psi$ between input and output voltage. A number n of such elements wired in a ladder network, with the last element connected to the first, generates the desired sinusoidal current distribution if driven with a frequency such that the total phase delay $n\Delta\Psi$ in the network equals 360° (constructive interference)

voltage, a high pass element consisting of a capacitor and an inductor (Fig. 9, for symmetry purposes two capacitors are used) if fed with an alternating input voltage provides an output voltage of equal magnitude with a phase difference of $\Delta\psi(\omega)$ dependent on the frequency ω. If a number N of such elements are connected in series, and the final element is reconnected to the first, a loop is created that supports currents at a very precisely defined frequency. Only for a total phase shift of $N\Delta\psi(\omega) = 2\pi$ does constructive interference (= resonance) occur. At frequencies that do not match the phase relationship the elements interact destructively. At the resonance frequency the currents J_n in the longitudinal filaments at position n are determined by:

$$J_n = J_0 Re(\exp(in\Delta\psi - \mathrm{i}\omega t)) = J_0 \cos(\varphi_n - \omega t), \tag{18}$$

where Re denotes the real part of the current derived in complex notation, and $n\Delta\psi$ describes the phase offset of the current in the nth filament with respect to the current phase of the first filament at $\varphi_0 = 0$. On resonance this phase offset is identical to the angular position φ_n of the nth filament: $n\Delta\psi = \varphi_n$. Increasing the number n of filaments yields an arbitrarily close approximation of the sinusoidal current distribution in Eq. (13). At a given time, ωt corresponds to an offset angle α, such that the sinusoidal current distribution according to Eq. (15) and the corresponding field direction is achieved. This sinusoidal current pattern rotates as a function of time around the cylinder axis (Fig. 10) and produces a homogeneous magnetic field which rotates in the xy-plane (circular polarization):

$$\bar{\boldsymbol{B}}_1 = B_1 \begin{pmatrix} \sin\omega t \\ \cos\omega t \\ 0 \end{pmatrix}, \tag{19}$$

(Compare Eq. (3)). If the proper orthogonal coupling to the driving amplifier and to the receiving preamplifier is provided to suppress counter-rotating fields [26], the advantages of this so-called "quadrature drive", including power reduction by 2 and SNR increase by $\sqrt{2}$, are obtained. Due to its characteristic shape, this design has been termed the "birdcage" resonator by its inventors. This structure gives rise to higher order resonances as well, as it supports all

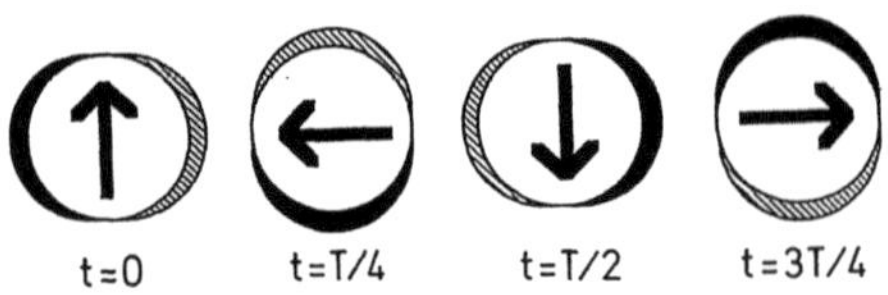

Fig. 10. The current distribution and the magnetic field on a circularly polarized birdcage resonator as a function of time. T denotes the time period for a full cycle at the Larmor frequency

frequencies in accordance with the phase relationship $N\Delta\psi(\omega) = k2\pi$. Under simplified assumptions the frequency spectrum is described by:

$$\omega_k = 1/(\sqrt{LC}2\sin(\pi k/N)) \qquad (k = 1 \ldots N/2) \tag{20}$$

where $k = 1$ corresponds to the desired dipolar field distribution, $k = 2$ describes the quadrupolar resonance, and $k = N/2$ the highest order resonance with every element in antiphase current to its neighbors. It has been proposed [27, 28] that this multipolar characteristics be used for spatial localization of the NMR signal (instead of gradient encoding). Construction and application experience with birdcage resonators have been described in [29–31]. A theoretical treatment of the birdcage field properties is found in Ref. [32]. Reference [33] contains a very elaborate description of construction and operation of a quadrature set of two orthogonal Alderman–Grant type resonators, a design that is virtually identical to a four-element birdcage probe.

The field homogeneity that can be achieved with the birdcage design depends of course on the number of filaments applied (see e.g. [34]), as a larger number of filaments yields a better approximation of the desired continuous sinusoidal current distribution. Figure 11 shows a plot of the field variation as a function of the normalized radial position in the central plane of the resonator, for different numbers of longitudinal conductors. While there is significant gain in homogeneity if the number of elements is increased from 4 to 8, the difference between a 16 and a 32-element resonator is marginal, and further increase in the number of elements only results in increased complexity and is in general not recommended.

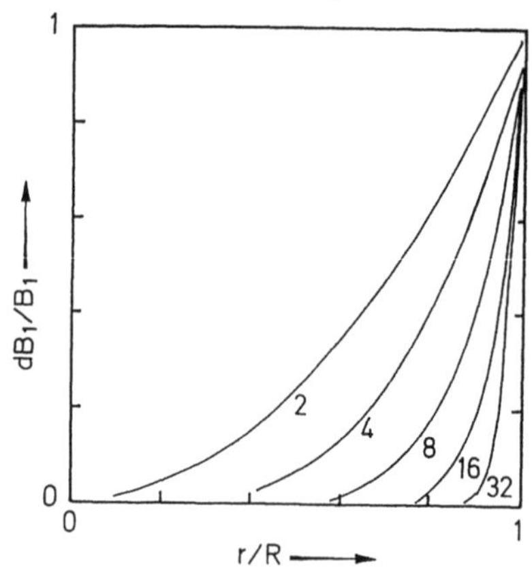

Fig. 11. Normalized field inhomogeneity in the central transversal plane of birdcage resonators with various numbers of longitudinal elements, as a function of the normalized radial distance from the resonator axis. B_1 denotes the desired dipolar field contribution, ΔB_1 denotes field variation, r the radial offset position and R the resonator radius

The lumped element view of this resonator is best suited for the description of its homogeneity behavior, but its frequency properties cannot be accurately determined in this view, as it is very difficult to quantify mutual inductive and capacitive coupling of the individual elements. A better approach (that is in fact applicable to all transversal field designs, including Aldermann–Grant and Slotted-Tube Resonator) is the microwave view, which visualizes the resonator as a conducting tube that supports longitudinal standing current waves of suitable wavelength $\lambda/2 = l$. If transversal currents are suppressed by slotting the tube longitudinally, (two narrow slots correspond to the STR, two large slots to the Alderman–Grant, and a number of slots to the birdcage resonator), one finds a longitudinal current and transversal field distribution on the open-ended tube as displayed in Fig. 12. The impedance Z of the tube as a function of longitudinal position z is related to the electric field E and to the current density j (and thus to the magnetic field B) as described in Ref. [35]. Carrying the thoughts in Ref. [35] a bit further [36], one arrives at the impedance function:

$$Z(z) = (-i\sqrt{\varepsilon/\mu}N \tan(2\pi z/\lambda))/4\pi, \tag{21}$$

where ε and μ are the free space permittivity and permeability, and N counts the number of longitudinal conductor elements. This function can now be readily applied to determine the shunt capacitances needed to achieve the desired operation frequency at the projected resonator length. At 100 MHz, a true $\lambda/2$ resonator is about 3 m long; thus a size reduction is necessary for biomedical applications. This is possible without changing the field behavior if the tube is cut off at both ends, and the end parts are replaced by lumped capacitances of equal impedance, as given by Eq. (21). It is desirable to "cut" the resonator symmetrically at both ends, in order to preserve the region of high and homogeneous B field in the tube center, while concentrating the high E fields at the ends into lumped capacitor elements. If a 16-element birdcage is to be

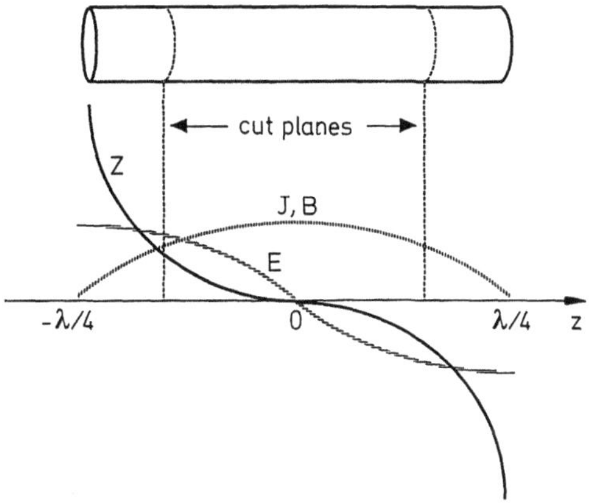

Fig. 12. Current, magnetic and electric field, and impedance on a $\lambda/2$ resonator as a function of position. Cutting at the marked positions and replacing the ends by the given impedance values at the cut planes preserves the current and field distribution on the central part

constructed at a length of 40 cm, each element has to be connected to ground on both sides by an impedance:

$$\begin{aligned} Z &= -i\sqrt{1.257 \times 10^{-6}\,\mathrm{Vs/Am}/8.854 \times 10^{-12}\,\mathrm{As/Vm}} \\ &\quad \cdot 16 \cdot \tan(2\pi \cdot 0.2\,\mathrm{m}/3.0\,\mathrm{m})/4\pi \\ &= 213\,\Omega \end{aligned} \tag{22}$$

corresponding to a capacitance of $C = 1/i\omega Z = 7.45\,\mathrm{pF}$.

The most general approach in NMR resonator design is to arrange a number of self-resonating half-wavelength antenna elements in an array of the desired symmetry and to feed them individually, imposing the desired amplitude and phase relationship. The birdcage resonator can be viewed as one realization of this concept, with the longitudinal filaments corresponding to the antenna resonators, and the feeding circuit being provided by capacitive or inductive end rings. The previously described phased array concept follows the same line of thought, using elementary loop antennas as basic building blocks.

4 Design Practicalities

While the described resonator designs are quite different with respect to symmetry and complexity, and therefore homogeneity and sensitivity, there are some general aspects of coil optimization that apply to all designs equally and are best discussed independently from a specific design.

4.1 *Shielding*

The purpose of shielding is to prevent the NMR resonator from behaving as an antenna, by partially or completely enclosing it in a more or less perfectly conducting "cage", with the aim to confine its field energy to some finite volume. This prevents radiation losses, sensitivity to outside noise sources (such as radio stations) and interaction with lossy or non-ideal outside materials. The perfect shield acts as a "mirror" (see e.g. [31]), such that all currents flowing in the antenna create oppositely flowing currents in the shield. Since the total external flux created by the currents flowing in the antenna structure thus cancels to zero (more precisely: the external dipole characteristics is transformed into quadrupolar characteristics with correspondingly steeper field decay), there is no effective field and no energy stored outside the shield. This effect is directly comparable to the working principle of gradient shielding. Well-designed rf shielding structures allow high-frequency currents to flow at minimal resistance, while low-frequency currents, such as those induced by switching gradients (eddy currents), should find a high-resistance path. Otherwise the probe interior would be shielded from the gradient fields as well. Therefore, if gradient fields have to penetrate, it is recommended, to not use plain copper foil, but coarse copper mesh or preferably "comb" structures that do not offer DC-current loops but capacitively connected loops only. A detailed description of this principle as applied to a phased array is given in Ref. [37].

Shielding may not only be appropriate towards the outside of the probe, but towards its inside as well. In a properly designed transversal field resonator high magnetic field prevails at the center of the structure, while high electric fields dominate the end areas [35]. To prevent undesirable dielectric interaction with the sample, it is recommended to shield the sample space from these high field regions by applying a "guard ring", that is, a ring-shaped equipotential surface, at both resonator ends (compare [18]). Preferably one would not use conducting rings, but capacitively connected half-rings, again to prevent low-frequency eddy-currents from being induced into the rings by z-gradients. These guard rings, an essential design element of the Alderman–Grant resonator, should be applied in the same manner in "birdcage" designs as well. Not only do they diminish electrically induced sample losses, but they also minimize the effect of sample change and movements on probe tuning. Also it is common practice to shield the tuning/matching capacitors of a surface coil by enclosing them in a conducting box, such that only the coil's fields penetrate the sample.

4.2 Coupling

The coupling mechanism is responsible for feeding transmitter power to the probe to continuously replace the energy lost in the sample and probe during a pulse, and to transfer signal energy picked up by the probe to the receiver. Energy can be transferred to the probe where it provides sensitivity. It can be coupled electrically (capacitively) to positions where the probe's internal electric fields are high (such as capacitances, whether as lumped elements or implicit in the probe structure). Alternatively, energy can be transferred by magnetic fields (inductively) at positions of high intrinsic magnetic field.

Capacitive coupling to a single current loop such as a surface coil is illustrated in Fig. 13. The driving voltage is applied to the probe's capacitor through an adjustable matching capacitor. The external matching and the internal tuning capacitor act as an imaginary "voltage divider", where C_T determines the

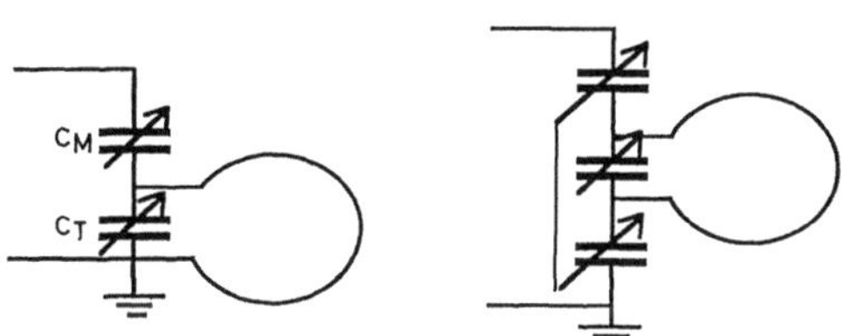

a b

Fig. 13 a, b. Capacitive tuning and matching, viewed as an imaginary voltage divider network (**a**). Symmetric matching capacities (**b**) uncouple the coil from feed ground and reduce capacitively induced sample losses resonance frequency, and the ratio $C_T^{-1}/(C_T^{-1}+C_M^{-1})$ determines the partial driving voltage and the coupling strength. More rigorous concepts of coupling as an impedance matching process have been developed in Refs. [38, 39].

It must be emphasized that tuning and matching are basically independent functions. In many cases, the "voltage divider" is set up with a variable matching capacitor which is connected to a fixed capacitor of the probe, while probe tuning is achieved by one or more variable capacities elsewhere in the probe.

In the coupling scheme of Fig. 13a one end of the coil is kept to feed ground, while the other end assumes maximum electric potential. The loss effect due to this coil potential can be strongly reduced if a balanced feeding scheme is employed (Fig. 13b). In this symmetrical scheme the coil is uncoupled from ground and assumes a virtual ground potential in its center, such that the electric potentials at both coil ends and the corresponding energy losses are reduced by a factor of two [39]. Virtual ground positions can easily be detected by touching the coil, connected to a sweep generator, with a finger tip. The higher the electric potential, the more pronounced is the shift and widening of the coil resonance in the wobble spectrum. Touching the coil at a virtual ground position barely effects the appearance of the resonance.

Inductive coupling is less frequently used in combination with surface coils, but it is instructive to demonstrate the principle with this simple example (Fig. 14). The primary surface coil resonator, tuned to the desired resonance, is approached by a coupling surface coil resonating in the vicinity of the main coil resonance. The primary coil is not connected to ground; therefore inductive coupling provides intrinsic balanced feeding. At some distance, the sensitive volumes of both coils barely overlap, and practically no energy is transferred from the driving coil to the main resonator (Fig. 14a). If the coupling coil approaches the main coil, the magnetic flux common to both coils is increased and the coupling becomes stronger. This behavior can be monitored if the coupling loop is connected to a sweep generator. Closer coupling corresponds to less power reflected from the coupling loop at the resonance frequency of the primary circuit, as the power is absorbed by the primary loop. Perfect match is achieved when all power is absorbed by the primary coil, and no power is reflected into the transmitter at the desired resonance frequency. Alternatively, coupling strength can be increased by bringing the resonance of the coupling coil closer to the main coil's resonance frequency (Fig. 14b), again increasing the flux from the coupling coil to the main coil in the desired frequency range.

In both cases of capacitive or inductive coupling, the coupling device can be viewed as a second resonating structure overlaid to, and providing the energy lost by, the main resonator. With respect to field homogeneity, this has no consequence if the symmetry of the coupling device matches the main coil symmetry. Probes that can be characterized by a single current loop (no parallel paths in the magnetic field producing conductors) are best coupled capacitively, leaving the symmetry of the current distribution in the inductors unaltered. Inductive coupling by a surface coil-type mechanism inevitably overlays the surface coil field characteristics onto the main coil field distribution. In low-loss

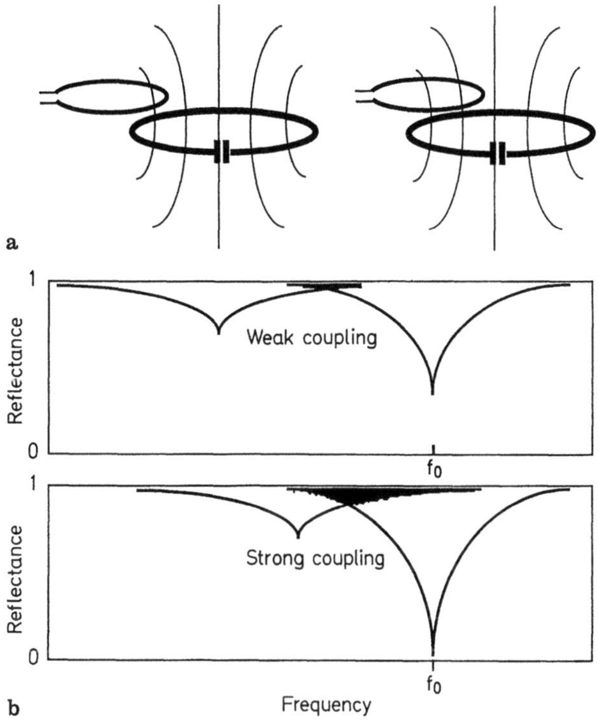

Fig. 14a, b. Inductive coupling. Coupling strength can be increased by spatially approaching the main resonator with the coupling coil, to increase the flux common to both coil (**a**), or alternatively by tuning the coupling coils resonance frequency towards the main coils' resonance (**b**) to increase the coupling coils' energy content at the desired frequency

probe/sample combinations only weak coupling is necessary; the coupling coil provides only a small percentage of the magnetic field energy stored in the probe. Thus, the symmetry break is negligible. High loss combinations require high energy flow through the coupling device, and thus result in higher inhomogeneity.

Example: If a coil shows a Q value of 200, the energy ΔE to be replaced per cycle amounts to $\Delta E/E = 2\pi/Q = 3\%$ (see Eq. (24)). This energy must be transported through the coupling device and increases field strengths locally by $\sqrt{(1+3\%)}-1=1.5\%$. A coil with $Q=50$ (high losses) requires an energy flow of 12% and thus a local inhomogeneity of $\sqrt{(1+12\%)}-1=6\%$.

For a multiloop coil (such as a birdcage resonator) no straightforward coupling scheme at matching symmetry behavior is available. In this case, it is recommended to use multiple coupling elements distributed symmetrically around the resonator.

Coupling to quadrature probes is more demanding: The transmit energy must be fed into two orthogonal feed positions, with a 90° phase shift to support

B_1 fields rotating in the desired direction, while suppressing B_1 fields rotating in the reversed direction. This can be achieved by a 90° phase delay line (e.g. $\lambda/4$ cable) introduced into one feed channel. If the same quadrature probe is to be used for reception, the phase relationship between the two channels must be inverted between transmit and receive mode to receive in-phase signal from both channels [26]. Several phase-splitting devices to perform this task have been described [26, 33, 40, 41]. In principle, both feed points can be tuned and matched individually, and can be connected to the phase splitter. In practice, however, the orthogonality of both channels required for independent tuning and matching, is difficult to achieve, and depends on probe loading conditions. Tuning and matching of both channels can become a tedious iterative process.

4.3 L–C Ratio

There exists an unlimited number of possibilities of constructing a 3 cm ^{31}P surface coil at 4.7 T: A single turn thick wire coil in combination with a fairly high capacitance of more than 50 pF will lead to the desired resonance at 81 MHz, as well as a 4 turn thin wire coil with a correspondingly lower capacitance of approximately 10 pF. Which design is preferable?

Two criteria lead to a narrower range of possibilities: In principle one wants the lowest possible inductivity of the coil in order to achieve high currents at low voltages, corresponding to high magnetic fields at low electric fields. This asks for high capacitance values, which is also favorable with respect to detuning due to changing the sample or patient movements. On the other hand, one wants to minimize losses in the coil's connection leads which cannot be arbitrarily shortened. If the connection leads length is 1/3 of the total length of the coil wire, roughly 1/3 of the magnetic energy does not contribute to the desired sample interaction, but is dissipated in the coil. Therefore one has to optimize towards short and thick probe wires, that are still substantially longer than the connection leads. It must be strongly emphasized that long connection leads (relative to the total coil wire length) lead to serious degradation of probe efficiency.

4.4 Trade-Off Homogeneity Vs. Sensitivity

A good resonator is characterized by both high sensitivity and high field homogeneity. Homogeneity is optimized by evenly distributed currents arranged with high symmetry. Examples are the solenoid and the birdcage resonator, with high longitudinal and pronounced rotational symmetry. The relationship between symmetry classes and field homogeneity properties for all kinds of magnetic field generating devices (applicable to the design of main magnets as well as gradient coils) was elaborated in a general framework by Romeo and Hoult [42]. Sensitivity is optimized for a given resonator and a given sample (assuming all electric loss factors have been eliminated) if the sensitive volume of the resonator is identical to the dimensions of the sample (or the region-of-interest to be measured).

If a sample of cylindrical dimensions, such as a solution in an NMR tube, is to be spectroscopically analyzed, the use of a matching solenoid results in almost perfect homogeneity across the sample at simultaneously maximum sensitivity. Conflicts between sensitivity and homogeneity arise, if the sample is much larger than the desired region of interest, e.g., if a transversal thin slice image of the sample is to be acquired. If the solenoid is much longer than the desired slice thickness, it shows good field homogeneity across the slice, but it will pick up noise from parts of the sample not contributing to the signal and degrade sensitivity. If, on the other hand, the solenoid is shortened to the desired slice thickness, its sensitivity is strongly enhanced at the cost of decreased homogeneity.

Bottomley et al. [31] have elaborated that the sensitivity of a birdcage resonator is increased by 2.6, if instead of the length/radius ratio of 4, which yields optimum homogeneity, a length/radius ratio of 1.4 is used. This decrease in homogeneity is less severe for imaging of transversal slices or spectroscopic voxel localization than for longitudinal slices. Hoult has exemplified [38] that the transformation of a two turn surface coil into a Helmholtz coil decreases sensitivity by $1-\sqrt{3}/2=13.4\%$ at a corresponding gain in longitudinal field homogeneity. In both cases this change in sensitivity is directly related to the change in the sensitive volume of the coil. It is therefore recommended to design resonator geometries in view of the desired samples, i.e., with sensitive/homogeneous regions not extending substantially outside the typically desired field-of-view dimensions. At this point one has to consider if employing a large transmitter probe with homogeneous transmit fields and a separate small reception probe yielding optimal sensitivity would be preferable. This choice, however, calls for measures of active or passive probe decoupling [3].

5 Performance Tests

The ultimate performance test of an NMR resonator is the desired experiment itself. However, if the experiment produces nonsatisfactory results, these results do not contain directly accessible information as to where to improve the probe design. Weaknesses can be revealed by using test samples and test protocols which shall be described in the following. Furthermore, there exists a wealth of tools to test the probe performance on the rf workbench during the construction process, without the need of a spectrometer.

5.1 On the RF Workbench

The basic rf workbench includes a sweep generator, a reflectance bridge, a diode demodulator, and an oscilloscope with at least two channels switchable to X–Y configuration, to display the reflected power vs frequency diagram of the NMR probe. On resonance the inductive and capacitive impedances of the probe cancel, and the circuit appears as a pure resistance. At this frequency the power absorbed by the circuit is maximized, and the reflected power is minimized: the resonator spectrum on the oscilloscope shows a minimum. At good impedance matching, providing that the resistance of the circuit matches the output impedance of the transmitter, the circuit does not reflect any power on resonance: The reflectance sweep spectrum hits the zero line.

Each closed current circuit resonates at one characteristic frequency and thus shows an absorption line in the reflectance spectrum. A simple solenoid is expected to show a single resonance line. If in such a circuit several resonances are present, an undesired parallel path is resonating in the cable and/or matching circuitry that dissipates energy and must be eliminated. A bird cage resonator consisting of 16 longitudinal elements (corresponding to 8 current loops), on the other hand, is expected to resonate at 8 different frequencies characterized by different multipolar current distribution.

If several resonances are present, a small pickup coil connected to an additional channel of the scope can serve to identify the origin of each resonance and its multipolar characteristics. Turning the coil 360° in the center of the resonator yields two signal maxima and minima for the desired dipolar mode, while the quadrupolar mode yields 4 maxima and minima, and so on. The pickup coil can further be employed to detect radiation at unexpected parts of the antenna and to roughly measure the field homogeneity inside the probe by scanning through its volume.

A regular pickup coil is sensitive to electric and magnetic fields in the resonator. To detect only magnetic fields one can build an electrically shielded coil according to the design in Ref. [43]. In many cases it is helpful to detect electric fields (which can create high losses if not shielded properly from the sample) to be able to eliminate them. A sensor to exclusively pick up electric fields is described in Ref. [35].

The bandwidth Δf (as determined by the spectral width at 1/2 reflected power) of a line of frequency f_0 in the wobble spectrum is directly related to the quality factor Q of a resonator by the relation:

$$Q = \frac{2\pi \text{ energy stored in the antenna}}{\text{energy lost per cycle}} = f_0/\Delta f. \tag{24}$$

The absolute value of Q is a problematic measure of probe quality as it is an integral value that does not account for inhomogeneous distribution of the stored B_1 field. A coil design that concentrates the field around very long but well shielded connection leads displays high Q but produces only poor signal. Q, however, is a very valuable quantity if evaluated as a function of different test loads.

To detail this matter, Q can be expressed in electrical quantities:

$$Q = \omega L/R \tag{25}$$

where L represents the field producing inductance, and R the resistance of the probe. R comprises several loss contributions:

$$R = R_{\Omega} + R_{m} + R_{e} \tag{26}$$

where R_{Ω} represents the ohmic resistance of the coil itself, R_{m} the effective resistance to currents magnetically induced in the sample, and R_{e} the effective resistance to electrically induced polarization currents in the sample. For the empty coil, R_{m} and R_{e} are zero, and:

$$Q_{e} = \omega L / R_{\Omega}. \tag{27}$$

Maximization of this value is desirable.

Introducing a magnetically lossy test sample (e.g., physiological salt solution) lowers the Q value to $Q = \omega L/(R_{\Omega} + R_{m})$. The lower Q drops upon introducing the lossy sample, the higher the magnetically induced sample currents, the more concentrated B_1 is in the sample space, and the better constructed is the coil. Magnetic sample losses cannot be eliminated [17].

If a dielectrically lossy nonconducting sample (e.g. a plastic polymer) is introduced, it couples exclusively to the electric fields in the probe. High Q drop corresponds to high electric fields, which do not contribute to spin excitation but to energy dissipation only. By proper electrical shielding these fields can be minimized in the sample space such that no losses are induced and Q does not drop noticeably [44]. To conclude, a good probe shows a high unloaded Q and a substantial Q drop upon introduction of a magnetically lossy sample, but no Q drop upon introduction of an exclusively dielectrically lossy sample.

Distribution of electric and magnetic fields can further be checked, even locally, by using metallic (high conductivity) respectively purely dielectric (high dielectric constant) non-lossy test samples. Metallic particles displace magnetic fields and thus increase the resonance frequency proportional to the displaced magnetic energy. Dielectric samples, on the other hand, increase stray capacitance and thus decrease frequency. A good test sample to check for electric fields is a bottle of distilled water, as it displays a very high dielectric constant [45] and negligible conductivity.

5.2 *In Situ Tests*

The essential quality of a resonator is a B_1 field with high magnitude $B_1 = |\bar{\boldsymbol{B}}_1|$ of the field strength and minimal changes of field magnitude or direction as a function of space. $B_1(r)$ at various positions of a coil can be determined directly by a spectroscopic NMR experiment. As the B_1 field nutates magnetization at the angular velocity $d\vartheta/dt = \gamma B_1$ (γ denotes the gyromagnetic ratio), a pulse of duration t achieves an angle $\vartheta = \gamma B_1 t$. If for a specific probe at a specified transmitter power a 90° pulse has duration τ the B_1 field associated with this pulse is:

$$B_1 = \pi/2\gamma\tau. \tag{28}$$

A 90° pulse time of 5 μs, for example, corresponds to a field strength $B_1 = 1.5$ mT. The 90° pulse length is a direct measure of probe sensitivity as well (principle of reciprocity). For comparison of the probe with other probes at different transmitter powers P and sensitive volumes V the following scaling relationship holds:

$$B_1 \sim \sqrt{P/V} \tag{29}$$

since the energy stored in the probe $E \sim \int B_1^2 dV \approx B_1^2 V$ is proportional to the loss power: $E = P \cdot Q/\omega$.

A small test sample can be used to determine local 90° pulse times. A B_1 map is then obtained by scanning through the whole volume of interest (see e.g. [40]). Using a full-size test sample or the target object yields the average B_1 which represents the sensitivity of the resonator under realistic conditions.

Reliable homogeneity mapping is achieved by performing 3D spin echo or 2D multislice spin echo series on a homogeneous phantom. In doing so one can reveal intensity variations close to conductors and boundaries, intensity gradients created by coupling elements, and rf holes in the interior of large lossy samples when rf penetration depth is limited. Spin echo techniques are favorized test sequences, as B_0 inhomogeneities are largely compensated and do not contribute to intensity variations, while B_1 inhomogeneities are enhanced by the application of two rf pulses. Alternatively, simple projections along all three main axes using a homogeneous cubic sample yield the desired homogeneity profile of the resonator under test. One should be aware, however, that system weaknesses can influence the results: 1) B_0 inhomogeneities lead to phase dispersion and to dark image regions, especially in gradient echo sequences. This effect is difficult to distinguish from B_1 inhomogeneities. 2) Nonlinear gradients result in scale distortions and over-intensified image areas. 3) Susceptibility boundaries lead to signal "displacement" and therefore to image inhomogeneities.

5.3 Probe Optimization: An Infinite Effort?

After deciding for a suitable coil geometry, and going through a set of construction compromises, one will finally have a newly designed working coil at hand. Performing some test procedures and the target experiment itself, some people are happy if the new coil performances better than from the one previously used. Others may be still unsatisfied, wondering how far their real design is from the unknown optimum performance. The following considerations are meant to give perfectionists a clue when to stop further optimization efforts.

For the following it is essential to select a sample that represents typical loading conditions of the probe in the target applications. At first one should verify that dielectric interaction is reduced to an absolute minimum, as described before, such that the electric loss resistance R_e can be neglected ($R_e = 0$). Loaded probe quality Q_l can then be represented by the effective inductance and the internal loss resistance R_Ω of the probe, and the resistance R_m of the purely magnetically coupled load

$$Q_l = L/(R_\Omega + R_m), \tag{30}$$

while the unloaded probe exhibits a quality factor:

$$Q_e = L/R_\Omega. \tag{31}$$

Now one can attribute a complementary, not directly measurable Q-factor Q_s to the sample contribution.

$$Q_s = L/R_m. \tag{32}$$

Coil sensitivity S is proportional to $\sqrt{Q}$ (see, e.g. [17]).

$$S = c \cdot \sqrt{Q_l} \tag{33}$$

For a given sample, sensitivity is maximized if an ideal coil of $R_\Omega = 0$ is employed:

$$S_0 = c \cdot \sqrt{L/R_m} = c \cdot \sqrt{Q_s} \tag{34}$$

For a given coil/sample combination, the ratio of true sensitivity S to maximal sensitivity S_0 can now be determined from the ratio of the loaded to the unloaded Q factors:

$$S/S_0 = \sqrt{Q_l/Q_s} = \sqrt{(1 - Q_l/Q_e)}. \tag{35}$$

The achieved sensitivity of a probe, scaled to the maximum as determined by the sample, is plotted as a function of the ratio loaded/unloaded Q in Fig. 15. If, for example, Q drops by a factor of 5 upon introduction of the sample, and one can exclude that this is due to dielectric losses, infinite efforts to further optimize the coil will only increase the SNR of the chosen sample by 10%.

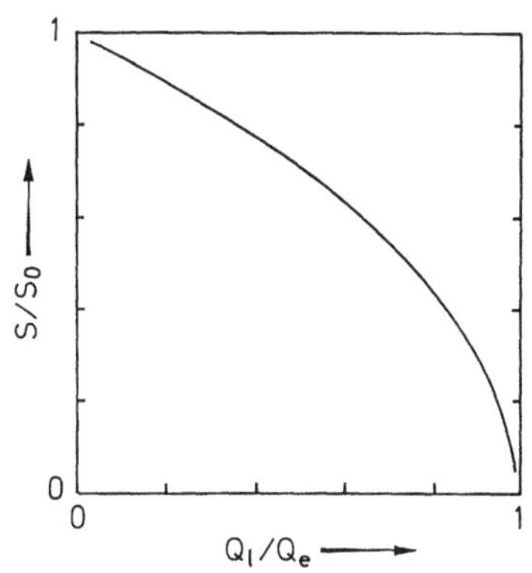

Fig. 15. Sensitivity S of a resonator-sample combination, normalized to the maximum sensitivity S_0 achievable with the sample if a lossless resonator were employed, as a function of the ratio of quality factors Q_l of the loaded probe and Q_e of the empty probe

In biological and medical environments typical samples generally dominate the coil losses, thus putting coil optimization on a relative position in the optimization protocol. This is reflected in several publications on the comparison of various probe designs (e.g. [8, 46]). While empty coils of similar volume may generally behave quite different, their performances are well comparable when loaded with patients. An elegant formulation of this fact has been casted by Hoult [47]: "A lossy sample dictates its own SNR".

Acknowledgements: The author wishes to thank B. Kohler, M. Rudin, M. Vaccaro, and P. Erhard for enlightening discussions, and valuable remarks and comments on this manuscript.

6 References

1. Terman FE (1943) Radio Engineer's Handbook. McGraw-Hill. New York
2. Chen CN, Hoult DI (1989) Biomedical magnetic resonance technology. Adam Hilger, Bristol, New York
3. Hyde JS, Kneeland JB (1988) in: Wehrli FW, Shaw D, Kneeland B (eds) Biomagnetic resonance imaging. Verlag Chemie, Weinheim, New York
4. Hoult DI, Richards RE (1976) J Magn Reson 24: 71
5. Hoult DI, Deslauriers R (1990) SMRM Book of Abstracts 203
6. Feynman R (1974) The Feyman Lectures on Physics II, 13-5, 6th edn. Addison–Wesley, Reading, Mass
7. Macalpine WW, Schildknecht RO (1959) Proc IRE 47: 2099
8. Arakawa M, Crooks LE, McCarten B, Henninger JC, Watts JC, Kaufmann L (1985) Radiology 154: 227
9. Roeder SBW, Fukushima E, Gibson AAV (1984) J Magn Reson 59: 307
10. Hardy WN, Whitehead LA (1981) Rev Sci Instrum 52: 213
11. Froncisz W, Hyde JS (1982) J Magn Reson 47: 515
12. Sobol WT (1986) Reviews of Magnetic Resonance in Medicine 1: 181
13. Mansfield P, Coxon R (1988) SMRM Work in Progress 156
14. Requardt H, Erhard P, Offermann J, Loeffler W (1988) SMRM Book of Abstracts 843; Requardt H, Offermann J, Erhard P (1990) Magn Reson Med 13: 385
15. Roemer PB, Edelstein WA, Souza SP, Hayes CE, Mueller OM (1988) SMRM Book of Abstracts 875; Hayes C, Hattes N, Roemer PB (1989) SMRM Book of Abstracts 175; Roemer PB, Edelstein WA, Hayes CE, Rohling KW (1989) SMRM Book of Abstracts 176; Roemer PB, Edelstein WA, Rohling KW, Darber WD (1990) SMRM Book of Abstracts 204; Roemer PB (1990) SMRM Book of Abstracts 255
16. Bolinger L, Prammer MG, Leigh JS (1988) J Magn Reson 81: 162
17. Hayes CE, Edelstein WA, Schenck JF, Mueller OM, Eash M (1985) J Magn Reson 63: 622
18. Alderman DW, Grant DM (1979) J Magn Reson 36: 447
19. Schneider HJ, Dullenkopf P (1977) Rev Sci Instrum 48: 68
20. Leroy-Willig A, Darrasse L, Taquin J, Sauzade M (1985) Magn Reson Med 2: 20
21. Cross TA, Müller S, Aue WP (1985) J Magn Reson 62: 87
22. Hong DH, Lee SW, Ra JW, Cho ZH (1987) Magn Reson Imag 5: 239
23. Kost GJ, Anderson SE, Matson GB, Conboy CB (1989) J Magn Reson 82: 238
24. Joseph PM, Fishman JE (1985) Med Phys 12: 679
25. Hinshaw WS, Gauss RC (1984) US Patent 4:439,733

26. Chen C-N, Hoult DI, Sank VJ (1983) J Magn Reson 54: 324
27. Carlson JW (1987) J Magn Reson 74: 376
28. Hutchinson M, Raff U (1987) Magn Reson Med 6: 87
29. Holcomb WG, Gore JC (1987) Mag Res Imag 5, supl 1
30. Joseph PM, Lu D (1989) IEEE Transactions on Medical Imaging, vol 8: 286
31. Bottomley PA, Charles HC, Roemer P, Flamig D, Engeseth H, Edelstein WA, Mueller OM (1988) Magn Reson Med 7: 319
32. Tropp J (1989) J Magn Reson 82: 51
33. Sank VJ, Chen C-N, Hoult DI (1986) J Magn Reson 69: 236
34. Sotgiu A, Hyde JS (1986) Magn Reson Med : 55
35. Joseph PM, Saadi-Elmandjra M (1987) J Magn Reson 75: 199
36. Link J (1989) Dissertation, Basel
37. Krause N (1983) Offenlegungsschrift DE 3133432 A1, Deutsches Patentant
38. Hoult DI (1979) The NMR Receiver; Progress in NMR Spectroscopy
39. Decorps M, Blondet P, Reutenauer H, Albrand JP (1985) J Magn Reson 65: 100
40. Hoult DI, Chen C-N, Sank VJ (1984) Magn Reson Med 1: 339
41. Mueller O, Edelstein WA (1988) SMRM Book of Abstracts 453
42. Romeo F, Hoult DI (1984) Magn Reson Med 1: 44
43. Chen C-N, Sank V, Cohen SM, Hoult DI (1986) Magn Reson Med 3: 722
44. Sergiadis G (1985) Magn Reson Med 2: 328
45. Grist TM, Hyde JS (1985) J Magn Reson 61: 571
46. Murphy MK, Fernandez EJ, Clark DS (1989) Magn Reson Med 12: 382
47. Hoult DI (1984) in: James TL, Margulis AR (eds) Biomedical magnetic resonance. S Francisco

Probes Tuned to Multiple Frequencies for In-Vivo NMR

Mitchell Schnall

Department of Radiology, Hospital of The University of Pennsylvania, 3400 Spruce St. Philadelphia, PA 19104, USA

Table of Contents

Multinuclear applications of NMR spectroscopy and imaging in-vivo demand high-performance, multiple tuned NMR probes. Much effort has been placed in the development of multiple-tuned probes that have minimal loss in sensitivity over the single-tuned probe. In this chapter the theory behind multiple tuned coils is presented in detail. Techniques using multiple pole tuning circuits and coupled tank circuits are discussed. Circuits for 2, 3 and 4 nuclear probes are presented. The actual construction of these probes is also discussed. A review of designs for multiple tuned volume coils is also presented.

List of Symbols

A magnetic vector potential
B magnetic field intensity (scalor)
B magnetic field vector
C capacitance
E Efficiency
emf electramotive force
F effective filling factor
I current
i $(-1)^{1/2}$
K coupling constant
k Boltzman constant
L inductance

M	magnetization vector
M	magnitude of magnetization vector
M	mutual inductance
Q	quality factor ($L\omega/r$)
r	resistance (ohms)
T	temperature (kelvin)
X	reactance (ohms)
Z	impedance (ohms)
β	**B**/I
Δf	frequency range
ω	frequency

1 Introduction

The growing interest in the application of NMR spectroscopy to intact biological systems has placed unique demands on instrumentation. Perhaps the most critical piece of instrumentation for the in-vivo spectroscopist is the NMR probe. The need to perform quick and reliable experiments, particularly in the clinical setting, makes probe sensitivity critical. Many studies involve the use of more than one nucleus. Typically, proton NMR is used to shim, obtain localizing images and/or spectra. Additional spectra may be acquired from ^{31}P, ^{23}Na, ^{13}C or other nuclei. This multinuclear approach to in-vivo NMR spectroscopy has placed a premium on multiple tuned NMR probes that perform with high sensitivity. There are many designs for double tuned probes in the NMR literature. Some designs for probes tuned to three and four nuclei have also been presented. All of these designs must sacrifice some sensitivity at one or more of the resonant frequencies at which they operate when compared to a similar single tuned coil. One common strategy of all of the multiple tuned probe designs is to minimize the loss in sensitivity. In most designs it is possible to arbitrarily distribute the loss in sensitivity between the different operating frequencies of the probe by proper choice of component values. Thus the evaluation of any multiple tuned probe can only be performed if the probe is optimized for a particular application. A second, less frequently discussed aspect of multiple tuned NMR probes is orthogonal tuning adjustment. In the in vivo setting it is very difficult to control loading of the probe. Therefore, the tuning must be adjusted with each application. Independent tuning of the individual resonances is important to ensure reliable performance by the probe. In this paper, I will discuss in detail many of the multiple tuned designs that have appeared in the NMR literature, paying particular attention to the advantages and disadvantages of each. I will focus on multiple tuned surface coils, but will also treat double tuned volume coils, which are of importance in the context of volume selective MRS techniques.

2 The NMR Probe

Before discussing multiple tuned probes, I would like to review some of the basic concepts of the NMR probe. The probe has two functions, radio frequency transmission (provide B_1 irradiation), and reception. These function may both be performed by a single structure, or by two different structures. As previously stated, the most important parameter of the NMR probe to the spectroscopist is the reception sensitivity. As it turns out, for structures that are used as transmitter and receiver coils, optimizing the reception sensitivity of the probe will also optimize its power efficiency as a transmitter by the principle of reciprocity [1]. For the purpose of brevity I will concentrate on the reception function of the probe.

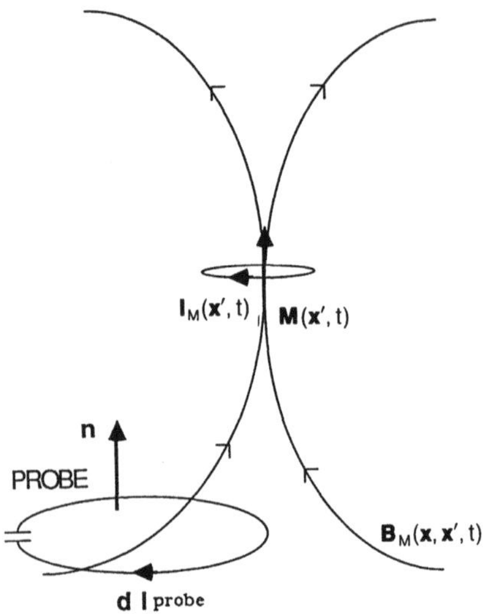

Fig. 1. A unit of magnetization, $\mathbf{M}(\mathbf{x}', \mathbf{x}, t)$, $\mathbf{x}'$ creates a magnetic field $\mathbf{B}_M(\mathbf{x}', \mathbf{x}, t)$ that induces an emf in the NMR probe

As a receiver, the probe detects the precessing transverse magnetization. The signal is detected as an emf induced in the probe coil by the time-varying magnetic field generated by the precessing transverse magnetization (Fig. 1). If we define $\mathbf{B}_M(\mathbf{x}, \mathbf{x}', t)$ as the magnetic field at the position $\mathbf{x}$ created by a spin precessing at a position $\mathbf{x}'$, then the emf induced in the coil can be calculated according to Faraday's law:

$$\mathrm{emf}(\mathbf{x}') = -(\partial/\partial t) \int_{\mathrm{probe}} \mathbf{B}_M(\mathbf{x}, \mathbf{x}', t) \cdot \mathbf{n} dS_{\mathrm{probe}} \tag{1}$$

where $\mathbf{n}$ is the unit vector normal to the surface of the coil, and dS_{probe} is an element of area on the surface of the coil. We can consider the magnetization dipole as an equivalent current loop with the current $I_M(\mathbf{x}') = M(\mathbf{x}')/a$, where M is the magnitude of the magnetic moment and a is the area of the equivalent current loop. The $\mathbf{B}_M$ field created by this equivalent current loop can be expressed in terms of the curl of the magnetic vector potential, $\mathbf{A}_M$. Use of Stokes' Law then allows us to express the emf in terms of a line integral of $\mathbf{A}_M$ along the probe coil:

$$\mathrm{emf}(\mathbf{x}') = -(\partial/\partial t) \int \mathrm{curl}\, \mathbf{A}_M \cdot \mathbf{n} dS_{\mathrm{probe}} = -(\partial/\partial t) \int \mathbf{A}_M \cdot d\mathbf{l}_{\mathrm{probe}}. \tag{2}$$

Noting that $d\mathbf{l}_{\mathrm{probe}}$ represents an element of wire along the probe loop. Expressing the magnetic vector potential as an integral along the equivalent current loop, the emf induced in the coil from magnetization at a point $\mathbf{x}'$ may be written:

$$\mathrm{emf}(\mathbf{x}') = (\partial/\partial t) \int \int (I_M(\mathbf{x}')/|\mathbf{x} - \mathbf{x}'|) d\mathbf{l}_M \cdot d\mathbf{l}_{\mathrm{probe}}, \tag{3}$$

with $\mathbf{l}_M$ being the equivalent current loop. If we reverse the order of integration and again apply Stokes' Law the emf can be expressed as:

$$\mathrm{emf}(\mathbf{x}') = (\partial/\partial t)\int (I_M(\mathbf{x}')\mathbf{B}, (\mathbf{x}', t)/I_{\text{probe}}(t)\cdot \mathbf{n}_M dS_M \tag{4}$$

where I_{probe} is the current in the probe and $\mathbf{n}_M$ is a unit vector normal to the surface of the equivalent current loop. Allowing the area of the equivalent current loop to approach zero, while maintaining the magnitude of the dipole moment (in other words allow the current loop to shrink to a point dipole), Eq. (4) may be written:

$$\mathrm{emf}(\mathbf{x}') = (\partial/\partial t)\{\beta_1(\mathbf{x}')\cdot \mathbf{M}(\mathbf{x}', t)\} \tag{5}$$

where we have defined $\beta_1(\mathbf{x}')$ as the magnetic field per unit current produced by the receiver probe and $\mathbf{M}(\mathbf{x}', t)$ is the magnetic moment vector at $\mathbf{x}'$ and time t. Explicitly expressing the time dependence of the magnetization vector results in:

$$\mathrm{emf}(\mathbf{x}') = i\omega\{\beta_1(\mathbf{x}')\cdot \mathbf{M}_{\text{tr}}(\mathbf{x}')\}e^{i\omega t}. \tag{6}$$

In this expression the subscript, tr, indicates the component of the magnetization transverse to the $\mathbf{B}_0$ field. Thus from Faraday's Law, it follows that the strength of the NMR signal from a unit of magnetization at position $\mathbf{x}'$ is proportional to the scalar product of the magnetization vector and the magnetic field that the receiver coil would make at $\mathbf{x}'$ if a unit current were passed in it. In order to obtain the total signal from any sample of interest, we need to integrate the above expression over the sample volume:

$$\mathrm{emf}_{\text{vol}} = \int i\omega\{\beta_1(\mathbf{x}')\cdot \mathbf{M}_{\text{tr}}(\mathbf{x}')\}e^{i\omega t} dV_{\text{sample}}. \tag{7}$$

There are several important features of the above result. First, and most important to our further discussion of multiple tuned coils is that the net signal intensity from any volume is proportional to the integral of the $\mathbf{B}_1$ field per unit current of the receiver coil (β_1) over the volume of the sample. A second feature is that the geometric relationship between $\mathbf{M}_{\text{tr}}$ and β_1 enters the expression for the total signal through the dot product. It is interesting to note that if the same probe is used for transmission and reception, there will be a constant relationship between the direction of the β_1 field and $\mathbf{M}_{\text{tr}}$, resulting in phase coherent addition of signal from all parts of the sample. Since $\mathbf{M}_{\text{tr}}$ is a function of the transmitter coil and the pulse sequence used, for the purposes of further discussion of double tuned probes we will assume that $\mathbf{M}_{\text{tr}}$ has a value of unity everywhere in space that contains sample, and that it is always parallel to the β_1 field. We will also ignore the time dependence of the NMR signal. This simplifies Eq. (7) to:

$$\mathrm{emf}_{\text{vol}} \alpha \omega \int \beta_1(\mathbf{x}') dV_{\text{sample}}. \tag{8}$$

It is important to realize, that in this expression, β_1 represents the magnitude of the β_1 field. In order to discuss the sensitivity of an NMR probe, in addition to the signal we must discuss the noise. The NMR probe coil (ignoring all tuning elements for the time being) can be modeled as an inductor in series with a resistor. The resistor, in the absence of loading, represents the resistance in the wires that make up the coil. Random thermal motion of electrons within the probe resistance is the major source of noise in the NMR experiment in the absence of sample loading. The RMS value of the resistive noise is well known:

$$N = (4kT\Delta f r)^{1/2} \tag{9}$$

where k is the Boltzman constant, Δf is the frequency range sampled, T is the temperature and r represents the coil resistance (2). It is assumed that the capacitors and leads contribute negligibly to the resistive losses. As stated above, in the absence of sample loading, the resistance term in the noise is dominated by the resistance in the wire of the probe coil (again neglecting tuning circuitry). The sample, when placed in proximity to the coil, effects its characteristics. This is particularly true in biological NMR since samples are lossy conductors, and coils are relatively large. The interaction between the coil and sample has two forms, inductive and capacitive (3). The inductive interaction consists of an inductive coupling between the probe coil and the multiple virtual lossy current loops that exist in a conducting sample. This interaction is mediated by the B field of the coil, and thus tends to be most pronounced in the locations in the sample that are subject top the highest B fields (usually close to the coil wires). The capacitive coupling between the probe and the sample occurs because electric fields that are generated by the coil cause current to flow in the sample. Since the electric fields are not essential to the measurement of NMR, it is generally desirable to eliminate as much of the stray electric fields as possible. This can be achieved in part by distributing the tuning capacitance, using a balanced drive, and by using Faraday shields (4, 5, 6, 7). The problem of coil loading is beyond the scope of this chapter, and thus I will not discuss it further. It is important to the assesment of multiple tuned probes to understand that in the in-vivo setting, the effect of sample loading is significant, and often is the dominant term in the loaded coil resistance.

Now that we have discussed the NMR signal and the noise, we are in a position to write down an expression for the signal to noise on the NMR probe. Dividing Eq. (8) by Eq. (9) yields:

$$S/N \propto \omega \int \beta_1(\mathbf{x}')dV'_{\text{sample}}/(4kT\Delta f r)^{1/2}. \tag{10}$$

The tuning network of a multiple tuned probe usually contains at least one inductor. Under these circumstances the β_1 (and thus the magnitude of the NMR signal) is unchanged relative to the same coil single tuned, but the resistance is altered due to the addition of inductors with finite loss. The signal to noise ratio efficiency of these probes can be defined as the S/N obtained with the multiple tuned coil relative to that obtained with the same coil single tuned. Since the noise is proportional to $(r)^{1/2}$, the efficiency, E, is equal to the square root of the ratio of the resistance of the single tuned coil (usually emanating from the sample inductor, r_s), to that of the same coil double tuned (includes the sample inductor, r_s, and the equivalent inductance of the tuning circuit,

$r_{t,\mathrm{equiv}}$):

$$E = (r_s/(r_s + r_{t,\mathrm{equiv}}))^{1/2}. \tag{11}$$

Interesting insight may be gained if we manipulate Eq. (10) further. The numerator and denominator may be multiplied by $(L)^{1/2}$. The relationships:

$$L = 2\int \beta_1(\mathbf{x}')\cdot\beta_1(\mathbf{x}')dV_{\text{all space}} \tag{12a}$$

$$L = Qr/\omega \tag{12b}$$

allow an expression for the S/N to be written:

$$S/N \propto (Q\omega/kT\Delta f)^{1/2}\{\int\beta_1(\mathbf{x}')dV'/(\int\beta_1(\mathbf{x}')\cdot\beta_1(\mathbf{x}')dV_{\text{all space}})^{1/2}\}. \tag{13}$$

From the standpoint of multiple tuned probes, there are two terms of this expression that are of interest. First is the dependence of the S/N on the Q of the probe. This is a measure of the net efficiency of the probe for producing a detecting β_1 field, relative to the resistive losses. The second is the bracketed expression which I will define as the effective filling factor, F. This term takes into account how efficiently the probe is used, i.e. how much of the useful β_1 of the probe is filled with sample. Thus to restate, the Q is a measure of the total β_1 produced relative to the resistance (noise) in the probe, and F determines how much of this β_1 is in the sample of interest. Note that if we were to place inductors in our probe that were far removed from the sample (as done in nearly all multiple tuned designs), the effect on the signal to noise of the probe would be manifested by a decrease in F. Thus despite an excellent Q, if the β_1 is not located in the sample, the S/N will be poor. This concept of the filling factor becomes useful when all the inductors in the multiple tuned coil are loaded with sample. In this situation Eq. (11) is not an accurate expression for the efficiency.

Thus far we have written an expression for the S/N of the NMR probe, and have not yet mentioned anything about tuning, even though it is part of the title of this chapter. We will define tuning as the elimination of the imaginary component of the impedance. The S/N of the probe is not directly related to tuning the probe. The purpose of tuning the NMR receiver coil is actually to optimize the noise figure of the preamplifier. Lets investigate this further. A generic circuit for a tuned probe is shown in Fig. 2. The voltage that would be measured across the probe is:

$$V = (\mathrm{emf})(L_s\omega)/(r_s + iL_s\omega + Z), \tag{14}$$

with L_s and r_s being the inductance and resistance of the sample coil respectively, and Z being the impedance of the tuning circuit. If $Z = -iL_s\omega$ (the condition of tuning), the voltage across the probe will be at a maximum. Maximizing V gives the preamplifier the best chance of amplifying the signal without further degradation by noise in the preamplifier. If we were able to build a preamplifier that made no noise at all, there would be no need to ever tune a probe.

In the case of a single tuned coil, the typical tuning element is a capacitor. It is useful to express the condition of tuning graphically. Figure 3 is plot of

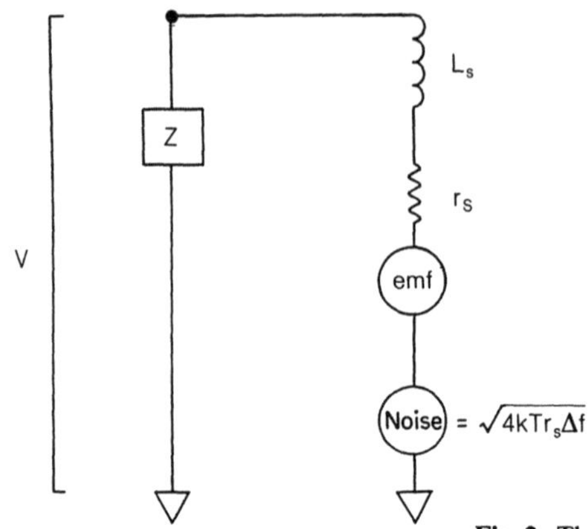

$$V = \frac{(\mathrm{emf})(L_s\omega)}{r_s + iL_s\omega + Z}$$

Fig. 2. The NMR receiver probe is represented by an inductor in series with the intrinsic resistance of the wires that constitute the inductor. The voltage measured across the probe is a function of the loading impedance, Z

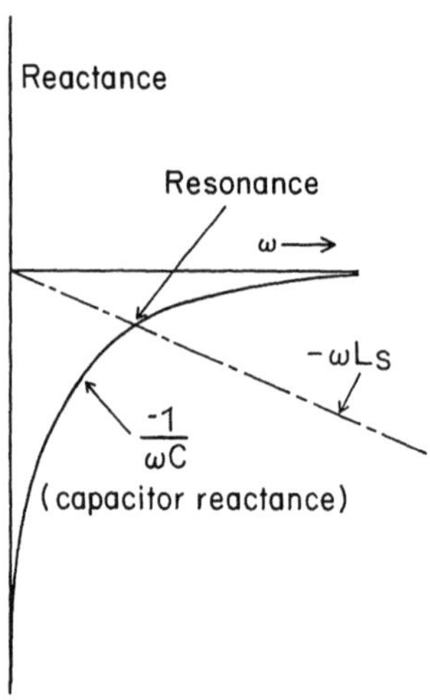

Fig. 3. The reactance of the tuning network (capacitor in this case) plotted along with the negative of the reactance of the probe coil ($-L_s w$). The intersection represents resonance

the negative of the coil reactance as a function of frequency. If the reactance of the tuning capacitor were plotted on the same graph, the point of intersection would represent the frequency at which $\omega L_s = -1/\omega c$, or resonance. Although this exercise seems relatively trivial in the single tuned case, it will be a valuable technique for studying multiple tuned coils.

Although it appears that tuning the probe to resonance should provide the largest signal and noise voltages to the preamplifier, and therefore optimize its noise figure, this is not usually the case. There are several issues to be considered.

First, the noise that the preamplifier itself generates is a function of the driving impedance. Many preamplifiers will not perform well at very high driving impedance. A second issue is that of a transmission line. In order to get the signal from the probe to the preamplifier a nonreactive transmission line is desirable. This condition is very hard to accomplish at very high impedances. Thus the impedance of the tuned probe is usually transformed to a lower impedance, typically 50 ohms, and run in a 50 ohm transmission line to the preamplifier. Depending on the characteristics of the preamplifier, the impedance may be adjusted again to optimize the noise figure of the preamplifier. As an aside, it is interesting to note that it is actually possible to build a field emission transistor preamplifier, that performs well at high impedance levels, directly on to the probe obviating the need for any impedance transformation.

The impedance match to 50 ohms can be accomplished in several ways. The parallel resonant probe may driven through its mutual inductance with a second coil in order to obtain a transformer match (8, 9, 10). If the coupling constant between the two coils is K, the impedance across the primary of the transformer is:

$$Z = K^{1/2} Q_p L_p \omega \tag{15}$$

where L_p and Q_p are the L and Q of the probe coil. Although there is some finite reactance and resistance in the primary coil, if the impedance transformation

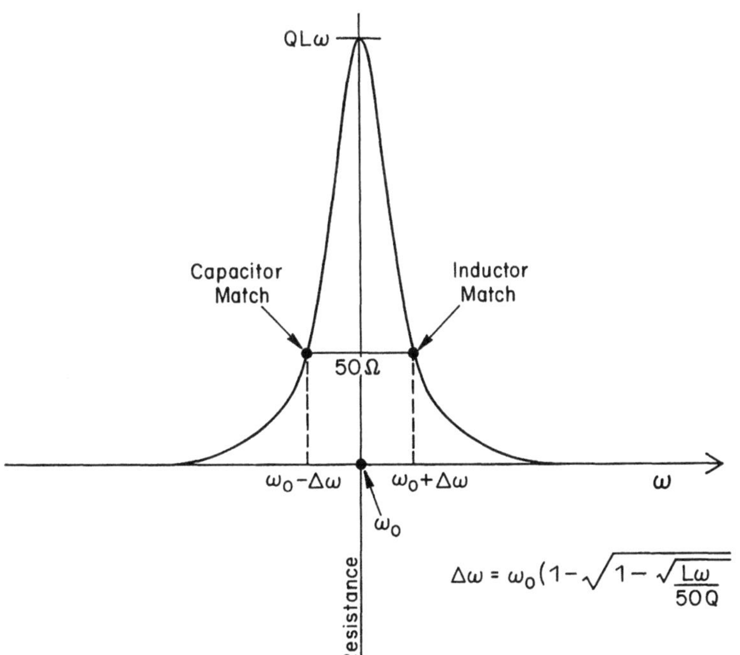

Fig. 4. The real part of the impedance of a probe tank circuit plotted as a function of frequency (ω_0 = resonant frequency). The probe can be matched by using it at 1 of the 50 ohm points

is large (as it usually is) it will only have a small effect on the net impedance and may be neglected. If desired, the primary coil may be series tuned in order to eliminate the reactance component of its impedance. In general, tuning the primary coil is only desirable when the geometry of the probe prevents tight enough coupling to achieve 50 ohms of impedance. Using the transformer matching technique, the impedance transformation is, to first order, independent of frequency, making it ideal to the application of single input multiple tuned coils. It also provides a balanced drive to the probe, reducing stray **E** fields and associated losses (5).

Alternately, the tuned probe can be used at a frequency other than its resonant frequency. Looking at a plot of the impedance of the tuned probe as a function of frequency (Fig. 4), it is always possible to find a frequency at which the real part of the impedance is 50 ohms. Unfortunately, by using the probe at this frequency a significant amount of reactance is introduced. This reactance may be cancelled by adding a reactive element of the appropriate value to the probe circuit. A matching inductor or capacitor may be used depending on which side of the resonance the probe is used. The required reactance to obtain a 50 ohm match is:

$$X_{\mathrm{M}} = (50 L_{\mathrm{p}} Q_{\mathrm{p}} \omega)^{1/2}. \tag{16}$$

This reactance may be split in order to drive the probe in a balanced configuration. As stated earlier, this has the effect of reducing stray electric fields, and thus loading effects. It is important to note that the reactance necessary to obtain a 50 ohm match in this way is a function of frequency. Thus for single input multiple tuned probes, a transformer match is most desirable.

3 Multiple Pole Multiple Resonant Probes

In order to obtain a multiple resonant probe, the normal tuning capacitor can be replaced with a tuning network that consists of a capacitor in series with a parallel *LC* circuit (11) (Fig. 5). A parallel *LC* circuit is often called a tank circuit. The effect of this tuning network can be represented graphically (Fig. 5b). A plot of the negative of the probe inductor reactance and the reactance of the tuning network as a function of frequency shows two points of intersection. This corresponds to two resonant modes of the resultant structure.

The current distribution in each resonant mode can be studied in detail. The low frequency mode is lower than the resonant frequency of the tank circuit in the tuning network. Thus in this mode, the tank circuit is inductive and contains current parallel to that of the main circuit (Fig. 6a). The higher frequency mode is at a frequency above that of the tank circuit in the tuning network. Thus at this frequency the tank circuit is capacitive and contains a current out of phase with the current in the main circuit (Fig. 6b).

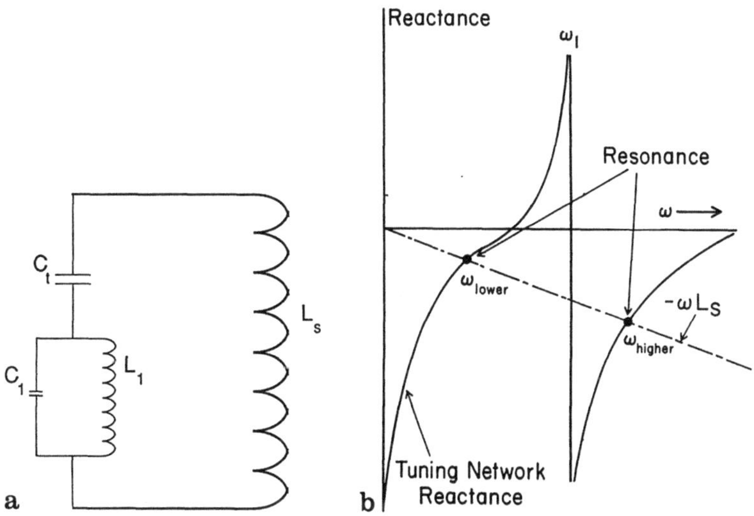

Fig. 5 a. Double tuned circuit using multiple pole design. **b.** Graphical analysis of the multiple pole tuning network. The 2 intersection points represent resonance. In reality there is a third resonance at $\omega = \omega_1$, but this corresponds to the tuning tank circuit resonance and is not useful for performing NMR with L_s

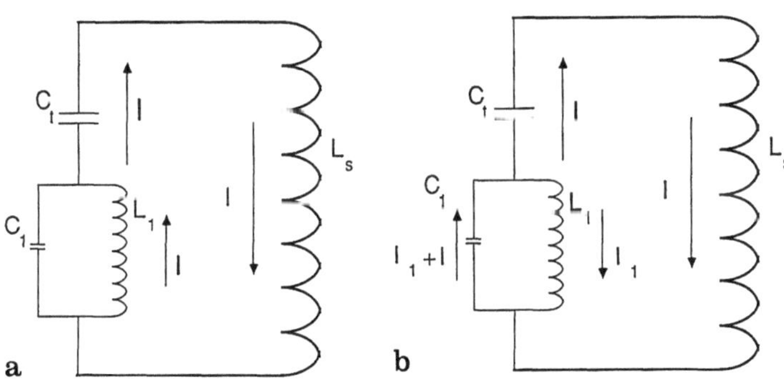

Fig. 6 a, b. Distribution of current in the low (**a**) and high (**b**) frequency modes of the multiple pole double tuned circuit

It is useful to quantitate the current distribution in the probe in order to assess its performance. The impedance of the tuning circuit may be written:

$$Z_t = r_1(1-\omega^2/\omega_1^{\,2})^2 + i\{L_1\omega/(1-\omega^2/\omega_1^{\,2}) - 1/\omega C_t\} + \text{terms in } r_1/Q_1 \quad (17)$$

where r_1 is the resistance of the tank inductor, L_1; C_t is the series tuning capacitor and ω_1 is the resonant frequency of the tank circuit. The resonant frequencies can be calculated by setting the reactance of the tuning circuit equal to the negative of the reactance of the probe inductor:

$$L_1\omega/(1-\omega^2/\omega_1{}^2)-1/\omega C_t=-L_s\omega \tag{18}$$

where L_s is the probe inductance. At the lower resonant frequency, if it is assumed that $\omega^2 \ll \omega_1{}^2$ (usually the case), then the resonant frequency is calculated to be:

$$\omega_l=1/(C_t(L_s+L_1))^{1/2}, \tag{19}$$

where ω_l is the lower resonant frequency. Note that the lower resonant frequency is dependent on the inductances of the circuit (which are usually fixed) and the capacitance of C_t, which may be adjusted to adjust the lower resonant frequency.

At the upper resonant frequency it is assumed that the reactance of the tank circuit dominates the reactance of the tuning network $(L_1\omega_h/(1-\omega_h{}^2/\omega_1{}^2)\gg 1/\omega c)$. Under these conditions:

$$L_1\omega_h/(1-\omega_h{}^2/\omega_1{}^2)=-L_s\omega_h \tag{20}$$

where ω_h is the higher resonant frequency. Solving for ω_h:

$$\omega_h=\omega_1(L_1/L_s+1)^{1/2}. \tag{21}$$

There is an important feature of the above results. Under the conditions described above, the lower resonant frequency depends on the value of C_t, while that of the higher resonant frequency is independent of this value and dependent on the value of C_1 through the ω_1 dependence. This means that the two resonances can be adjusted independently, greatly facilitating construction and tuning adjustment for loading.

At this point it is important to consider the sensitivity of the two probe resonances. The efficiency of the multiple tuned coil is defined as the sensitivity of the multiple tuned coil relative to the sensitivity of the same coil single tuned. If is assumed that the inductor of the tuning circuit is placed remote from the sample then the signal will be the same for the double and single tuned case since this quantity only depends on the $\mathbf{B}_1$ field per unit current in the sample coil. Assuming that lead and capacitor losses are negligible, in the single tuned case the resistive noise emanates only from the probe coil, L_s. Noise contributions from the probe coil and the resistance in the tuning circuit must be considered for the double tuned case. The efficiency of the probe is then the square root of the ratio of the probe coil to that of the sum of the resistance of the tuning network and the probe coil (see Fig. 7, Eq. (10)):

$$E=\{r_s/(r_s+r_{t,\mathrm{equiv}})\}^{1/2}$$

where $r_{t,\mathrm{equiv}}$ is the equivalent resistance of the tuning network. In the low frequency mode, the current in the L_1 is nearly equal to the current in L_s (if we assume $\omega_l \ll \omega_1$). Expressing r in terms of Q and L, the efficiency for the lower frequency mode is:

$$E_l=1/(1+L_1Q_s/L_sQ_1)^{1/2}. \tag{22}$$

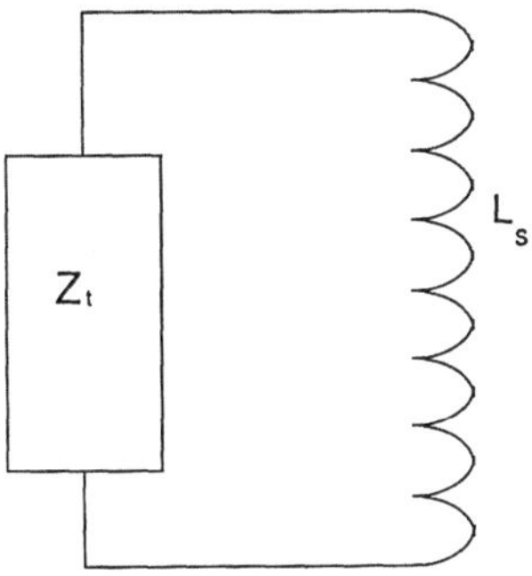

Fig. 7. The tuning circuit is represented by an equivalent impedance $Z_t = r_t + iX_t$

In the higher frequency mode, we must calculate the equivalent resistance of the tuning network. The tank circuit reactance is typically much higher than that of the tuning capacitor, C_t. The condition of tuning can therefore be expressed as:

$$L_1\omega/(1-\omega^2/\omega_1{}^2) = -L_s\omega. \tag{23}$$

This allows us to write:

$$1-\omega_h{}^2/\omega_1{}^2 = L_1/L_s \tag{24}$$

where ω_h is the higher resonant frequency. This can be substituted into the real part of the tuning circuit impedance in Eq. (17) to express $r_{eq} = r_1(L_s/L_1)$ at the higher frequency resonance. Substituting into Eq. (10) the efficiency of the high frequency mode is expressed:

$$E_h = 1/(1+L_sQ_s/L_1Q_1)^{1/2} \tag{25}$$

There are several important aspects of the efficiency equations ((22) and (25)). First is that the ratio of Q_s/Q_1 enters both expressions. If the Q of the tuning inductor is high compared to that of the sample coil, the efficiencies of both modes will improve. This is a quantitative statement of the fact that if the resistive losses in the sample coil are high, the contribution of resistive losses from the tuning circuit become less important. In in-vivo NMR, $Q_s/Q_1 < 1$ for two reasons. First, the sample inductor geometry is fixed by the required sensitivity profile, while that of the tuning inductor is not constrained, and can always be made to have an optimal Q. Second, the sample coil is loaded with tissue while the tuning coil is not. The loading of the sample coil often dominates the resistive loss in the sample coil, making the losses incurred in the high Q tuning circuit less significant. Thus to accurately assess the performance of the probe it must be loaded with sample.

The efficiencies of the double tuned probe also depend on the ratio of the tuning to sample inductances. Making the tuning inductance small compared to the sample inductor will improve the low frequency efficiency and degrade the high frequency efficiency. Increasing the tuning inductor will have the opposite effect. This effect can be demonstrated graphically. Decreasing the tuning inductance will move the resonant frequency of the tuning tank circuit closer to high frequency resonance. The closer the frequency of operation of the probe is to ω_1, the higher will be the circulating current in the tuning tank circuit, and thus the higher its contribution to noise. Thus, by moving the resonant frequency of the tuning tank circuit, the relative efficiencies of the two resonant modes my be adjusted. This design is typically used to tune a coil to ^{1}H and ^{31}P, ^{23}Na or ^{13}C. In this application it is most useful to use $L_t/L_s \ll 1$. There are two reasons for this. First, the sensitivity of the lower frequency nucleus is much poorer, and loss in sensitivity is less well tolerated. Second, the sample coil will load more at the higher frequency mode, making the losses in the tuning circuit less significant in this mode.

It is important to remember that in our theoretical treatment of this double tuned design, we have assumed $\omega_l \ll \omega_1$. This is usually the case as described above. If we were to use this design to tune closely spaced resonances, such as ^{19}F and ^{1}H, or ^{23}Na and ^{31}P, this treatment would no longer be valid. The efficiencies of the closely spaced resonances would be poor, because they would be closely spaced around ω_1, resulting in high circulating currents in the tuning tank circuit at both resonances.

In order to use this probe, its impedance must be matched to its load, typically 50 ohms. As previously discussed, matching may be accomplished by using a transformer drive, or by a reactive type match. The large advantage of a transformer drive is the match only depends on the coupling between the primary and secondary of the transformer, and therefore is largely independent of frequency (10). In reality, two effects will cause the transformer to have some frequency dependence. First is the inductance of the primary drive inductor. The second, is that the contribution of the sample coil (typically the secondary of the transformer drive) to the total inductance of the tuned probe is different for the two modes. This will result in the coupling constant ($K = M/(L_{\text{primary}} L_{\text{secondary}})^{1/2}$) being different for the two modes of operation. Note that M represents the mutual inductance between the two inductors. In my experience, it is usually possible to obtain a satisfactory match at both frequencies with a straight transformer drive (Fig. 8a). However, the match may be improved by placing a capacitor in the primary, and using it to adjust the relative impedance of the two modes.

It is possible, although less desirable to use a reactive match to drive this double tuned probe. As shown earlier, if the probe is driven as a parallel tank circuit, the required series matching reactance is dependent on the inductance and Q of the probe (these are different for the two modes) and the frequency. Since the required reactance will be very different for the two frequencies of operation, at least two reactive elements, 1 of which must be an inductor, must be used to achieve a perfect match (11). The use of an inductor in the matching network will add additional loss in signal to noise. For this reason, I have used this probe primarily in the transformer drive mode. If a reactive match is used, it has been shown that balancing the drive will improve performance (Fig. 8b, c) (12).

The actual construction of a surface coil tuned to ^{31}P and ^{1}H frequencies with the multiple pole design is very simple. The first step is to decide on the

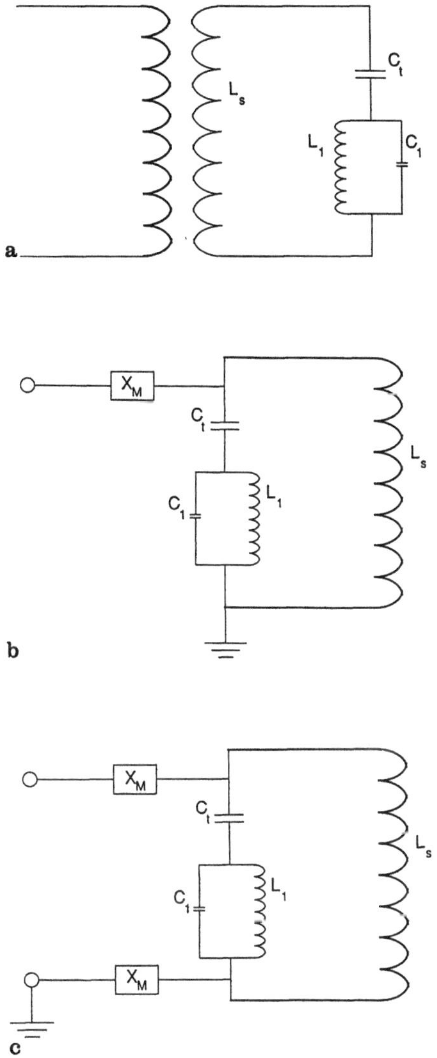

Fig. 8 a–c. Circuit for transformer match (**a**), reactive match (**b**), and balanced reactive match (**c**) drive configurations of the multiple pole double tuned circuit

required sample coil. Once this is determined, the appropriate tuning inductor is made, keeping in mind the efficiency requirements of the two modes. I find an L_s/L_t ratio of 5 to be useful for 10 cm diameter surface coils used for ^{1}H and ^{31}P spectroscopy. The higher frequency may be tuned by adding C_1 to the circuit and adjusting its value to the desired resonant frequency. C_t may then be added to circuit and adjusted so that the lower frequency resonance is at the desired frequency. I find it most useful to place the tuning circuit across the sample coil, and test the resonant frequency with a small loop (pick-up coil) inductively coupled to the probe coil. This not only provides information about the position of the resonant frequencies, but also about the current distribution of the probe. The coupling between the pick-up coil and the probe, will depend on the relative currents in the sample and tuning inductors (assuming that the pick-up is coupled to the sample inductor). This information helps one to be sure that the currents in each of the modes are as planned. Once tuned, the probe can be matched by fashioning a proper primary loop to drive the probe.

In addition to the particular configuration of the multipole tuning network that is described above, the dual of this network may also be used effectively to yield a double tuned probe (see Fig. 9a). In the interest of brevity I will not discuss this mode in detail, but it is useful to point out that in this mode, C_1 tunes the high frequency mode and C_t tunes the low frequency mode. This particular design was actually one of the first multiple tuned designs to be described (2). Hoult describes a version of this design, in which the two resonances are driven by separate inputs (Fig. 9b). It is, in general, possible to drive any of the multiple pole coils at two different points as long as the drive point for each resonance is placed so that the probe is driven across that part of the tuning circuit that provides most of the tuning reactance of the particular mode (see Fig. 9c).

An interesting variation on the multiple pole design has been proposed. It is possible to use one turn of a multiple turn surface coil as a tuning tank inductor by shunting across this loop with a capacitor (see Fig. 10) (13). Thus the tuning inductor is also loaded with sample and will detect an NMR signal. Although in the low frequency mode, the current in the tank and sample inductor are in phase, they are nearly 180 degrees out of phase at the high frequency mode. Thus the NMR signals add coherently in the low frequency mode yielding an efficiency of one. At the high frequency mode, signal from the sample loop will be canceled to some extent by that in the tuning inductor loop, degrading efficiency. Thus this design, allows one to further optimize the lower frequency at the expense of performance at the higher frequency. When treating this design quantitatively it is also necessary to take into account the mutual inductance between the two coils.

Before leaving this design, it is proper to address the problem of decoupling a double tuned probe from the transmit coil. It is often the case that the requirement of a homogeneous transmitter field and high sensitivity surface coil detection make the use of separate coils for transmitting and receiving necessary. The problem of blocking current from flowing in the surface coil during transmitting has been treated in detail for the single tuned case (14, 15). The solutions are almost identical for the double tuned case. Prammer et al. have proposed a circuit that allows simultaneous decoupling of a double tuned surface coil from the transmitter field (16) (Fig. 11). Briefly considering a solution to the problem of decoupling, it customary to use a PIN diode switch to activate an inductor which, in combination with the tuning capacitor forms a resonant high impedance tank circuit that blocks the flow of current during application of the transmitter field. This diode may be activated passively (by the emf

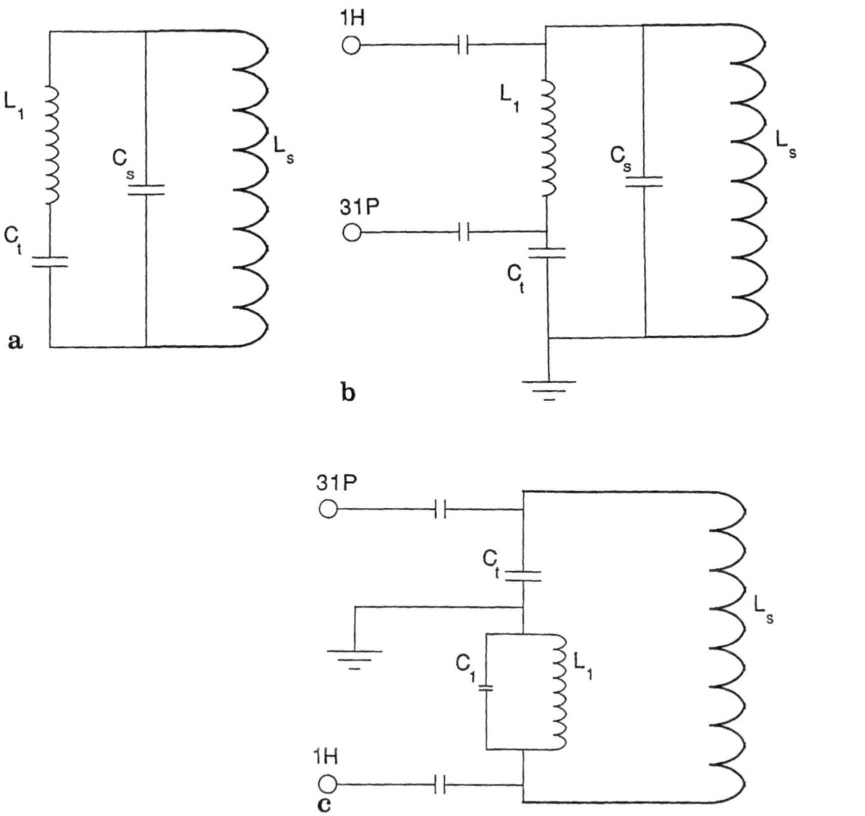

Fig. 9 a. Alternate configuration of the multiple pole double tuned circuit. **b.** Two input configuration of the circuit shhown in a. **c.** Two input configuration of the multiple pole network shown in Fig. 5

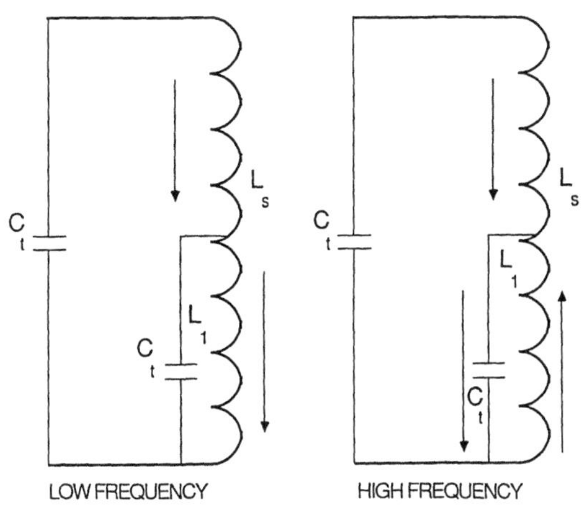

Fig. 10. Circuit using part of the coil inductor to form the tank circuit in the multiple tuning circuit. The current distribution in the two modes demonstrates out of phase in the sample inductor in the high frequency mode

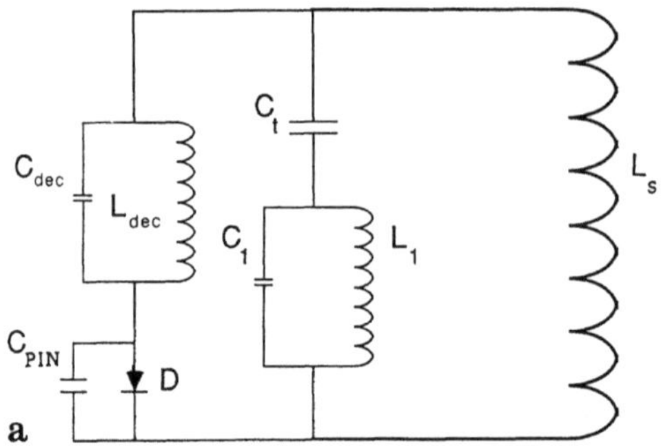

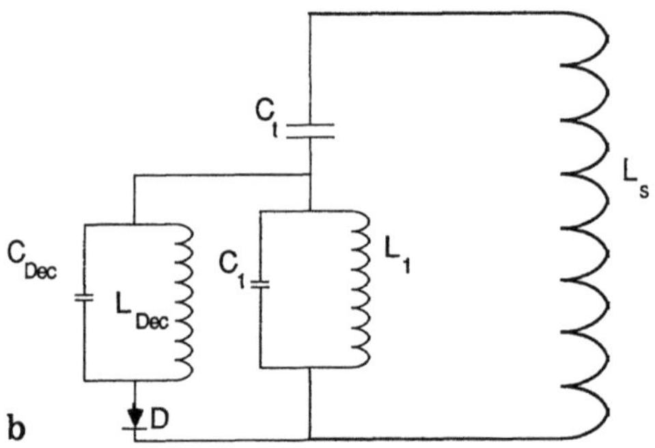

Fig. 11 a. Circuit for decoupling the double tuned circuit from a transmiter coil at both modes. (D = diode, C_{PIN} = capacitance of the PIN diode). **b.** Circuit for decoupling only the high frequency mode from a transmiter coil

induced by the transmitter field) or actively (by DC passed down the probe cable). If a single lumped element tuning capacitor is used, the resonant condition of the blocking network is satisfied if L_{dec} is equal to the inductance of the sample inductor (both will be tuned by the same tuning capacitance). This is also true for the double tuned case. If $L_{dec} = L_s$ then both the decoupling tank circuit and the sample coil will be tuned at both resonant frequencies by the same tuning circuit. Because of the finite, but small, capacitance of the PIN diode in the off position, Prammer (16) suggests placing a capacitor, $C_{dec} = C_{PIN}$, in parallel with the tuning inductor to compensate.

It is also possible to decouple the probe from the transmit field at the higher (^1H) frequency, and use the surface coil for transmitting and receiving at the lower (^{31}P) frequency. This is accomplished by placing a decoupling network across only the tank circuit of the tuning circuit (Fig. 11b). Since this part of the tuning circuit carries most of the tuning reactance at ^1H, it is clear that it will perform at this frequency identical to the single tuned case. Two issues can be considered at ^{31}P. First, no DC current is used to force the diode on, the voltage across the tuning tank circuit (and thus the diode) will be small and the diode will not conduct very well. This makes the influence of the decoupling network small. Second, the double tuned probe for ^1H and ^{31}P is usually built with $L_1 \ll L_s$. Thus even if the diode were to conduct, the addition of L_{dec} ($L_{dec} = L_s$) in parallel to L_1 will only minimally effect the overall probe resonance.

The lower resonant frequency may be decoupled independent of the higher resonant frequency by a similar argument.

4 Overcoupled Tank Circuits in Multiple Tuned NMR Probes

The creation of a double resonant structure by coupling two resonant coils together, is the basis behind several double tuned designs. If the two resonant

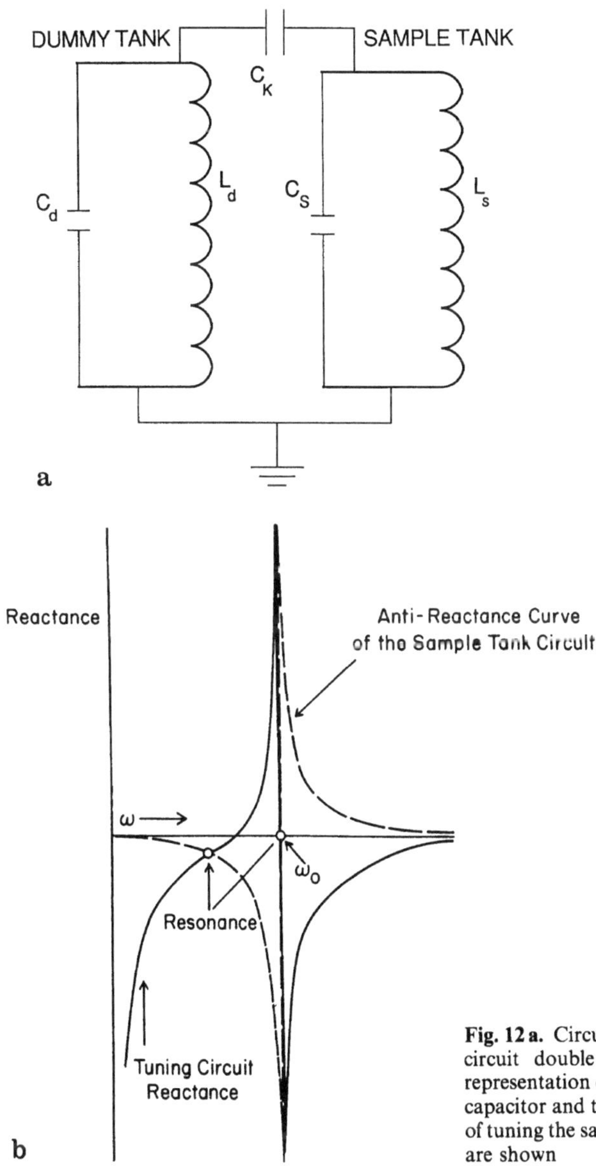

Fig. 12 a. Circuit for capacitively coupled tank circuit double tuned circuit. **b.** Graphical representation of the circuit in a. The coupling capacitor and the dummy tank are thought as of tuning the sample tank. The two resonances are shown

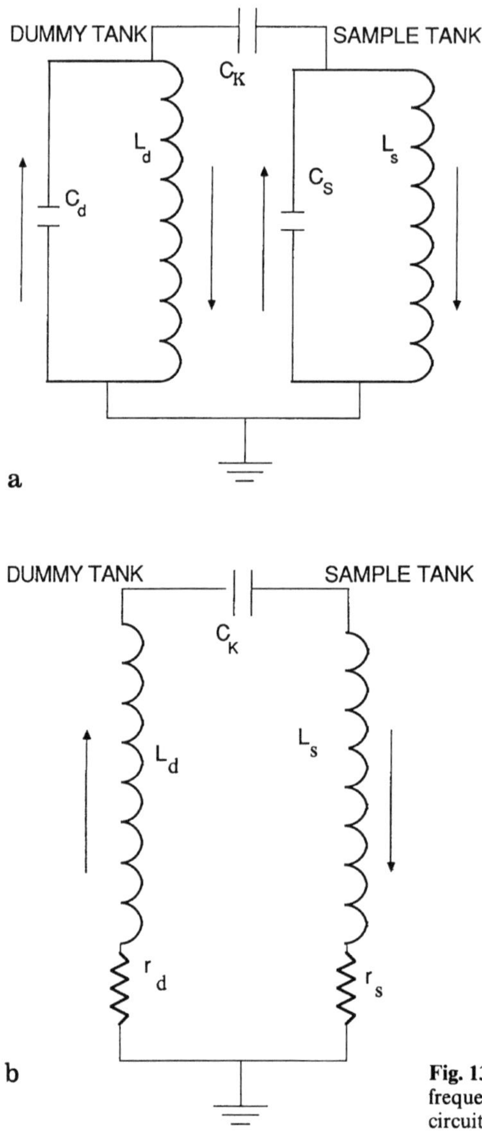

Fig. 13 a, b. Current distribution in the high frequency (**a**) and low frequency (**b**) modes of the circuit shown in 12

coils are tuned to the same frequency, coupling them together will result in two distinct resonant modes if $KQ > 1$ (called overcoupled) (17, 18). In general the coupling between the two resonators may either be capacitive, inductive or a combination of both. The case of capacitive coupling is perhaps the easiest to understand, therefore we will discuss it first.

A schematic of two capacitively coupled tank circuits is shown in Fig. 12 (19). In this design, a "dummy" tank circuit is used to double tune the sample coil by capacitively overcoupling it to the tank circuit which contains the probe coil. The coupling constant is defined as:

$$K = C_K/(C_s C_d)^{1/2}, \tag{26}$$

where C_K is the coupling capacitor, C_s is the capacitor on the tank circuit of the sample coil, and C_d is the capacitor on the dummy tank circuit (20). Before trying to understand the resultant resonent modes, it is useful to consider this design graphically. Considering the coupling capacitor and dummy tank circuit as the tuning network acting on the sample tank circuit, a graphical solution to the resonant modes is shown in Fig. 12b.

The high frequency mode of operation is at the resonant frequency of the overcoupled tank circuits. This mode corresponds to parallel current in each of the tank circuits (Fig. 13a). In this mode the impedance of the two resonant tank circuits is so high that of the coupling capacitor can be neglected. This is true is $KQ \gg 1$, which is the case in grossly overcoupled tank circuits. In this mode, the current will distribute itself between the two tank circuits according to their impedance. At the higher resonant frequency (again the resonant frequency of each tank circuit) the impedance of each tank circuit is:

$$Z_{d/s} = L_{d/s}\omega Q_{d/s}, \tag{27}$$

and is real (*d/s* refers to the dummy or sample tank). Again, the square root of the ratio of the net resistance of the sample coil to that of the double tuned coil is the efficiency of the probe. In this case, the efficiency is calculated to be:

$$E_h = (1/1 + L_s Q_s/L_d Q_d)^{1/2}. \tag{28}$$

The second resonant mode is at a lower frequency than that of the two coupled tank circuits. This mode corresponds to the resonance created by the coupling capacitor tuning the two equivalent inductances of the tank circuits operated below their resonant frequency (Fig. 13b). We can thus calculate an expression for the lower resonant frequency by setting the reactance of the coupling capacitor equal in magnitude to the sum of the reactance of the two tank circuits,

$$L_s\omega_l/(1 - \omega_l^2/\omega_0^2) + L_d\omega_L/(1 - \omega_L^2/\omega_0^2) = 1/C_K\omega_L, \tag{29}$$

where ω_0 is the resonant frequency of the two coupled tank circuits. Rearranging Eq. (29) yields:

$$\omega_l = \omega_0/(1 + C_K(L_d + L_s)\omega_0^2)^{1/2}. \tag{30}$$

The resonant frequency of the lower resonant mode is dependent on the value of the coupling capacitor. Since the frequency of the higher resonant mode is independent of this quantity, the probe resonant frequencies can be independently adjusted.

It is straightforward to calculate the efficiency of the low frequency mode. In this mode the current in each of the equivalent resistors is equal, thus the efficiency is just the square root of the equivalent resistance of sample inductor

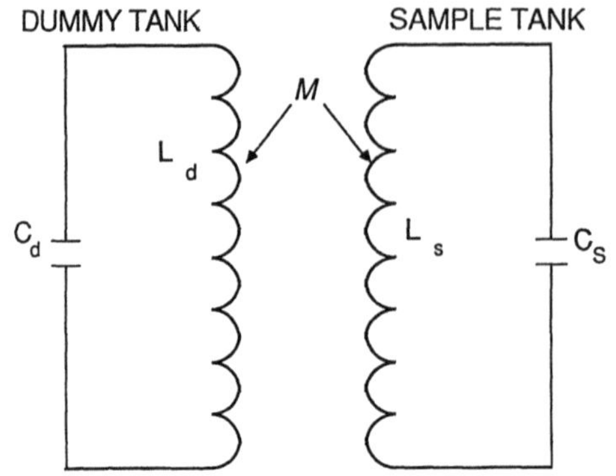

Fig. 14. Inductively overcoupled tank circuit double tuned probe

to the sum of the equivalent resistances of the sample and dummy inductors:

$$E_1 = (1/(1 + L_d Q_s / Q_d L_s))^{1/2}. \tag{31}$$

The efficiency expressions for this probe are identical to the multiple pole case. Thus again the efficiency is improved by the higher relative Q of the dummy coil due to the loading effects on the sample coil. The relative efficiency of the two modes can again be adjusted by the proper choice of L_d relative to L_s.

In order to construct this probe, the first step is to choose the required sample coil. The inductance of the dummy inductor may be chosen according to the efficiency requirements of the two resonances. Each tank circuit is tuned to the frequency of the high frequency mode. The coils are than coupled together. As the coupling capacitor is increased, the lower frequency mode will move down in frequency, but the upper frequency mode will not move. The lower frequency mode is adjusted to the proper frequency by adjustment of the coupling capacitor. We use a transformer drive to match the coil, however, it is possible to use a reactive match as described above.

It is possible to use inductive coupling between the resonant sample and dummy coils to double create a double tuned NMR probe. A diagram of such a probe is shown in Fig. 14. Again if the two coils are overcoupled,

$$KQ = (M/L_s L_d)Q > 1, \tag{32}$$

there will be two distinct resonances. Note that K represents the coupling constant, $M/L_s L_d$, in the case of inductive coupling. The resonant modes correspond to parallel, and antiparallel current in the two resonant coils. If the two coils are both tuned to the same resonant frequency, ω_0, then the frequencies of the two resonant modes will be:

$$\omega_h = \omega_0(1 + K), \quad \omega_l = \omega_0(1 - K). \tag{33}$$

It is more difficult to calculate the efficiency of this type of double tuned probe. The calculation of the noise made in the coupled probe is complicated by the noise correlation between the two coils introduced by the inductive coupling. The fraction of noise that is correlated between the two loops is equal to the coupling constant between the loops. Thus if the two coils are remote from each other (as they are in the capacitively coupled case), there will be no significant noise correlation between the two coupled loops. If there is significant coupling between the two resonant circuits, there will also be significant noise correlation. The correlated and uncorrelated noise components will sum differently when the total noise made in the probe is calculated. In the case of the correlate noise, the noise voltages add, just as the signal voltages add. In the case of uncorrelated noise, the noise powers add. The implication of this is that the sum from uncorrelated noise components is relatively less than that of noise components of similar values that are correlated. For the probe described above (two coils tuned to the identical resonant frequency inductively overcoupled with a coupling constant of K) the total noise in each mode is:

$$N = N_{\mathrm{corr}} + N_{\mathrm{uncorr}} = K(r_{\mathrm{s}}^{1/2} + r_{\mathrm{d}}^{1/2}) + (1 - K)(r_{\mathrm{s}} + r_{\mathrm{d}})^{1/2}. \tag{34}$$

The calculation of the signal is also more complicated than in the case of the probes discussed earlier. In our previous discussions, only the sample coil interacted with the NMR sample, and thus in the different modes of coil operation, although noise sources were different, the same detection field was used, resulting in the identical signal for all modes according to Eq. (9). In the inductively overcoupled case, the requirement of the two coils being tightly coupled, necessarily leads to the $\mathbf{B}_1$ field of both coils being loaded with sample. Thus in order to calculate the signal from each mode relative to the single tuned case it is necessary to calculate the integral of $\mathbf{B}_1 \cdot \mathbf{M}$ over the volume of the sample. The $\mathbf{B}_1$ field will be different for the different modes, resulting in a different signal intensity. In particular, in the high frequency mode the $\mathbf{B}_1$ from the dummy coil will partially cancel that of the sample coil (antiparallel currents run in this mode), resulting in a lower signal. The $\mathbf{B}_1$ fields will be predominantly in the same direction in the lower frequency mode resulting in a relatively higher NMR signal. In addition to a different integrated signal intensity for the two modes, the spatial distribution of the acquired signal is also different. Since it is often helpful to compare the data acquired from the different nuclei, it is useful to have the same spatial distribution of the signal from the two modes. Thus the different net $\mathbf{B}_1$ distributions of the two modes is a major drawback of this type of double tuned probe. It is interesting to note that if we were consider the efficiency of the double tuned probe on a small sample volume, the efficiency would depend on where in space that volume is chosen. Thus it is often useful to consider the efficiency of probe at a point in space instead of over all space. This makes it possible to calculate the efficiencies in closed form. Some typical calculated efficiencies of the probe described above are shown in Table 1.

When using inductively overcoupled coils to create a double tuned probe, it is not necessary for both coils to be at the same resonant frequency. Fitzsimmons et al. have described the use of two tightly inductively coupled coils that are at different resonant frequencies to obtain a double resonant probe

Table 1. Efficiency of inductively coupled double tuned probe (E_h/E_L)*

	$R=0$	$R=a$
$D=0.2a$	0.4/1.2	0.2/1.1
$D=0.5a$	0.2/1.1	0.36/1.0
$D=a$	0.45/0.95	0.53/0.89
$D=1.5a$	0.59/83	0.58/81

* These efficiency calculations assume no loading. The efficiencies for the low frequency mode are greater than 1 because the decrease in noise correlation dominates over the loss of B_1 strength, making the two turned parallel current (lower frequency mode) coil better than the single turned single tuned coil used for comparison. R, a, and D are defined in the diagram.

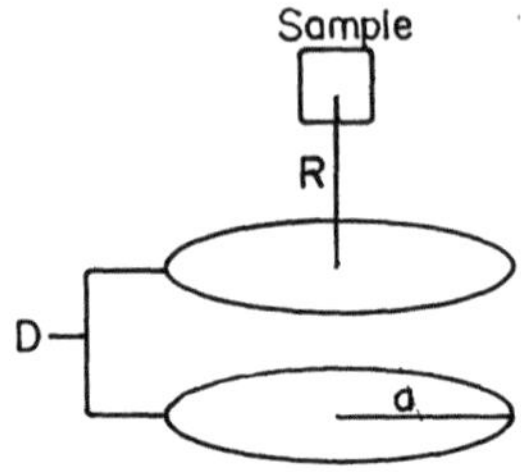

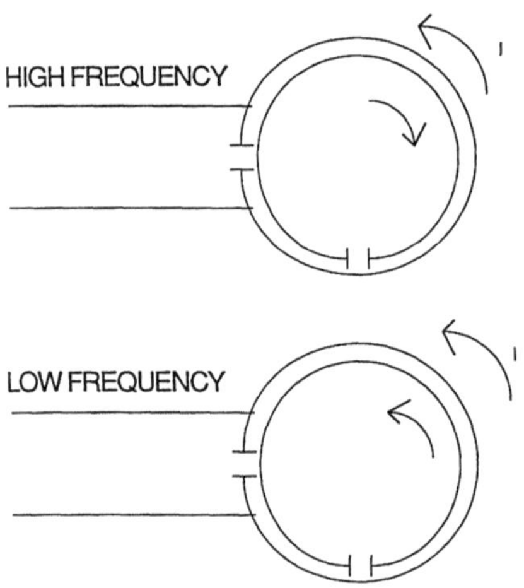

Fig. 15. Cartoon of the resonance inductively coupled double tuned probe suggested by Fitzimons et al. The current distributions in the low and high frequency modes are shown

(21) (Fig. 15). In this case the two resonances still correspond to parallel and antiparallel current in two coupled coils, but since the coils are at different resonant frequencies, the current in two coils will be different. If the coils are placed in near degenerate positions, the net B_1 field distribution will be nearly the same at the two resonant frequencies. As expected, the efficiency of this probe is nearly 1 at the lower frequency mode since the currents in the two coils are parallel in this mode. The measured efficiency of this probe at higher frequency mode was significantly lower (14%). Again, this was due to the antiparallel current in the two coils.

It is also possible to use a combination of capacitive coupling and inductive coupling to between two coils to make a double resonance probe. Grist et al. have describes a modification of the CRC loop gap resonator that uses this technique to obtain a probe tuned to ^{31}P and ^{1}H (22). As expected, the performance of the lower resonant mode is better than that of the higher resonant mode, with the sensitivity distributions of the two modes being different.

The actual construction of the inductively coupled double tuned probes is more difficult than the capacitively coupled type. This is because the tuning of the individual modes is interactive. The final tuning adjustment of the probe involves iterative adjustment of the individual resonant frequencies and the coupling of the coils. While doing this adjustment it is important to keep track of the resonant frequencies of the individual tank circuits in order to ensure the desired relative efficiencies. If the resonant frequencies need adjustment when the probe is loaded with different samples, this exercise needs to be repeated with each application of the probe. Because of the cumbersome tuning adjustment, and the different sensitivity distribution of the two modes, I prefer to use the capacitively coupled design, although the inductively coupled design can be used with good results.

5 Three and Four Resonance Probes

There are many application in in vivo NMR where probes tuned to more than two frequencies are required. This is best achieved by using combinations of the tuning circuits described above. It is possible to build three and four frequency probes that perform adequately at all frequencies. Of course, as more

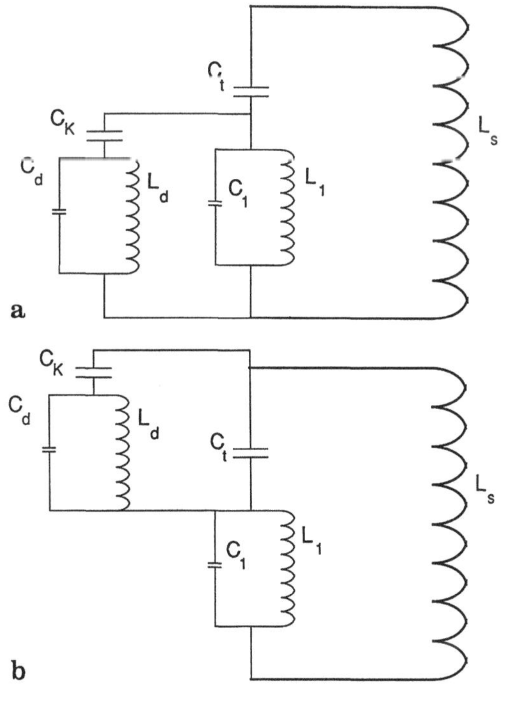

Fig. 16 a, b. Triple tuned NMR probes by splitting the high (**a**) and low (**b**) frequency resonances of the multiple pole double tuned probe backbone

resonances are added to the probe, the tuning circuits become more complex, and care must be taken to distribute the relative efficiencies between the resonances as the planed experiment demands.

In my experience, most multiple tuned coils applied to in vivo NMR spectroscopy require a ^{1}H and ^{31}P resonance. This is most conveniently made by using the two pole design described above. In order to create additional resonant modes, each of the individual resonances can be split by using it as part of an overcoupled tank circuit (19). This is easily accomplished by using a capacitive coupling across that part of the tuning network that provides the dominant reactance at the resonance that is to be split. This would be the tank circuit for the ^{1}H resonance and the capacitor, C_t, for the lower resonant frequency. Figure 16 contains diagrams of three frequency probes that use a dummy tank circuit capacitively coupled to the high frequency mode (a) and the low frequency mode (b). The circuit shown in (a) is useful for ^{19}F, ^{1}H, ^{31}P NMR spectroscopy, while that shown in (b) is useful for ^{1}H, ^{31}P and ^{13}C, ^{23}Na or ^{7}Li. The efficiency and frequency of the non split mode of the ^{1}H/^{31}P will not be significantly perturbed by the addition of the coupling to the other resonance. It is important to consider the required efficiency of all of the resonances prior to construction of the initial ^{1}H/^{31}P probe. This is because the efficiency of the split resonance is dependent on the efficiency of the resonance from which it is split and the efficiency of the splitting. For example, considering a probe tuned to ^{1}H/^{19}f/^{31}P, the efficiencies for the three resonances of this design can be calculated:

$$E_{^{31}\mathrm{P}} = 1/(1 + Q_s L_1/Q_1 L_s)^{1/2} \tag{35A}$$

$$E_{^{1}\mathrm{H}} = 1/(1 + Q_s L_s/Q_1 L_1 + Q_s Q_1 L_s/Q_1 Q_d L_d)^{1/2} \tag{35B}$$

$$E_{^{19}\mathrm{F}} = 1/(1 + Q_s L_s/Q_1 L_1 + Q_s Q_1 L_s L_d/Q_1 Q_d L_1^2)^{1/2}. \tag{35C}$$

As stated above, the efficiency of the unsplit resonance, ^{31}P in this case, is not significantly altered. The denominator of the efficiency expressions for the ^{19}F and ^{1}H resonance each have a term from noise made in the multiple pole tank, and an additional term from the noise made in the dummy tank circuit. Since the efficiency of the two split resonances each contain the noise term from the higher resonance of the multiple pole design, this sets the upper limit of the efficiencies of the two higher resonant frequencies of the triple tuned coil. It is essential to keep this in mind when building the ^{31}P/^{1}H backbone to a triple tuned coil.

A four resonant probe can be constructed by combining the two approaches described above (23). A circuit for such a probe is shown in Fig. 17. The efficiencies of this probe can be calculated in a similar fashion to those of the triple nuclear coil shown above. Since splitting the lower resonant frequency does not influence the efficiency of the high resonance, and thus the efficiencies of the ^{1}H and ^{19}F resonances are identical to the expressions in Eqs.(35B) and (35C). The efficiencies for the ^{23}Na and ^{31}P resonance of the four frequency

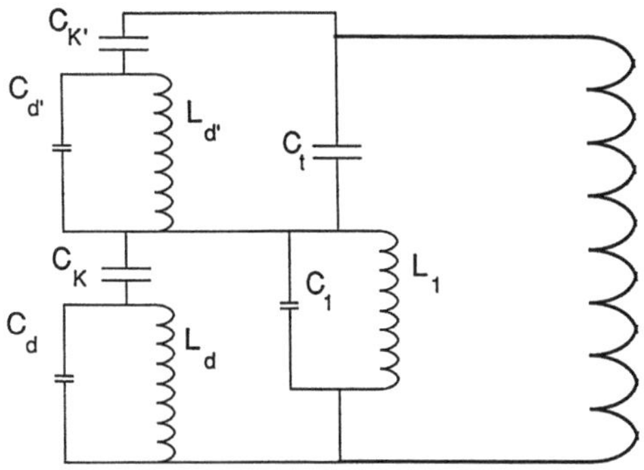

Fig. 17. Quadruple tuned probe by splitting low and high frequency resonances of the multiple pole double tuned probe

probe are:

$$E_{^{31}P} = 1/(1 + Q_s L_1/Q_1 L_s + Q_s Q_1 L_1^2/Q_1 Q_{d'} L_s L_{d'})^{1/2} \tag{36A}$$

$$E_{^{23}Na} = 1/(1 + Q_s L_1/Q_1 L_s + Q_s Q_1 L_1 L_{d'}/Q_1 Q_{d'} L_s L_1)^{1/2}. \tag{36B}$$

We have successfully used this four nuclear probe to perform physiological experiments on cat brain metabolism with good results (23). The construction of such a probe involves first building a $^{31}P/^{1}H$ backbone, being sure to consider all of the efficiency considerations discussed above. The dummy tank circuits are tuned remote from the probe, and then added to the probe circuit with the accompanying coupling capacitor. The coupling capacitor is adjusted so that the lower resonant mode of the split resonance is at the desired frequency. Minimal iterative adjustment of the higher frequency resonance of a split mode may be required if the dummy tank is not exactly on resonance (this is usually the case, since placing the dummy tank onto the circuit board slightly alters its resonant frequency).

It is obvious from the efficiency equations above, that as more resonances are added to the probe, the efficiency must deteriorate. As stated many times in this chapter, the relatively low Q of the sample inductor due to loading effects always serves to improve the efficiency of these probes. It is this fact, which makes these probe viable at all. It is possible to further improve the relative Q of the tuning circuit inductors and lower the noise they make by lowering the temperature of the tuning circuit (24). Reducing the temperature of the tuning circuit inductors has two effects (25). First, the conductivity of the inductors is improved, increasing the Q of the coil. Second, the noise is reduced from the decrease in the temperature (remember noise $\propto T^{1/2}$). Cooling the probe coil to liquid nitrogen temperature has been described, but in the in vivo setting, this will not have a large effect because losses in the sample dominate. Since the tuning inductors are not loaded with sample, the noise that is made in this part of the circuit can be further reduced by cooling them down to liquid nitrogen temperature. If this were done, the Q_s/Q_t term in the denominator of all of the efficiency expressions would be replaced by $Q_s/3.8Q_t^c$, where the 3.8 accounts for the difference in noise made at the different temperature, and Q_t^c is the Q

of the cooled tuning inductor. Q_t^c will in general be larger than Q_t by an amount that depends on the composition of the particular coil. Thus if the losses in the tuning inductors are debilitating, this technique could significantly improve the efficiency of a multiple tuned coil.

6 Double Tuned Volume Coils

Thus far multiple tuned coils have been discussed in the context of small surface coils. There are many applications that require volume coils that provide a homogeneous $\mathbf{B_1}$ field. There are two approaches that have been used to obtain a double tuned volume coil. The first uses a crossed coil design, with each of the coils tuned to a different frequency (26,27). Many such designs have been proposed. If the isolation between the two coils is sufficient, each resonator will behave as if the other were not present. Thus the efficiencies of these probes are high. These devices are also relatively easy to build and adjust. However, the use of this probe does limit the potential geometry of the resonator, and make quadrature operation difficult.

A second approach to multiple tuned volume coils is to apply a multiple tuning circuit to an already existing structure. This can be applied to any resonant structure that contains discrete tuning capacitors that account for a significant amount of the tuning reactance. Two such designs have been described in the literature. Bolinger et al. have proposed a coil consisting of multiple longitudinal struts along the surface of a cylinder, with a sinusoidal density of the struts along the circumference (28) (Fig. 18). This

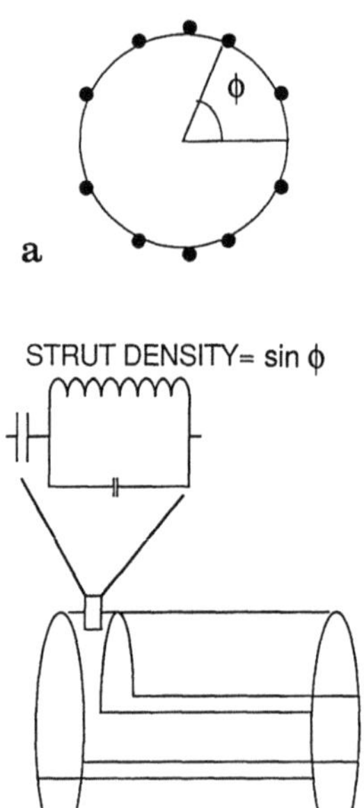

Fig. 18 a, b. Multiple tuned coil suggested by Bolinger et al. The distribution of the struts is according to sin ϕ (**a**). The coil is tuned with a multiple pole network (**b**)

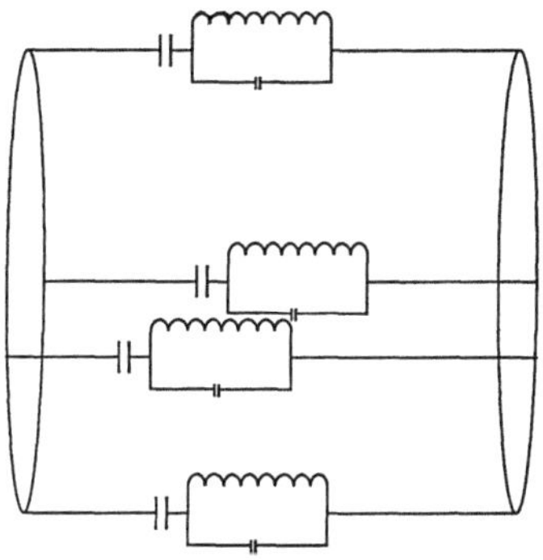

Fig. 19. The capacitors in a low pass birdcage resonator may be replaced with a multiple pole tuning network in order to create a double tuned resonator

current distribution provides a perfectly homogeneous $\mathbf{B}_1$ field inside the cylinder (29). Neglecting mutual inductance between the struts, this design provides a homogeneous $\mathbf{B}_1$ field everywhere inside the cylinder. The coil is tuned to resonance by a single discrete capacitor across the input. It is also possible to distribute this capacitance across both sides of the cylinder to reduce stray **E** fields. In order to multiple tune this structure, the tuning capacitor or capacitors are replaced by the appropriate double tuning circuit. In order to tune this probe to ^{1}H and ^{31}P, the multiple pole design described above work very well. The expressions for the resonant frequencies and the efficiencies hold.

A similar tactic has been used in our laboratory to create a double tuned birdcage coil. A birdcage resonator can be thought of as a full wavelength lumped element transmission line placed on the outer surface of a cylinder, providing a sinusoidal current distribution (30). In the low pass design, there is a single capacitor on each of the struts. This capacitor (in combination with the strut inductance) provides the required capacitance to tune the lumped element transmission line to a full wavelength. It is possible to replace each of these capacitors with a multiple pole type multiple tuning circuit (31, 32). The resultant structure is then resonant at two frequencies, retaining a sinusoidal current distribution (Fig. 19). Because of the intrinsic symmetry of the birdcage coil, this structure can also be driven in quadrature. Since loading losses often dominate in these probes, the addition of the double tuned circuitry does not have a larger effect on the probe performance. However, it is quite cumbersome to build this coil since it requires each of the tuning circuits to be identical in order that the two quadrature modes are tuned to the same frequencies and remain isolated.

Recently, Fitzimons has shown that a quadrature double tuned birdcage can be built by off resonance coupling of two birdcage coils (33). The two birdcage resonators are placed in a near degenerate position, similar to the surface coil design described earlier. As expected, again the low frequency resonance has an efficiency near 1, but that of the lower frequency is less than 0.5 due to out of phase currents in this mode. Another double tuned birdcage design involves tuning each of the orthogonal modes of the resonator to different frequencies (34). In theory, this is identical to using a crossed coil configuration. If the frequencies are relatively close (i.e. ^{19}F and ^{1}H) high B_1 homogeneity of the birdcage resonator is retained. This design provides an efficiency of 1 for each mode, but does not allow quadrature operation.

7 Conclusions

Many methods for obtaining multiple tuned RF coils for nearly every application have been presented. All of these designs have one unifying concept. They take advantage of the combination of two or more resonant circuits to create a multi-resonant structure. For double tuned coils, the resultant resonant modes correspond to parallel and anti-parallel current in the two resonant circuits. Although it had not been presented as such, the multiple pole design actually involves off resonance coupling of a parallel tank circuit and a series resonant sample coil. Since all of the double tuned circuits presented are based on the concept of coupled resonators, the performance of all of these circuits can be made identical with the proper choice of components. The only caveat to the previous statement is those probes in which both resonant circuits in the multi-resonant structure interact with the sample. This situation is more complicated, resulting in different sensitivity profiles for the different resonant modes, and efficiencies that are a function of location. The net efficiency values, however, are similar to the other coils. Thus it is best to build the probe that is the most convenient for a particular application, and the easiest to construct.

8 References

1. Hoult D, Richards RE (1976) J Magn Reson 24: 71
2. Hoult DI (1976) Progress in NMR Spectroscopy 12: 41
3. Bottomly PA, Andrew RA (1976) Phys Med Biol 23: 630
4. Murphy-Boesch J, Koretzky AP (1983) J Magn Reson 54: 26
5. Decorps M (1985) J Magn Reson 65: 100
6. Misc G (1987) Radiology 165: 344
7. Hoult DI, Chen CN (1988) In: Proc. seventh annual meeting Soc Magn Reson Med, 20–26 Aug. San Fransico
8. Froncisz W, Jesmanowicz A, Hyde JS (1956) J Magn Reson 66: 135
9. Kuhns PL, Lizak MJ, Lee S, Conradi MS (1988) J Magn Reson 78: 69
10. Schnall MD, Barlow C, Subramanian VH, Leigh JS (1986) J Magn Reson 68: 161
11. Schnall MD, Subramanian VH, Leigh JS, Chance B (1985) J Magn Reson 65: 122
12. Chew WM, Taveres N, Auffermann W, Higgans CB (1988) Mag Res Med 7: 321
13. Rajan SS, Wehrle JP, Glickson JD (1987) J Magn Reson 74: 147
14. Edelstein WA, Hardy CJ, Meuller OM (1986) J Magn Reson 67: 156
15. Bendall MR (1988) J Magn Reson 8: 380
16. Prammer MG, Vogele KE, Leigh JS (1989) In: Proc. eighth annual meeting Soc Magn Reson Med, 12–18 Aug. Amsterdam, The Netherlands

17. Purington ES (1930) Proc I.R.E. 18: 983
18. Aiken CB (1937) Proc I.R.E. 25: 230
19. Schnall MD, Subramanian VH, Leigh JS (1986) J Magn Reson 67: 129
20. Terman FE (1943) Radio Engineer's Handbook, 1st edn, McGraw Hill, New York p 151
21. Fitzimmons JR, Brooker HR, Beck B (1987) Magn Reson Med 5: 471
22. Grist TM, Kneeland JB, Jesmanowicz A, Francisz W, Hyde JS (1988) Mag Res Med 6: 253
23. Eleff SE, Schnall MD, Ligetti L, Osbakken M, Subramanian VH, Chance B, Leigh JS (1988) Mag Res Med 7: 412
24. Styles P, Soffe NF, Scott CA, Cragg DA, Row F, White DJ, White PCJ (1984) J Magn Reson 60: 397
25. Damico LA, Summers JJ, Schnall MD (1987) In: Proc. sixth annual meeting Soc Magn Reson Med, 17–21 Aug. New York
26. Boesch C, Martin E (1988) Radiology 168: 481
27. Cross TA, Mueller S, Aue WP (1985) J Magn Reson 62: 87
28. Bolinger L, Prammer M, Leigh JS (1989) J Magn Reson 81: 162
29. Hayes CE, Edelstein WA, Schenk JF, Mueller OM, Eash M (1985) J Magn Reson 63: 662
30. Mascart E, Joubert J (1886) Lecons sur l'electricite et le magnetisme 1: 357
31. Rath A (1989) In: Proc. eigths annual meeting Soc Magn Reson Med, 12–18 Aug. Amsterdam, The Netherlands
32. Isac G, Schnall MD, Lenkinski RL, Vogele K (1990) J Magn Reson 89: 41
33. Fitzimmons JR, Beck B (1990) In: Proc. ninth annual meeting Soc Magn Reson Med, 18–24 Aug. New York
34. Joseph PM, Lu D (19899) IEEE Trans Med Imaging 8: 286

RF Pulses

A short radio-frequency (RF) pulse can be considered as a multifrequency source allowing the excitation of a large range of resonance frequencies simultaneously. The frequency response of a rectangular pulse of duration T is given by $\sin(\pi fT)/\pi f$ yielding a bandwith of $-1/T < f < 1/T$, which should be considerably larger than the width of the spectral region of interest in order to guarantee homogeneous RF excitation of all resonances. This basic type of pulse is extensively used not only in high-resolution NMR but also in a large number of in vivo MRS studies.

Many in vivo applications, however, require more sophisticated RF pulses. RF probes with inhomogeneous RF field distribution (surface coils) and tissue heterogeneity do not allow homogeneous excitation throughout the sample with simple rectangular pulses. Other in vivo studies require frequency selective pulses, e.g. for the selection of a volume of interest and for the elimination of solvent signals. In this section, RF pulse design for in vivo MRS is discussed in more detail.

The problem of RF field heterogeneity prompted the development of B_1 insensitive pulses. They are based on the adiabatic principle, stating that as long as the condition $|B^e(d\alpha/dt)^{-1}| \ll 1$ is fulfilled (i.e. the rate of change in the direction of the effective magnetic field is small compared to the strength of this field), the magnetization will remain aligned along B^e. Since B^e is the vector sum of the static field B_0 and the RF field B_1, this implies that under adiabatic conditions the RF pulse becomes insensitive to B_1 variations. The design of such amplitude and phase modulated pulses is discussed by Garwood and Ugurbil.

Frequency selective pulses such as Gaussian or sinc pulses are used with great success for slice selection in MRI or for the localization step in MRS. Such pulses, however, are not ideal with respects to their excitation profile. A new generation of phase compensated frequency selective RF pulses is described by Morris.

Multi-frequency selective pulses offer the possibility of studying several volumes-of-interest simultaneously. The development of the such pulses is discussed by Müller.

Finally, an important issue in in vivo proton MRS is the elimination of the intense water signal. A number of techniques such as the use of selective water presaturation or the use of binomial pulses has been developed. Strategies to overcome problems associated with solvent suppression, such as magnetic and RF field inhomogeneity, short relaxation times, sample motion, or power deposition are discussed by van Zijl and Moonen. Special attention is given to the techniques involving magnetic field gradients.

Solvent Suppression Strategies for In Vivo Magnetic Resonance Spectroscopy

Peter C. M. van Zijl[1] and Chrit T. W. Moonen[2]

[1] Georgetown University Medical School, Dept. of Pharmacology, 4 Research Court, Rockville, Md 20850, USA
[2] In Vivo NMR Research Center, Biomedical Engineering and Instrumentation Program, NCRR, National Institutes of Health, Bldg 10, Rm B1D-125 Bethesda, Md 20892, USA

Table of Contents

The presence of dominant water and lipid resonances is one of the major obstacles for the use of the sensitive proton nucleus in in vivo NMR metabolite studies. The suppression of these signals is complicated by problems like B_0 and B_1 inhomogeneity, motion, low concentrations, short transverse relaxation times and heat deposition by rf fields. The strategies that can be employed to overcome these difficulties are the topic of this review. Conventional approaches are discussed, but special attention is given to techniques employing magnetic field gradients. Recent advances have shown that the availability of these gradients may revolutionize water suppression approaches. In these techniques excellent water suppression can be attained in a single scan, without the need for phase cycling. This is especially important for the motion-sensitive in vivo situation. However, eliminating phase cycling is also very advantageous for multidimensional solution NMR. In view of the present availability of sophisticated gradient technology for high resolution instruments it is therefore expected that these gradient techniques will soon become accepted in in vitro NMR.

NMR Basic Principles and Progress, Vol. 26

1 Introduction

Despite its obvious potential for assessing in vivo metabolism, NMR spectroscopy has yet to acquite a significant place in clinical practice. This situation can be partly ascribed to the technique's inherently low sensitivity and to the low metabolite concentrations. Thus, compared to the microliter resolution of clinical MRI, metabolite spectroscopy suffers from low spatial resolution. When acquiring larger volumes (1–100 ml) to attain sufficient signal, specificity may be lost due to tissue heterogeneity. This greatly complicates the reproducibility of clinical trials, especially in cases where the size (or position) of the lesion changes. For the purposes of sensitivity and access to many metabolites, it is advantageous to use proton NMR. However, proton spectroscopy is hampered by its limited spectral resolution and by the need to effectively suppress the dominating water and lipid resonances. The strategies that can be employed to attain suppression of these and other unwanted resonances are the topic of this review.

Most solvent suppression techniques [1–4] for NMR were designed for in vitro studies. However, the application of these methods to biomedicine is far from straightforward due to the specific requirements and problems encountered when working with patients and living animals (in vivo), or when studying perfused organs or cells (ex vivo). The major issues to be addressed are increased B_0- and B_1-inhomogeneity. Other complications are motion, heat deposition, low concentrations and short transverse relaxation times T_2.

As far as water suppression is concerned, the most important in vivo problem is the attainable homogeneity in the static magnetic field B_0. This may be a consequence of the limited homogeneous volume of the magnet, but, with present high quality equipment, the major cause is variation in magnetic susceptibilities over the object. As an illustration of the problem, consider the acquisition of a spectrum of the cerebral cortex. When using a whole-head resonator, signal detection sensitivity is approximately similar for all regions. If, due to susceptibility differences, a small volume in the sinuses has the water resonance at a different frequency (for example, at 2.0 instead of at the normal 4.7 ppm) than a similarly-sized volume in the cerebral cortex, this signal will interfere with metabolite detection. Lipid signals from the superficial structure will also distort the spectrum. These perturbing signals have proton concentrations that are orders of magnitude (10^3–10^4) higher than those of the metabolites and may give artifacts throughout the spectrum. Thus, suppression techniques based on differences in precession frequency between metabolites and interfering resonances will only work if they can be combined with high quality spatial localization procedures [5–13].

Spatial inhomogeneities in the applied radiofrequency (rf) field B_1 may also reduce suppression efficiency, since not all the water to be suppressed can be equally excited with a single pulse. This especially poses a problem with surface coils, which are often used to improve localization and sensitivity. Although this difficulty can be partly alleviated by the use of localization and/or adiabatic rf pulses [14–19], the situation generally remains much more complicated than for the well defined in vitro experiment.

In addition to these problems, in vivo methods are often hampered by macroscopic motions like the cardiac and respiratory cycles, fluid flow, and muscle spasms. Microscopic motions like diffusion or capillary perfusion generally do not appreciably change the detectable water magnetization in the region of interest within each acquisition. Although most of the macroscopic motions are coherent, and can therefore in principle be corrected for, this is not easy in practice since the combined motions may appear incoherent over the volume of interest. As a consequence, the popular phase cycling and spectral difference techniques are often unsuccessful, and single-scan suppression techniques are necessary for optimum results in vivo.

Heat deposition due to rf pulses is a function of the dielectric constant and conductance of the solvent, and both in vitro studies on acqueous solutions and in vivo experiments are sensitive to rf heating. However, the higher ionic strength in the in vivo situation strongly increases this sensitivity. Also, due to the coil size, much higher powers are demanded than for in vitro experiments and extreme care has to be taken to avoid overheating at certain hotspots near the coil. Thus, some conventional suppression techniques may not be applicable in vivo, and rf power requirements have to be carefully analyzed for each sequence.

Living systems are not very tolerant towards biochemical or physiological perturbations. Although ex vivo studies have been performed in which water was replaced by D_2O, changing the system is generally not desired. As a consequence, biomedical studies often require the detection of low concentration metabolities in their natural environment. The concentrations of metabolites studied by in vivo NMR are in the millimolar range, but studies of perfused organs and cells may suffer from much lower concentrations. This is mainly due to the presence of a large extracellular volume. For instance, for perfused cells, the concentration ratio of metabolites to the solvent may be as small as 10^{-6}, putting high demands on the efficiency of suppression methods.

At first glance, attaining sufficient suppression in a single scan does not seem very difficult, since it effectively means that one just needs to solve the dynamic range problem of detecting small metabolite signals in addition to the solvent signal. For instruments equipped with a 16-bit analog to digital (ADC) converter this requirement need not be very demanding. However, since the enormous water signal may also distort the baseline (e.g. when zero-order phasing is insufficient to accomplish an in-phase spectrum), water suppression generally has to be much more effective than expected from this theoretical requirement.

The suppression of a particular resonance in a NMR spectrum requires a difference in a nuclear property between the protons in the molecule under investigation and those in the compound interfering with detection. In this respect it is important to note that this property need not be a magnetic one, but that it must be reflected in a NMR-observable parameter. Thus, in addition to chemical shifts, scalar coupling and relaxation, properties like diffusion and exchange can be exploited. Section 3 is subdivided into paragraphs treating these different properties that are presently used in solvent suppression techniques. The methods will be analyzed with respect to the criteria mentioned above and, in addition, in terms of: 1) insensitivity to rf pulse imperfections; 2) insensitivity to relaxation effects; 3) ease of phasing the spectra, and the possibilities of 4) studying exchangeable protons and 5) detecting resonances at the water frequency. The most important aspects will be summarized in Sect. 4. Since many methods have recently been reviewed in detail [1], we will concentrate on those applicable in the biomedical research environment.

2 Experimental Considerations

Most of the modern techniques to be discussed here consist of a series of rf pulses and gradient pulses of variable length and strength, after which the resulting signal is acquired. Since phase cycling can generally not be optimally exploited to correct for system instabilities, high performance hardware is necessary to make most methods practically applicable. For instance, the application of long pulses requires excellent stability of the rf and gradient amplifiers. Other prerequisites for obtaining high quality spectra are more specific.

2.1 Radiofrequency Pulse Generation

Many water suppression sequences employ frequency selective $\pi/2$- or π-pulses. When accurate metabolite concentrations are to be determined, only the solvent line should be suppressed. Thus, the frequency response resulting from a certain time-domain rf pulse should preferably be rectangular (Sect. 3.1). To be able to optimize frequency response, hardware and software capabilities should include the generation of shaped (tailored) pulses [14–28]. Also, for the suppression to be as good as possible, it is necessary that these pulses are effective over the whole region of interest, independent of B_1 inhomogeneity (adiabatic). This can sometimes be accomplished with the adiabatic pulses that have been developed in the last half of this decade [14–19]. Since adiabatic pulses are phase- and amplitude modulated, they also require a high resolution (preferably millidegree) phase shifter. When they are frequency modulated, a high resolution frequency synthesizer is required. Disadvantages of adiabatic pulses are their higher rf power deposition and increased length compared to shaped pulses like sincs, Gaussians or Hermitians. These latter pulses may also provide good results, especially when localized spectroscopy techniques provide small areas with good B_1 homogeneity or when a series of pulses with different angles is used.

Composite long pulses consisting of a series of short rectangular pulses with shifted phases [2, 29, 30] in principle may reduce B_1-inhomogeneity effects, but these pulses can not be implemented easily in localization schemes.

2.2 *Pulsed Magnetic Field Gradients*

Many recent water suppression techniques (Sect. 3) use magnetic field gradients to spoil the water magnetization [31]. These methods should perform localization and generate a clean signal in the required dynamic range in a single scan. To accomplish this, it is necessary to apply strong gradient pulses (crushers) of different strength, length or direction at various places in the sequence. The result can only be optimal if residual gradient effects (eddy currents) are negligible when acquiring the signal. This requirement is more stringent for spectroscopy than for imaging, since linewidths in the order of a few Hz (circa 0.001 G) are necessary. Ideally, the gradient coils should to be self-shielded [32–34]. When shielding is not completely sufficient, additional compensation can be applied by preemphasis methods [35, 36]. To avoid induction of eddy currents in the rf coil itself, it is very useful to split the coil in several sections and to reduce the thickness of the metal as much as possible.

2.3 *Data Acquisition and Signal Processing*

Limited dynamic range of the ADC can create data manipulation problems like Fourier-transform roundoff errors or baseline curvature. There are several methods dealing with dynamic range problems after signal detection [1, 2]. Hardware methods, e.g. the tunable notch filter [37] and differential sampling [38] reduce analog dynamic range before the signal reaches the ADC. Oversampling [39] can be used to increase ADC resolution. Software methods [1, 2, 40–49] may be used to lower the digital dynamic range and/or to correct spectral phase errors.

Even when dynamic range problems are solved, the "tails" of the intense solvent resonance may still cause phasing problems for spectra with a linear (first order) or higher order phase dependence. Another reason for this phase behaviour may be delayed signal detection, since signal cannot be detected until the rf pulse (and a certain dead time) are finished. This induces an approximately linear phase shift as a function of off-resonance frequency. Many binomial suppression sequences ([2], Sect. 3.1) also cause a linear spectral phase shift. Since first order phase correction may not work in the case of overlapping lines [50, 51], good magnetic field homogeneity is an experimental prerequisite. A simple solution to some of the above problems is the use of echo spectroscopy, where the frequency evolution due to field inhomogeneities and chemical shifts is refocussed, and phasing can in principle be accomplished by zero-order correction. When not using echoes, a solution may be found in the so-called self-refocussing rf pulses [21, 52–54]. However, when used for selective excitation, these pulse are long [50–100 ms at 270 MHz) with respect to T_2 (T_{1p}) of many metabolities and will have to be shortened to be practical for in vivo spectroscopy. Even when echoes or self-refocussed pulses are used, the suppression method itself may still cause a linear or higher-order phase dependence. Recently [43–46], deconvolution signal processing techniques were proposed that are able to remove large solvent resonances and their tails from the metabolite spectra, thus improving phasing. In one of these methods [46], the acquired time domain signal is not Fourier transformed, but processed with singular-value decomposition methods [55–57]. These techniques provide amplitudes, frequencies and damping factors of the signal components, which can then be used to reconstruct the frequency spectrum. When a large solvent line with a tail can be described by the sum of exponentially-damped sinusoids with frequencies different from those of the metabolites, the solvent (and its tail!) can be subtracted out [46]. An interesting aspect of this method and the data shift accumulation methods [38, 41–45] is that, although they are chemical shift selective, they can remove tails of the solvent peak that extend beyond the selective frequency region.

Although these processing methods may be very helpful, they are often based on assumptions (e.g. Lorentzian lineshapes) and many of them may not work in some practical situations. It is therefore more advantageous to reduce the water signal prior to acquisition in order to directly avoid dynamic range problems, water line broadening due to radiation damping [58] and errors due to scan to scan variation of phase. The latter may destructively interfere with some of the signal processing techniques for the correction of spectral phase. Some of the processing methods are also time consuming, which may be disadvantageous in the reconstruction of multidimensional datasets.

3 Suppression Techniques as a Function of Nuclear Properties

3.1 Chemical Shift

Suppression techniques using chemical shift differences between solvent and solute are the most frequently employed ones. However, these chemical shift methods have the inevitable disadvantage of being unable to study solute resonances at and closely around the water frequency. They also become more difficult for studies in inhomogeneous static magnetic fields, which often require a wider frequency region to be suppressed. When appropriate shimming cannot be accomplished, the wings of broadened signals may cause baseline distortions and thus phasing problems in the regions of interest.

Single Frequency Selective Soft Pulses

Figure 1 shows the generalized pulse sequence for frequency selective suppression experiments using a non-composite soft pulse. It consists of a soft rf pulse P1 of a certain shape, a gradient (B_0 or B_1) homospoil pulse G1 of length DG1 [31, 59–64], and a delay D1. This combination is generally placed in front of the sequence used for detecting the metabolites, but can in principle be used at any instance in this sequence where the water magnetization is longitudinal (e.g. in the TM period of the stimulated echo sequence [63–66]). In the discussion we always assume that the gradient pulse is of sufficient length and strength to dephase all transverse magnetization present after P1. The gradient parameters can be calculated, but it is most accurate and efficient to determine them experimentally for the localized volume. In order to avoid reduction in T_1 due to radiation damping effects [58], the gradient pulse should be applied directly after P1.

The terminology "soft" for P1 is generally used for pulses with a duration DP1 longer than 0.2 ms. These pulses are frequency selective and have to be optimized with respect to frequency response. Linear response theory predicts that the frequency profile resulting from a selective pulse of a certain shape is the Fourier transform of this shape. For instance, for a rectangular rf pulse this is a sinc ($\sin \nu/\nu$) function and for a sinc-shaped rf pulse the frequency profile is expected to be rectangular. However, linear response is only valid for small flip angles ($\sin \theta \approx \theta$) and on resonance. Also, pulse time has to remain short with respect to relaxation times T_2 and T_1 ($T_{1\rho}$), and shaped pulses often have to be truncated. This truncation again leads to extension of the suppressed region in the frequency response. Thus, it is difficult to predict optimum pulse shapes based on linear response theory. Attempts have therefore been made to predict optimum pulse response from the Bloch equations. However, next to the rectangular pulse, there is only one pulse shape that has been solved analytically for all offsets and amplitudes, namely the hyperbolic secant inversion pulse [14–17].

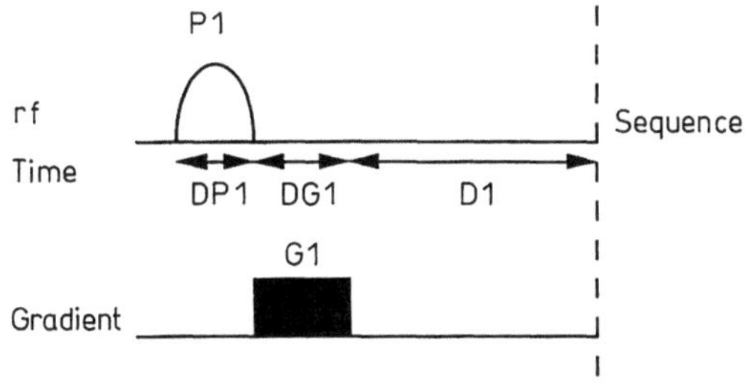

Fig. 1. Generalized sequence for frequency selective suppression. After a soft rf pulse of length DP1 and strength P1, a gradient G1 of length DG1 spoils all residual transverse magnetization. The delay D1 is as short as possible when the rf pulse has ($\pi/2$)-character (CHESS) and optimized to null the solvent magnetization for an initial inversion pulse. For presaturation DP1 is of the order of T_1 or longer

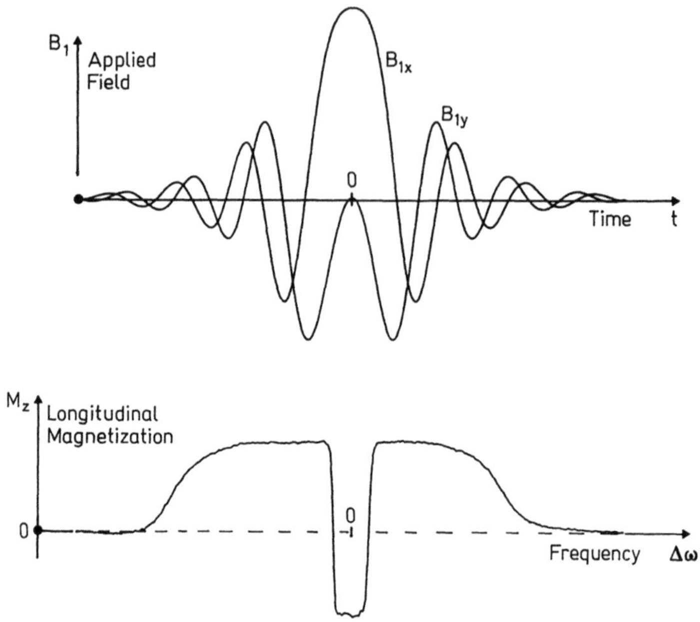

Fig. 2. Time domain shape and resulting frequency profile of a complex hyperbolic secant pulse. The frequency profile illustrates the highly selective population inversion attained. Reproduced, with permission, from Silver MS, Joseph RI, Hoult DI (1984) J Magn Reson 59: 347

The frequency profile of this pulse is almost perfectly rectangular (Fig. 2). Since theoretical prediction of optimally shaped pulses is difficult, practical approaches are used which numerically optimize the pulse shape by looking at the frequency response for all pulse angles [18–25]. At present, a family of shaped pulses is available that give approximately rectangular frequency profiles. Regretfully, an adiabatic $\pi/2$ pulses with a rectangular frequency profile has not yet been realized, but this is an active area of research [18, 19]. An adiabatic, slice selective $\pi/2$ pulse has been reported (Conolly et al. (1990) 9th Ann Mtg Soc Magn Reson Med, p. 419). However, this pulse is not useful for water suppression, since it requires two acquisitions. Adiabatic $\pi/2$- and π-pulses [14–19] may become the most important pulses for in vivo solvent-suppression studies, especially when using surface coils. A disadvantage of adiabatic pulses is that they have to be made rather long to produces narrow transition zones (e.g. see Murdoch et al. (1987) J Magn Reson 74: 226).

When P1 is a $\pi/2$ pulse [60–64], the sequence in Fig. 1 has been dubbed CHESS (CHEmical Shift Selective [60]) or SUBMERGE (SUppression By Mistimed Echo and Repetitive Gradient Episodes [61, 62]). We use the CHESS nomenclature. A $\pi/2$ pulse should be followed by an as-short-as-possible period DG1 + D1 in order to avoid T_1 relaxation. For the same reason, DP1 should be short also. Alternatively, the power of P1 can be slightly increased to account for T_1 relaxation. This is what effectively happens when optimizing P1 experimentally. Since DP1 is determined by the desired suppression width in Hz, shorter pulses can be used at higher fields to attain a similar suppression width in ppm. As mentioned, suppression efficiency will depend on the ability of the rf pulse to generate transverse magnetization for all water spins in the sample. Also, the quality of shimming is crucial for the dephasing efficiency of the gradient [63]. Thus, for optimum B_0 and B_1 homogeneity, a single CHESS sequence should suffice. In practice, however, experiments with a surface coil and a localized volume give a suppression factor of only about 15 for a single CHESS [63]. For birdcage coils, a factor of about 100–500 can be attained. However, a series of CHESS sequences gives much better results [61–67]. For instance, for a surface coil, application of three CHESS sequences before the stimulated echo sequence (STEAM: STimulated Echo Acquisition Mode [12, 13]) gives a suppression factor of about 500–800 [63, 64]. STEAM with 3 CHESS before and 2–3 in the middle of the sequence (TM-period) can give suppression factors of over 15,000 [63], even at echo times as short as 12 ms and without using phase cycling (Fig. 3, [63, 67]). With a birdcage coil large

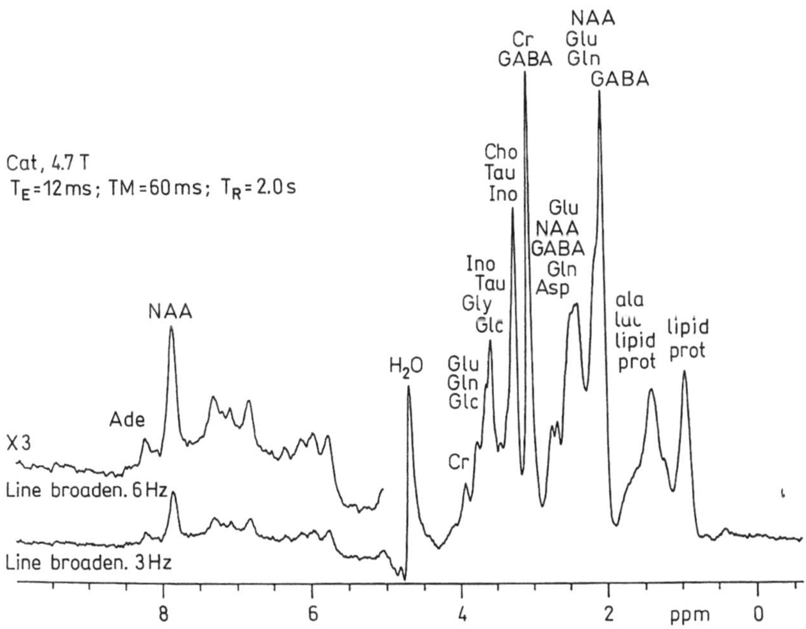

Fig. 3. In vivo proton spectrum of a 700 μl nominal volume localized in the cerebral cortex of cat brain in parts of the parietal and frontal lobes. The sequence used was a 3,3-DRYSTEAM (2048 scans). The CHESS pulse length was 5 ms (one-lobe sinc), partly suppressing the creatine CH_2-group at 4.01 ppm (assignments are from Ref. [89]). The only processing applied was a 2-Hz exponential broadening. Abbreviations are: NAA: *N*-acetylaspartate; Cho: cholines; Cr: creatines, Lac: lactate; Ino: inositols; Glu: glutamate; Gln: glutamine; Glc: glucose derivatives GABA: γ-aminobutyric acid; Asp: aspartate; Gly: glycine; Ala: alanine; Tau: taurine; Ade: adenosines; Prot: protein

suppression factors can be achieved with a single CHESS sequence before STEAM and one in TM. The flip angles of the rf pulses in each CHESS can be adjusted individually for optimum performance [61–64], but experiment shows that this is not really necessary [63]. However, proper results can only be attained if care is taken to avoid unwanted echoes that can arise when a series of rf and gradient pulses is applied [61–63, 68]. A detailed practical and theoretical approach to avoid these unwanted echoes by strategical placement of gradient homospoil pulses (crushers) was recently published [63]. Thus, the availability of sufficient gradient strength makes phase cycling superfluous [65, 66]. From a practical point of view, this removal of unwanted echoes requires the availability of strong gradient pulses to avoid unnecessary lengthening of the sequence. When these gradient pulses can be applied without causing residual gradient effects and when properly shaped rf pulses are available, spectral phasing can be accomplished with a zero-order correction.

Combinations of CHESS sequences with STEAM have been named DRYSTEAM [63] and VAPOR (Volume localized solvent Attenuated PrOton nmR [64]). Combination with the double echo localized spectroscopy sequence was named DRYPRESS [63, 66]. In order to denote the number of CHESS sequences, a prefix can be added to the name: e.g. a 3-DRYSTEAM describes a STEAM sequence with 3 CHESS in front. A second prefix can be attached to describe the number of CHESS sequences in TM (e.g. 3, 3-DRYSTEAM above). Many successful applications have been reported using the STEAM [13, 63–67] or other sequences [66] combined with CHESS pulses.

When P1 is a π pulse, the name "WEFT" (Water Eliminated Fourier Transforms [69–76]) applies to both nonselective (Sect. 3.3, [69–72]) and selective [73–76] inversion pulses. For selective excitation the name "Modified WEFT" [73] is in use. After the π pulse, DG1 + D1 is chosen as the time for the solvent magnetization to null during T_1-relaxation. This time period will be a function of the length of the relaxation delay between scans and can be optimized for rapid acquisition (Variable nutation WEFT [73], Super WEFT [72] and VASE [76]). Depending on the type of rf pulse used, suppression factors of about 1000 can be obtained. The selective adiabatic inversion pulse is probably the most suitable due to its frequency profile. An example of the experimental success of this pulse is seen in Fig. 4 [77]. The features of this spectrum differ from that of the cat brain in Fig. 3 due to the longer echo time ($T_E = 136$ ms) used. Of course, this T_2-weighting helps in the water suppression also. Nonselective relaxation weighting (T_1 and T_2) is discussed in more detail in Sect. 3.3. One disadvantage (Fig. 4) of a single WEFT sequence is that it can only be optimized for one T_1. Thus, when large tissue regions are to be studied with multiple water-T_1 components, complete suppression cannot be attained. Recently, a double-WEFT has been proposed to partly avoid this problem: Duyn JH, Matson GB, Maudsley AA, Hugg JW, Weiner MW, Radiology, in press).

When P1 is a very long low-power rf pulse (in the order of T_1 or longer), the term (pre)saturation [78–84] is used. The definition of this term is somewhat ambiguous. In most NMR text books saturation refers to a state in which the

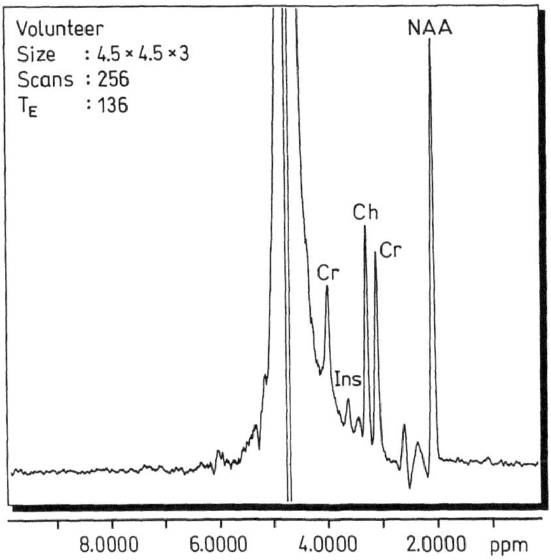

Fig. 4. In vivo proton spectrum of a 60 ml nominal volume localized in the brain cortex of a human volunteer. The sequence used was a volume-selective double echo with an adiabatic inversion pulse preceding the first slice selection. A delay D1 of 350 ms was used for optimum suppression, but complete suppression cannot be attained due to the presence of multiple T_1 components: the broad component is already positive, while the narrow one is still inverted. The selective character of the inversion pulse is illustrated by the presence and positive phase of the creatine CH_2-group. Since T_1 of this signal is longer than that of the narrow water component, attenuation would be much stronger if the signal were inverted by the adiabatic pulse. The time-domain signal was processed by a Lorentzian/Gaussian multiplication ($-2/+3$ Hz). Courtesy of Drs. PR Luyten, JA den Hollander, Philips Medical Systems BV, The Netherlands

populations of the different spin states of a certain nuclcus are equalized. In the imaging literature a rather broad definition is often used, referring to a situation where the total longitudinal magnetization is zero (e.g. in a certain spatial location). This would include the two suppression techniques just discussed. In this paragraph we will therefore refer to rf presaturation when a rf field is continuously applied at the solvent frequency in order to reach the hypothetical situation of equal populations or, classically speaking, of complete dephasing of magnetization in a plane perpendicular to the excitation axis [81]. Since saturation is competing with T_1 relaxation, complete saturation cannot be achieved. For the best result, irradiation should be maintained as long as possible, e.g. for a time in the order of 2-3T_1 [81]. In practice a certain time DP1 is chosen and the pulse power P1 and repetition time T_R have to be optimized with respect to suppression of the solvent and to signal to noise (S/N) for the signals of interest. The exact optimization conditions have been described in detail by Hoult [81]. With sufficient rf power available, suppression factors of about one thousand can be reached for DP1 in the order of T_1. If it is necessary to obtain a reproducible state of saturation at the beginning of each scan, dummy scans must be applied. This may for instance be the case when exchangeable protons are studied and the irradiation time is long with respect to the inverse of the exchange rate (saturation transfer), or when intermolecular relaxation between water and the signals of interest is important (cross-relaxation) [80]. Thus, for exchangeable protons, signal intensities will be a function of repetition time and pulse length and, for the case of rapid exchange, optimum signal intensities are reached for shorter irradiation periods. This obviously competes with the requirement for long irradiation to achieve better saturation and a narrow frequency bandwidth. Thus this suppression technique is not very suitable when studying exchange.

Presaturation methods have the advantage that the pulse bandwidth need not extend over the complete water resonance. Suppression of nonexcited water may occur through saturation transfer due to chemical exchange and cross-relaxation. When presaturation of the water resonance simultaneously saturates some resonances of molecules with a long correlation time, intramolecular cross-relaxation may also suppress the non-irradiated protons in the same molecule. This principle has been cleverly applied to suppress the dominating hemoglobine resonances in the proton spectra of erythrocyte suspensions [82].

Irradiation can be applied with an additional decoupling coil or with the same coil as for transmission and acquisition [83]. It is preferable to continue saturation (gated) during detection, but this approach may cause S/N losses due to receiver blanking and due to possible system instabilities and leakage between decoupler and receiver. Leakage induces a residual rf spike at the irradiation frequency, which can be removed by phase synchronization between transmitter (decoupler) and receiver, by a two-step phase cycle, or by unsynchronized rf irradiation and averaging [84]. Another serious problem is possible heat deposition (hyperthermia risk). Also, since long (selective) pulses are used, the strength of the irradiation field B_1 drops off rapidly off-resonance. Thus, rf saturation is more difficult in inhomogeneous magnetic fields B_0. Inhomogeneities in the irradiation field B_1, on the contrary, may be helpful in suppression [81, 85–88].

Despite the many potential problems, rf presaturation techniques have been very popular due to their ease of implementation and the sufficient suppression

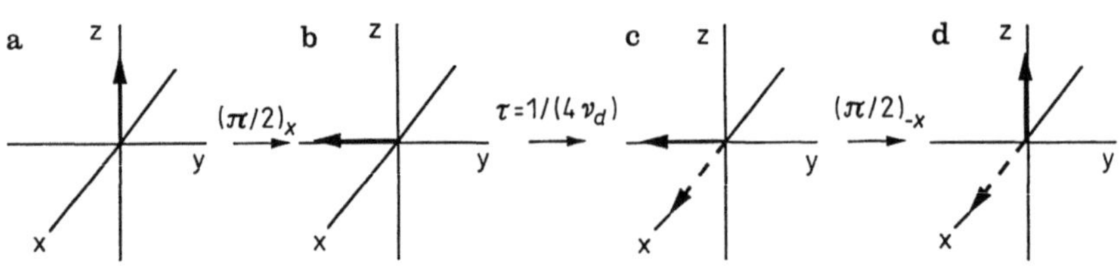

Fig. 5 a–d. Principle of the jump-return sequence. All magnetization is excited by a hard $(\pi/2)_x$ pulse with water on resonance (**a, b**). After evolution for a period $\tau = 1/(4\nu_d)$ the magnetization to be detected (*dashed arrow*) has evolved over a phase angle of 90°, while the water spins are still along the *y*-axis (**c**). A final $(\pi/2)_{-x}$ pulse makes the water magnetization longitudinal, while the signal of interest remains transverse and can be detected (**d**)

attained with respect to dynamic range. As a matter of fact, the first in vivo proton spectra [89] were obtained with the help of this suppression technique.

Frequency Selective Hard Pulses/Composite Soft Pulses

An alternative method of attaining frequency selective suppression is by using evolution of transverse magnetization to exploit the frequency (chemical shift) differences between the solvent and the resonances of interest. The most simple example of this principle is the so-called jump-and-return (JR) sequence of Plateau and Gueron [90]. All magnetization is excited with a short $\pi/2$ pulse and allowed to evolve for a period τ after which a final $\pi/2$ pulse of opposite phase ($-\pi/2$) is applied (Fig. 5). When the transmitter offset is on the solvent, no evolution occurs for this compound and, theoretically, all magnetization is returned to the positive z-axis by the second rf pulse. For a resonance with frequency difference $\nu_d = 1/(4\tau)$ from the offset, evolution has occurred over a phase angle ($\phi = 2\pi\nu_d\tau$) of 90° and the magnetization remains transverse. Of course, the frequency profile of this sequence depends on this frequency difference. The theoretical Fourier transform of two infinitely short square pulses (short with respect to τ) separated by a period τ is $\sin(\pi\nu_d\tau)$. The experimental profile closely resembles this shape [90], indicating that simple Fourier transform theory can be used to design a series of frequency selective hard pulses. Hore [2] and Turner [91] noted that the frequency profiles of a series of $n+1$ square pulses separated by τ, and with relative intensities given by the binomial coefficients of n (Table 1), are very useful for suppression purposes. There are two possible series using these coefficients, each producing a characteristic profile. Series with a 180° phase shift for alternating pulses (e.g. 1-$\underline{1}$, 1-$\underline{2}$-1, [92], 1-$\underline{3}$-3-$\underline{1}$, etc.) give a $\sin^n(\pi\nu_d\tau)$ profile (Fig. 6), while series of hard pulses with constant phase (1-1, 1-2-1, etc.) have a $\cos^n(\pi\nu_d\tau)$ appearance. The purpose is to have the sum of pulse angles to be 90° for the spins of interest and zero for the solvent. For the alternating series, the approach is similar to the jump-return case: transmitter offset on the solvent and the spins of interest evolve. However, contrary to JR, evolution is over a phase angle of π. Thus for a 1-$\underline{1}$, the flip angle is $45° + (-45°) = 0°$ for the solvent, while it is 90° for spins at $\nu_d = 1/(2\tau)$.

Table 1. Binomial coefficients for integer *n*

n	Binomial coefficients (relative pulse intensities)
1	1:1
2	1:2:1
3	1:3:3:1
4	1:4:6:4:1
5	1:5:10:10:5:1

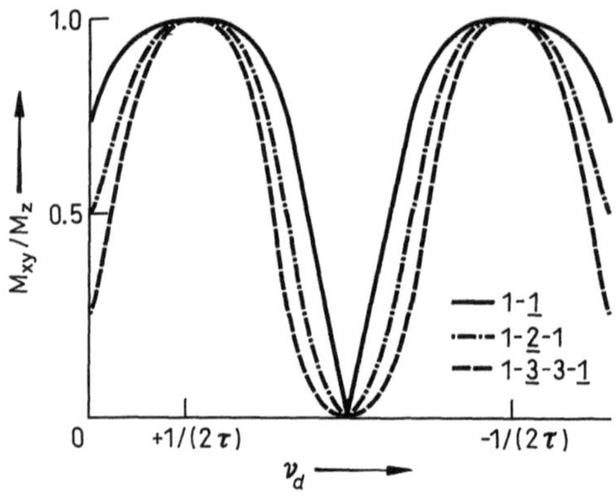

Fig. 6. Frequency response for the first three pulse combinations ($n = 1$–3) of the alternating-phase binomial series. Note the increasingly Gaussian character of the suppression profiles for higher n, reducing the detectable range, but improving suppression of broader water lines. Reproduced, with permission, from Blondet P (1988) thesis

For the cosine series, the transmitter is placed on the spins of interest and the solvent magnetization is allowed to evolve over phase π. In practice the sine series is preferred over the cosine series, since the null always appears on the carrier frequency (solvent), even if the delay times deviate or if the excitation width of the hard pulse is too small. Due to off-resonance effects the null produced by the cosine series is never as flat as for the corresponding sine series [2]. Sequences that have the transmitter offset on the solvent are also less sensitive to B_1 and B_0 inhomogeneities [2] and require a smaller sweep width due to the position of water in the center of the spectrum. Although cosine sequences are less susceptible to errors in pulse phase, the sine methods are more popular based on the previous considerations and the availability of phase cycling (EXORCYCLE [93]) to correct for pulse-phase errors. A final practical point is that sequences with an even number of pulses are less prone to deviations in the pulse angle, since they are symmetric: the first pulses are always compensated by similarly deviating pulses in the second half of the sequence. However, the odd sequences can be improved by splitting the central pulse [94]. The calculated frequency profiles for the most-used binomial sequences are in Fig. 6. Addition of extra pulses (OBTUSE [95]) and new sequences that are not binomial [96] have been proposed to improve these frequency profiles. Trial-and-error approaches [97] and computer optimization [98] have also delivered sequences that give better profiles.

Although reasonable frequency profiles can be attained, the binomial sequences (except for the Jump-Return) and many related variations suffer from an expected linear (and experimentally often higher-order) phase dependence in the spectrum. Several investigators [99–102] suggested to remove the linear phase dependence by refocussing the chemical shift evolution with an echo sequence. The proposed refocussing pulses vary from a hard π pulse [99] to binomial ones with the pulse height [101] or length [100, 102] doubled. The use of binomial pulses for refocussing [103] or for both the excitation and refocussing inside an echo sequence was already suggested earlier [104, 105]. The important advance in the later papers was the use the EXORCYCLE phase cycling. This procedure not only improves the suppression factor of 50 for a single scan to circa 5,000–10,000 in four scans, but also removes most of the linear phase roll. Similar results should be attainable if phase cycling is substituted by a pair of gradient pulses around the refocussing pulse. Other pulse combinations have been designed by optimizing frequency and phase response using Bloch equation analysis (NERO family [106–108]), but in practice these sequences still show phase errors and are also susceptible to B_1 and B_0 inhomogeneities. Recently a refocussed sequence with the excitation and refocussing pulses consisting of 8 and 12 hard pulses, respectively, was calculated using an optimization procedure. The experimental results show good phase and frequency response [109]. A detailed discussion of performance of the different sequences has been given in a recent review by Hore [2]. Practical aspects have been reviewed by Hull [110].

However, the addition of more pulses is not very suitable when B_1 inhomogeneities are a problem, e.g. when using surface coils for in vivo studies. An excellent evaluation of the use of binomial sequences in combination with surface coils has been given by Blondet et al. [101]. They show that, by using EXORCYCLE, the occurrence of a chemical shift dependence for the position

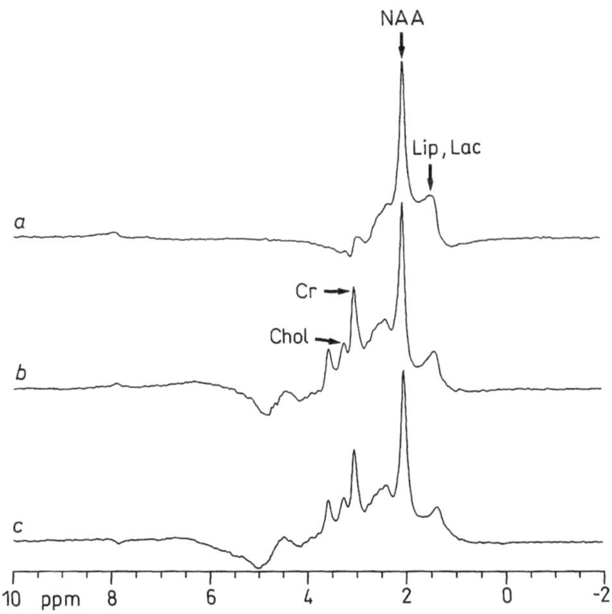

Fig. 7 a–c. In vivo proton NMR spectra of cat brain obtained with the 1-$\underline{3}$-3-$\underline{1}$/2-$\underline{6}$-6-$\underline{2}$ (**a**), the 1-$\underline{1}$/2-$\underline{2}$ (**b**) and the 1-$\underline{1}$/1-$\underline{1}$ (**c**) binomial echo sequences. To attain sufficient localization, the skull was exposed and a small surface coil (16 mm) was used. Each spectrum is the result of 16 scans using EXORCYCLE/CYCLOPS phase-cycling. $T_R = 5$ s and $T_E = 136$ ms were used. Exponential broadening (5 Hz) and zero-order phase correction were applied. Reproduced, with permission, from Alger JR, Brunetti A, Nagashima G, Hossman KA (1989) Magn Reson Med 11: 73

and shape of the sensitive volume of the coil is avoided for the binomial refocussing pulses and reduced for the excitation pulse. This approach was based on the successful application of EXORCYCLED refocussing pulses (depth pulses) for the surface coil study of deep seated regions by Bendall et al. [111]. Recently, the binomial echo principle has been combined with adiabatic pulses to provide an efficient water suppression sequence [112].

When sufficient localization can be attained with a surface coil, the binomial pulse sequences can be very useful for in vivo applications (Fig. 7 [101–105, 113]). Also, since water is not saturated, exchangeable protons can be studied. However, the presence of the large water magnetization may cause water line broadening due to radiation damping effects. Another disadvantage is that hard pulse sequences are not easy to combine with many of the popular B_0-gradient localization schemes. Also, when motion is a factor (e.g. in the clinic), phase cycling to attain improved suppression may not be applicable anymore.

The first composite soft pulse that used to be very popular in older NMR literature is the so-called "Redfield 214 Pulse" [4, 114]. Redfield et al. noticed that this pulse was much better for suppression purposes than a single selective soft pulse [115, 116] of the same total length. In modern notation, the 214-pulse would be written as $2\underline{1}4\underline{1}2$. No delays occur between the five pulses. For this composite pulse, the carrier is placed on the resonance of interest and the bandwidth is determined by the shortest component. The frequency and phase profiles of the pulse are asymmetric and only spectral parts above or below the water frequency can be studied. The suppression in this part is comparable to the 1-$\underline{3}$-3-$\underline{1}$ sequence. However, the resulting water component on the other side may complicate phasing. Although adjustments can be made to reduce this contribution, it is a tedious procedure and the Redfield pulse is not very popular anymore. Improvements and another well-performing soft pulse ($1\underline{2}1$ [117, 118]) have been proposed, but they are not as attractive as the binomial echo variations that have a much better phase behavior and are suitable for surface coil use. Recently some soft-hard composite pulses were proposed [119] that have excellent frequency profiles. Most of these techniques have not yet been used in vivo. In principle, the simplest soft-hard combination can be considered equal to a CHESS sequence followed by a hard excitation pulse when a gradient pulse is inserted in between the two pulses.

Another group of composite pulses proposed by different groups [85–88] and discussed in detail by Blondet et al. [85] consists of a hard excitation pulse followed by a delay τ and one or more spin-lock pulses (of only a few ms) separated by additional delays τ. Again the delay is optimized for evolution of the spins of interest. During the spin locking pulse, B_1 inhomogeneity (e.g. when using a surface coil) dephases all components (solvent and resonances close to it) that remained in the plane perpendicular to the excitation axis. It should be noticed that this is not dephasing in the transverse plane, but in a plane orthogonal to the axis along which the rf field is applied. Higher order suppression can be attained by addition of more periods τ and spin locking pulses along perpendicular axes [85]. Although some perturbing echoes may occur in these series of pulses, they can be removed with short gradient pulses. Phasing problems can be alleviated by addition of a binomial refocussing pulse. This dephasing in a longitudinal plane using B_1 inhomogeneity [81] can be seen as the classical description of rf saturation. As with continuous saturation, care has to be taken that a reproducible situation is generated from scan to scan. Thus long repetition times (T_R) should be used or a continuous state of saturation should be achieved so that T_R can be optimized for optimum signal to noise. Excellent in vivo results were obtained using this sequence [85]. When long T_Rs without continuous saturation are used, exchangeable protons can also be studied, since the total length of the composite sequence is only in the order of 5–10 ms. Furthermore, the rf power deposition will be low compared to continuous rf presaturation, although local heating effects may still be important. An extra advantage is that the decoupler is not used, making it available for other purposes, e.g. spectral editing.

Continuous Wave (CW) and Rapid Scan Acquisition

CW and rapid scan [120–122] methods are in principle very suitable for water suppression, since regions of the spectra can be excited selectively without perturbing the water. As a matter of fact, one of the first biological application using ^{1}H NMR [122] was performed using rapid scan NMR for the study of metabolism of bacteria. However, the CW approach by itself cannot be combined with localization and (even rapid scan) has a lower S/N per unit time than Fourier transform NMR, making it less suitable for in vivo NMR. Also, the methods are difficult to combine with relaxation time measurements.

3.2 Scalar Coupling

Many editing techniques [103–105, 123–165] use the evolution of scalar coupled nuclei to suppress the water signal. Since the two water protons have equal chemical shift, their scalar coupling (J) is not effective for evolution. Thus, complete water suppression should be attainable when selecting coupling nuclei. However, this simple statement is only true if editing can be achieved in a single scan. When difference techniques or phase cycling are necessary to accomplish editing, the large single-scan water signal will still reach the receiver and often additional suppression has to be added to the sequence. This is the case for many of the conventional multiple quantum techniques [123–125], polarization transfer methods [123, 141–148] and echo-difference editing schemes [103–105, 156–163]. As we will demonstrate, the availability of static field gradients allows the transformation of many multiscan suppression techniques into single scan alternatives. For this review we restrict ourselves to these gradient-enhanced methods, where editing is accomplished by gradient selection of certain coherences for the coupled spins, while the other signals are dephased. However, it should be kept in mind that this dephasing is a coherent, and therefore reversible, process. Thus care should be taken to avoid unwanted gradient-recalled echoes for the solvent and other interfering signals [63, 68].

Although a detailed treatment of the use of gradients for coherence selection [126, 166–169] is beyond the scope of this paper, we will review a few basic principles which are necessary to understand the sequences. We use the operator formalism with standard rotation conventions (right-handed frame [170]), neglecting relaxation. Although many techniques apply to AX_n systems, we will confine our treatment in this chapter to AX-systems. Proton magnetization will be identified by the symbol I, magnetization of other nuclei by S. In the homonuclear case, I_1 will be the X-spin, I_2 the A-spin. In all examples the first $\pi/2$ pulse will be given along the x-axis. Under the influence of a gradient G of length τ, magnetization of coherence order p will obtain a phase $p\gamma Gr\tau$.

$$I_p \xrightarrow{\gamma Gr\tau} I_p e^{-ip\gamma Gr\tau}. \tag{1}$$

Thus, for single quantum coherences ($p = \pm 1$), this can be expressed as:

$$I_- \xrightarrow{\gamma Gr\tau} I_- e^{i\gamma Gr\tau} \quad \text{and} \quad I_+ \xrightarrow{\gamma Gr\tau} I_+ e^{-\gamma Gr\tau}. \tag{1a, b}$$

Using the relations $I_\pm = I_x \pm iI_y$ and $e^{\pm i\phi} = \cos\phi \pm i \sin\phi$, one can derive the evolution of magnetization expressed in cartesian coordinates:

$$I_x \xrightarrow{\gamma Gr\tau} I_x \cos(\gamma Gr\tau) + I_y \sin(\gamma Gr\tau). \tag{2}$$

Thus the evolution at a single point r in space can be compared to that of a chemical shift. However, for the total magnetization, the phases of all points in space are summed and, when the gradient is strong enough, the random distribution of phases causes the residual signal to be negligible. In the following discussion we assume that this is always the case when there is a residual gradient term in the expression for the magnetization at the start of acquisition.

An important point that should be noticed is that the classical definition of spin-up and spin-down operators used above differs from that in the coherence transfer formalism [171, 172]. In the latter formalism, normalization factors are added to make the operators correspond to the spherical harmonics. Of course, spin evolution patterns will be equivalent in both notations.

From the above equations, a few simple rules of thumb can be obtained. First (Eq. (1)), spins of a certain coherence order ($p \geqq 1$) get a phase proportional to this order when evolving under the influence of a gradient, and can be selectively rephased. Second (Eq. (1a, b)), coherences of opposite sign are phase encoded with opposite sign and can be selectively refocussed. In 2D-spectroscopy this means that, when dephasing is performed in t_1 and rephasing in t_2, quadrature detection in the t_1-domain can be attained [138, 150–152]. This principle can of course be extended to more dimensions. Third (Eq. (2)), when signal is dephased, the phase of the first rf pulse after this gradient is not important anymore, since signal is equally distributed in all directions (of course a subsequent rf pulse may have to have a certain phase). This effect gives extra editing possibilities, especially when the gradient is followed by a $\pi/2$ pulse. For instance, applying a $(\pi/2)_x$ pulse to the magnetization in Eq. (2) gives:

$$I_x \cos(\gamma Gr\tau) + I_y \sin(\gamma Gr\tau) \xrightarrow{(\pi/2)_x} I_x \cos(\gamma Gr\tau) + I_z \sin(\gamma Gr\tau). \tag{3}$$

Thus evolution continues for one component, and, under the assumption of equal distribution of phases, a factor of two is lost in signal if no further rf pulses are applied. For coupled spins, using the general preparation period of $1/(2J)$ with a refocussing pulse for chemical shifts and inhomogeneity dephasing (Fig. 8), one has:

$$\begin{aligned} &2I_{1x}I_{2z}\cos(\gamma Gr\tau) + 2I_{1y}I_{2z}\sin(\gamma Gr\tau) \\ &\xrightarrow{(\pi/2)_x} -2I_{1x}I_{2y}\cos(\gamma Gr\tau) - 2I_{1z}I_{2y}\sin(\gamma Gr\tau). \end{aligned} \tag{4}$$

Thus both multiple-quantum excitation and polarization transfer are attained. These opportunities to simultaneously obtain different coherence pathways (Eqs. (3, 4)) have been used to edit for coupled spins in the so-called split-pathway echo techniques [136, 164, 165].

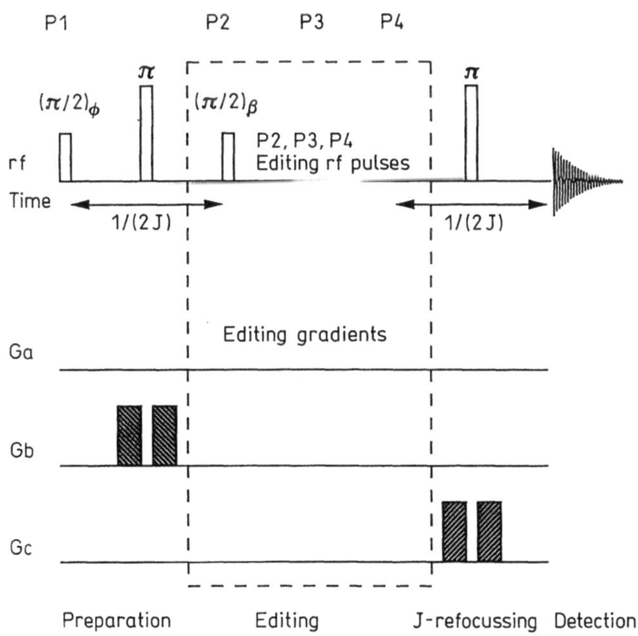

Fig. 8. General pulse sequence for gradient-enhanced scalar coupling editing. The different editing schemes (*dashed block*) consist of 1–3 rf pulses and a combination of gradient pulses (Ga) and are given in subsequent figures. Phases of the rf pulses will be given with respect to that of P1 (ϕ). Preparation and J-refocussing is the same for most of these sequences. The gradient pairs around the π pulses assure that new signal excited due to B_1 imperfections is dephased

A final fact to know is that a symmetric gradient pair around a π pulse gives complete rephasing:

$$I_x \xrightarrow{\gamma Gr\tau} I_x \cos(\gamma Gr\tau) + I_y \sin(\gamma Gr\tau) \xrightarrow{\pi_x} I_x \cos(\gamma Gr\tau) - I_y \sin(\gamma Gr\tau)$$

$$\xrightarrow{\gamma Gr\tau} \{I_x \cos(\gamma Gr\tau) + I_y \sin(\gamma Gr\tau)\} \cos(\gamma Gr\tau) + \{-I_y \cos(\gamma Gr\tau)$$

$$+ I_x \sin(\gamma Gr\tau)\} \sin(\gamma Gr\tau) = I_x. \tag{5}$$

New signal excited due to pulse imperfections is dephased.

In the following paragraphs, we discuss different gradient-enhanced editing schemes based on the general sequence structure in Fig. 8. After the $\pi/2$ excitation pulse P1 of phase ϕ ($\phi = 0$ for an x-pulse, 90 for a y-pulse), the coupled spins are prepared by evolution over $1/(2J)$ with refocussing of chemical shifts and inhomogeneities. The spins in an AX-system have then reached the antiphase situation ($2I_{1x}I_{2z}$ for $\phi = 0$ or $2I_{1y}I_{2z}$ for $\phi = 90$). For "non-coupled" spins, neglecting relaxation, the initially-excited magnetization is regained. When the second $\pi/2$ pulse P2 has phase $\beta = \phi$, pure multiple quantum coherence is excited. When $\beta = \phi \pm 90$, complete polarization transfer is accomplished. With a gradient pulse before P2, both coherences are created (Eq. (4)). The editing possibilities arising from these three preparations will be discussed in separate paragraphs. Since preparation, J-refocussing and detection are identical for most of the techniques, we will only discuss the actual editing scheme of one to three pulses (P2–P4) in the block in Fig. 8. All gradient pulses are assumed to have the same length τ. In the editing we will only vary the amplitude of these pulses. In the figures, selective rf pulses are dashed and their length, shape or kind (e.g. binomial) is not specified.

Gradient-Enhanced Multiple-Quantum Editing [126–140]

Figure 9a shows the general sequence for homonuclear multiple quantum editing using gradients [126–135]. Two $\pi/2$ mixing pulses are necessary (called P2 and P4). The phase of P2 has to be equal to that of P1, while the phase of P4 is irrelevant. Magnetization of coherence order p attains phase $\gamma pG\tau$ from evolution through gradient G of length τ during t. After conversion to single quantum magnetization by P4, refocussing of this multiple-quantum-evolved magnetization occurs with gradient of strength pG. Since the water magnetization was longitudinal during t, it was not phase encoded and will be dephased by the final gradient pG. However, when $p = 0$ (zero quantum), no automatic water suppression is attained and from this point of view it is always more convenient, in these gradient-enhanced homonuclear experiments, to edit for higher-order coherences ($p \geqq 2$).

When editing the X-spins in AX_n systems ($n = 1 - 3$), a factor of two in signal is lost due to selecting double quantum or zero quantum coherences only. An additional factor two is lost due to different phase encoding for positive

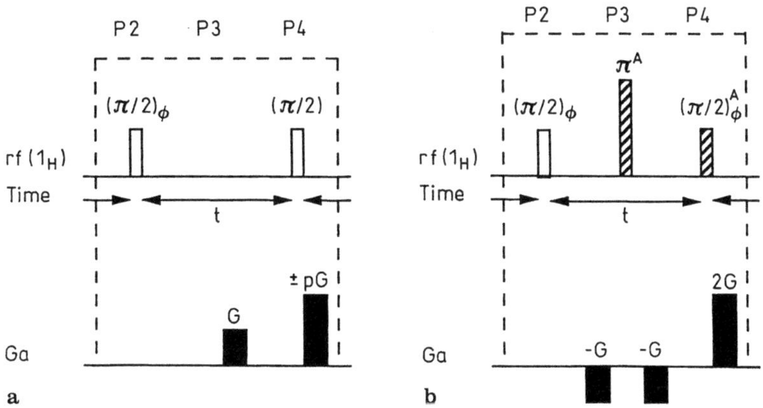

Fig. 9 a, b. Gradient-enhanced homonuclear multiple quantum experiment: **a**) nonselective sequence; the phase of P2 has to be equal to that of P1 (Fig. 8), the phase of P4 is irrelevant. **b**) Selective sequence for combined detection of positive and negative zero- and double-quantum-filtered signals for X-spins in AX_n systems (e.g. methyl group of lactate); the sign of the refocussing gradient depends on the sign of the dephasing gradient and the phase of P4 should be equal to that of P2, P1. For $p \geqq 2$, water suppression is automatic for both (**a**) and (**b**). For (**b**), lipid suppression is also attained

and negative coherences during t. For AX_n systems, when detecting the X-spins, these signal losses can be avoided. Both positive and negative coherences are rephased when the read pulse P4 is selective for A-spins only [128, 134]. It should be noticed that, in this case, the phase of P4 is still irrelevant, but that the sign of the refocussing gradient depends on the sign of the dephasing gradient. Notice that this is not important for a nonselective read pulse. An A-spin-selective π pulse P3, converts zero quantum coherence into double quantum and vice versa. When P3 is applied in the middle of t, the chemical shift of A is refocussed and the experiment is pseudo single quantum for X-spins [134]. An additional advantage is refocussing of the scalar coupling for the X-spins in any AX_n-system. When using both a selective refocussing and a selective read pulse (Fig. 9b), the phase of P4 has to be equal to that of P2 and the detection gradient has opposite sign with respect to the dephasing gradients. Selective editing schemes have the additional advantage that lipid signals are also dephased. It is not obvious from the pulse sequence in Fig. 12b why suppression should be attained, since residual transverse water magnetization that is still present after P2 does not experience any dephasing. However, the sequence was reported to be effective for lactate detection [134], which may perhaps be a consequence of water excitation/inversion during the selective A-pulses, which were centered at the lactate CH at 4.0 ppm.

An alternative method to avoid lipids, which are also scalar coupled, uses the specific multiple-quantum evolution frequency of the metabolite of interest to select in-phase positive and negative signals in subsequent scans. Subtraction then gives a clean signal [127, 131], which is illustrated for in vivo detection of

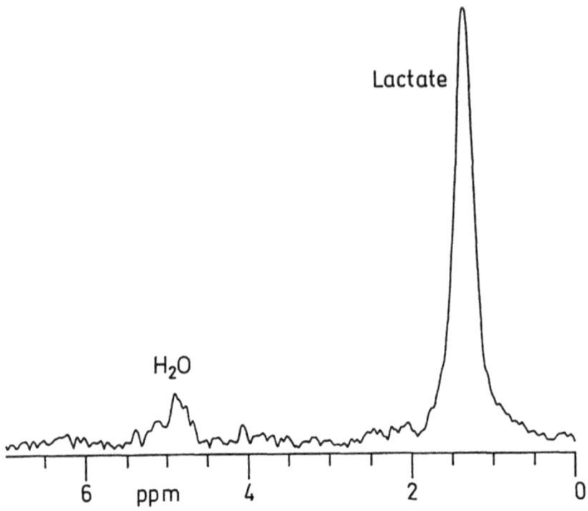

Fig. 10. In vivo, water suppressed, gradient-selected double-quantum-filtered proton spectrum of a nominal volume of 0.5 ml localized in a RIF-1 tumor subcutaneously implanted on the lower back of a C_3H mouse. The sequence in Fig. 9a was used, with lipid suppression attained by subtracting two data blocks (128 scans each) with $t = 8$ and 10 ms, respectively [131]. Field strength 2.0 T, $T_R = 3$ s, preparation time 68 ms ($1/(2J)$). Courtesy of C Sotak (Worcester Polytechnic Institute, Worcester, MA, USA) and RF Kallman (Stanford University, Stanford, CA, USA)

lactate in a mouse tumor in Fig. 10. Another approach to separate lipid signals from lactate is the use of 2D-NMR [127, 128, 131]. Notice that, although the latter two lipid suppression methods are not single-scan, the water resonance is suppressed in each scan.

Figure 11 shows the general gradient-enhanced scheme for heteronuclear multiple-quantum coherence (HMQC) editing [136–139]. The heteronuclear method has the direct advantage that the refocussing pulse P3 for the nucleus to be detected ($I = {}^1H$) is automatically selective. The ${}^1H$ inversion pulse converts double and zero quantum terms into zero and double quantum terms, resp. Proton chemical shifts refocus during t and the experiment is pseudo single quantum for ${}^{13}C$ spins. The sequence in Fig. 11 has two disadvantages. First, the refocussing gradient for HMQC editing depends on the ratio of gyromagnetic ratios of the coupled spins. In proton-detected HMQC editing, this means that only a small refocussing gradient is required, namely $p\gamma^* G = p(\gamma_x/\gamma_H)G$. Depending on the coupling nucleus, the available maximum gradient strength and the B_0-homogeneity, this may not be sufficient for complete water dephasing. A second disadvantage is loss of a factor of two in signal due to selective detection of only positive or negative coherences. Although this feature can be used in 2D-NMR to attain quadrature detection in the $f1$-domain [138, 150, 152], the signal loss is disadvantageous in the 1D case.

A final important fact is that in any 1D-HMQC sequence, signal may be lost due to evolution of the S_y terms into S_x components, or vice versa, during t (I-terms are refocussed), while the detection pulse has only a certain phase:

$$2I_xS_y \xrightarrow{t} 2I_xS_y\cos(\omega_s t) - 2I_xS_x\sin(\omega_s t). \qquad (6)$$

For instance, if $\omega_s t$ is equal to a multiple of $\pi/2$, $\cos(\omega_s t) = 0$ and a $(\pi/2)_x$ pulse would not excite any observable magnetization. In the above example, this problem can be solved by applying a $(\pi/2)_y$ pulse or by setting the offset of the

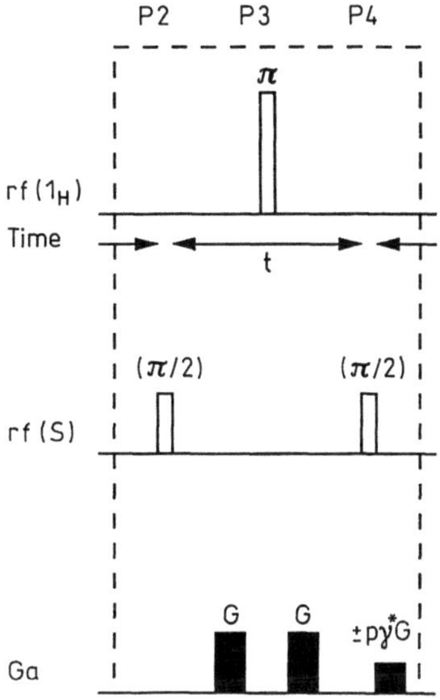

Fig. 11. Scheme for editing of proton-detected heteronuclear multiple-quantum-filtered signals. Positive and negative chemical shift evolutions are encoded with opposite phases, which can be used for quadrature detection in the $f1$-domain in 2D experiments. Gradient refocussing depends on $p\gamma^* = p\gamma_s/\gamma_H$. The phases of all pulses are irrelevant and both water- and lipid-suppression are attained when enriched compounds are used

S-nucleus on resonance and applying a $(\pi/2)_x$ pulse. It should be noticed that when gradients are present during evolution, both components are always present:

$$2I_xS_y \xrightarrow{t} 2I_xS_y\cos(\omega_s t + \gamma Gr\tau) - 2I_xS_x\sin(\omega_s t + \gamma Gr\tau) \tag{7}$$

Even with $\omega_s t$ chosen properly, the final editing pulse P4 can now only convert one of these two. However, to detect exactly half the signal, it is still necessary to choose $\omega_s t$ appropriately. In homonuclear experiments analogous considerations have to be kept in mind.

The use of gradients to select multiple quantum coherences may also benefit polarization transfer methods like inverse DEPT [141] and inverse POMMIE [144], which use multiple-quantum evolution periods.

Gradient-Enhanced Polarization Transfer

The simplest way to attain gradient-enhanced polarization transfer is with a single rf pulse P2 enclosed by two gradients (Eq. (4) [150–154]). We will not discuss this simplest case, since the homonuclear situation [150–153] does not provide automatic water suppression, and work on the heteronuclear proton-detected version is still in progress [139]. Preliminary work on carbon-detected gradient-enhanced heteronuclear polarization transfer has been reported [154] but is not of concern in the context of water suppression.

Knüttel et al. have reported multiple-pulse gradient-enhanced polarization transfer sequences with inherent water suppression for homonuclear [155] and heteronuclear application [136] to AX_n systems. The homonuclear sequences (Fig. 12) use frequency selective pulses and therefore are only able to edit for certain metabolites at a single time. However, they have the inherent advantage that lipid resonances are simultaneously suppressed.

In the sequence of Fig. 12a, the selective $\pi/2$ pulse on the X-spins produces longitudinal scalar order for the spins of interest, while the water and lipid magnetizations remain transverse and are dephased. The selective $\pi/2$ pulse on the A-spins completes the polarization transfer:

$$2I_{1x}I_{2z} \xrightarrow{(\pi/2)_y^X} -2I_{1z}I_{2z}(\text{water, lipid dephase}) \xrightarrow{(\pi/2)_y^A} -2I_{1z}I_{2x}. \quad (8)$$

Notice that the second $\pi/2$ pulse can have any phase. This sequence has the disadvantage that when P4 is not selective enough, new water magnetization may be excited.

In the sequence in Fig. 12b only one selective pulse is used, which may enable simultaneous editing of a few metabolites. If the transmitter offset is on the water resonance, P2 does not influence the water magnetization, which is subsequently dephased by both gradient pulses. For the coupled spins, a so-called cyclic polarization transfer is accomplished:

$$2I_{1x}I_{2z} \xrightarrow{(\pi/2)_y} -2I_{1z}I_{2x} \xrightarrow{(t/2,\pi^A,t/2)} -2I_{1z}I_{2x} \xrightarrow{(\pi/2)_y} 2I_{1x}I_{2z}. \quad (9)$$

The heteronuclear variant of this cyclic polarization transfer works similarly (Fig. 13). However, both sequences have the disadvantage that no dephasing

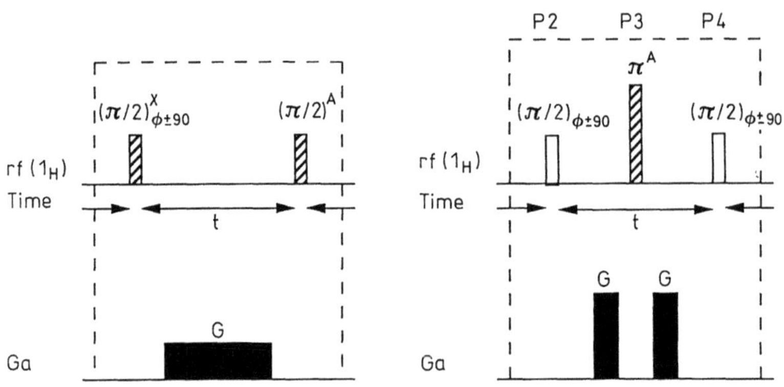

Fig. 12 a, b. Homonuclear polarization transfer methods for selective detection of X-spins in AX_n systems with inherent water and lipid suppression. **a**) Single polarization transfer needing selective pulses on X- and A-spins. The phase of P4 is irrelevant. **b**) Double (cyclic) polarization transfer, needing only a A-selective pulse. The phase of P4 is equal to that of P2, which is perpendicular to P1. The transmitter has to be on the water resonance

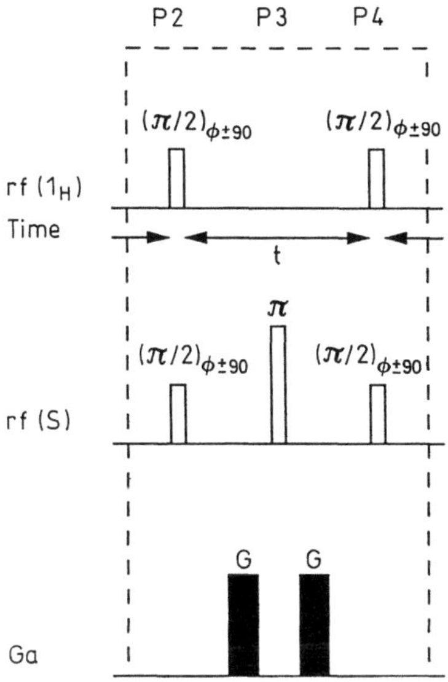

Fig. 13. Heteronuclear cyclic polarization transfer sequence with inherent water and lipid suppression. The phases of the P4 pulses are equal to those of P2, which are perpendicular to P1. The transmitter has to be on the water resonance

gradient is present after P4. Since the read pulse may excite longitudinal water magnetization (e.g. that has relaxed with T_1) somewhere in the sample, the water suppression efficiency of these sequences is not good. To overcome this, an alternative sequence has been proposed with additional gradients and evolution times [136]. However, compared to the original sequences a factor of four in signal is lost.

Split-Pathway Echoes [164, 165]

Figure 14 displays the homonuclear scalar-coupling editing sequence using split-pathway echoes (For introduction see Eqns. 3, 4 and ref. 164). The principle

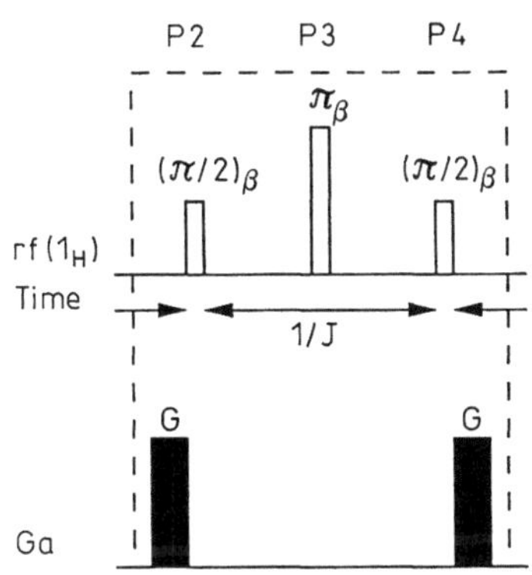

Fig. 14. Homonuclear split-pathway-echo sequence. Notice the mandatory evolution period of $1/(2J)$ between P2, P3 and P3, P4. The phase of P2 is irrelevant, but the phases of P3 and P4 are determined with respect to P2. The simplest case is one in which all phases are equal to the one of P2. Complete water suppression is attained

of the editing is based on the fact that the result of the chemical shift evolution scalar coupled spins is the same as that of a single echo sequence (Eq. (5)):

$$2I_{1x}I_{2z}\cos(\gamma Gr\tau) + 2I_{1y}I_{2z}\sin(\gamma Gr\tau)$$

$$\xrightarrow{(\pi/2)_x} \xrightarrow{1/(2J)} \xrightarrow{\pi_x} \xrightarrow{1/(2J)} \xrightarrow{(\pi/2)_x}$$

$$2I_{1x}I_{2z}\cos(\gamma Gr\tau) - 2I_{1y}I_{2z}\sin(\gamma Gr\tau). \quad (10)$$

Thus evolution through a final gradient G of equal sign refocusses all original signal (Eq. (5)). The water magnetization also undergoes split-pathway evolution (Eq. (3)), but the result is comparable to that of an effective 2π pulse (no sign change for the γ-component), and the signal is further dephased by the second gradient:

$$I_{1x}\cos(\gamma Gr\tau) + I_{1y}\sin(\gamma Gr\tau) \xrightarrow{(\pi/2)_x} \xrightarrow{1/(2J)} \xrightarrow{\pi_x} \xrightarrow{1/(2J)} \xrightarrow{(\pi/2)_x}$$
$$I_{1x}\cos(\gamma Gr\tau) + I_{1y}\sin(\gamma Gr\tau). \quad (11)$$

As demonstrated in Eq. (4), the magnetization of the coupled spins is split into pathways for multiple quantum and single quantum coherence. In order to accomplish the correct result, the phases of pulses P3 and P4 are determined with respect to P2, but the phase chosen for P2 is irrelevant. Several combinations are possible [164], of which the most simple one is with the phase

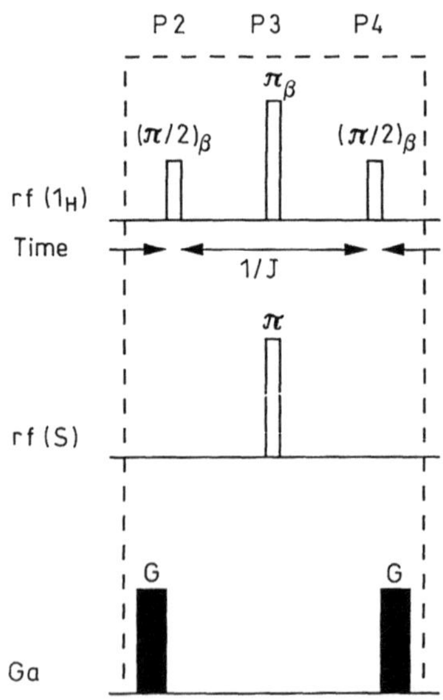

Fig. 15. Heteronuclear split-pathway-echo sequence. Notice the mandatory evolution period of 1/(2J) between P2, P3 and P3, P4. The phase of P2 is irrelevant, but the phases of P3 (for the proton pulse) and P4 are determined with respect to P2. The simplest case is with all phases equal to that of P2. Complete water and lipid suppression are attained

of all three editing pulses equal (e.g. Eq. (10)). Editing procedures with split pathways were first applied in the so-called BIRD pulses [173].

In the heteronuclear alternative (Fig. 15, [165]), the principle of Eq. (12) applies again, but the split pathways for the coupled spins are now for single quantum coherence and longitudinal scalar order:

$$2I_xS_z\cos(\gamma Gr\tau) + 2I_yS_z\sin(\gamma Gr\tau) \xrightarrow{(\pi/2)_x^I} 2I_xS_z\cos(\gamma Gr\tau) + 2I_zS_z\sin(\gamma Gr\tau). \qquad (12)$$

The phases of the proton rf pulses P2–P4 are the same as in the homonuclear case. The phase of the heteronuclear π pulse is irrelevant.

General Remarks for Scalar-Coupling Editing

A great advantage of sequences using scalar coupling evolution (J-modulation) is that water suppression is frequency independent, allowing the study of protons at the water frequency. When diffusion losses (Sect. 3.4) are avoided, exchangeable protons can be studied. Also, rf pulse imperfections will not influence the quality of the suppression, although signal losses may occur.

A specific in vivo disadvantage is that the optimum signal can only be achieved after a certain preparation time, in general equal to $1/(2J)$. Thus echo detection in most techniques occurs at $1/J$, while it is even $2/J$ for the split-pathway method. For the metabolites in the in vivo proton spectrum ($J \approx 7.3$ Hz) this corresponds to detection at $T_E = 136$ ms ($1/J$) or at $T_E = 272$ ms ($2/J$). Thus, only compounds with a long T_2 (e.g. lactate) can be edited efficiently. However, when heteronuclear techniques with proton detection are used, this problem is alleviated. For instance, ^{13}C—H couplings in C—H bonds are in the order of 120–170 Hz, allowing T_Es of 9–6 ms ($1/J$) and even the split-pathway method offers good possibilities. When sufficient gradient strength is available, all water is suppressed efficiently due to the gradient pulse after P4.

When lipid signals are to be suppressed, it is often necessary to apply selective pulses on the coupled nuclei of interest. As a matter of fact, without using gradients, von Kienlin et al. [149], reported successful in vivo lactate editing with a single-scan polarization transfer technique in which both water and lipids were suppressed using selective binomial pulses on the lactate- methyl and methine groups. For natural abundance heteronuclear studies, selective pulses on the nuclei of interest may also be necessary. When enriched compounds are used, lipid suppression should be inherent. Also, excellent sensitivity can be attained, which has enabled detection of glutamate fluxes in vivo in brain using conventional difference techniques [161]. The methods discussed in this section should drastically improve the detection of enriched metabolites and allow quick spectroscopic imaging over large volumes with effective water and lipid suppression.

One possible disadvantage that has to be mentioned is that many methods discussed in this section use gradient pairs separated by a certain length of time. Thus, diffusion losses (Sect. 3.4) and increased motion sensitivity may occur. However, in some cases diffusion losses may be beneficial. For 2D-NMR, this effectively means that diffusion weighting should be kept constant while incrementing t (t_1), which can be accomplished by slightly adjusting G for each increment.

3.3 Relaxation

A suppression approach that has been successful for many in vivo applications is the exploitation of differences in relaxation times T_1 and T_2 between solvent and solutes. However, when nonselective excitation is applied, this approach requires significant differences in relaxation time in order to avoid unacceptable signal losses for the metabolites of interest. In practice, the only cases that have been applied are those in which T_1 of the solute or T_2 of the solvent were as short as possible. The latter is easier to achieve, and has been applied in vivo as well as ex vivo. Solute T_1s that are much shorter than T_1 for water can generally only be found in macromolecules. An important advantage of nonselective relaxation techniques is that solute signals at the water resonance can in principle be studied. Also, relaxation methods do not require any phase cycling for suppression purposes.

T_1 Methods [69–77, 174–177]

The general pulse sequence for the T_1-based inversion-recovery method was already given in Fig. 1 and discussed in Sect. 3.1. Of course, when a selective inversion pulse on the solvent is used, the disadvantage of having to attain large differences in relaxation times does not apply, but the drawbacks of chemical-shift-selective techniques are found.

One T_1-based sequence that has not yet been discussed is the so-called Driven-Equilibrium Fourier Transform method (DEFT, Fig. 16 [174–177]). In this sequence magnetization excited by P1 is refocussed by P2 and returned to the z-axis by P3. Water magnetization inverted by P2 is nulled by correct choice of D1. In the nonselective case, faster-relaxing solutes are supposed to have relaxed back during 2D1. The advantage of DEFT over conventional WEFT is that choice of D1 is not so critical, since it is only necessary to remove water that relaxed back during the first D1. Phase cycling may also improve suppression.

Inversion recovery methods have the disadvantage of lengthy adjustment times (dependence on T_R, optimization for only one T_1 value) and the possibility of loosing signal of exchangeable protons. Although nonselective approaches

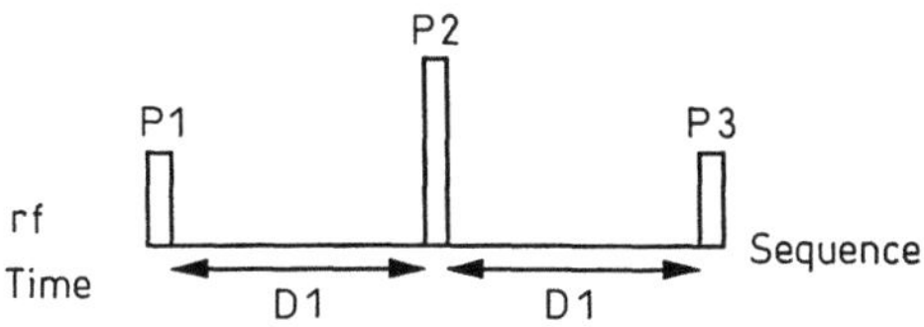

Fig. 16. Driven-Equilibrium Fourier Transform (DEFT) sequence. P1 and P3 are $\pi/2$ pulses of equal phase, P2 is a π pulse. P1 and P2 can be selective if desired. Water magnetization excited by P1 is refocussed and returned to the z-axis by P3, which is the first pulse of the actual sequence to detect metabolites. Magnetization excited by P2 is nulled by judicious choice of D1

may be useful in specific cases, the selective methods do not seem to be very advantageous anymore when high performance gradients are available. It is simpler to apply the sequence in Fig. 1 with a pulse angle slightly larger than 90°, and optimize rapidly for minimum water signal. For this approach, all different T_1 components are suppressed simultaneously. A slight T_1-optimization is also included in this procedure [63], but D1 is so short that the settings apply for each T_R. Of course for very short T_Rs unwanted interscan echoes have to be avoided. A reason for using WEFT or DEFT with a selective inversion pulse is the excellent frequency profile that can be attained with the hyperbolic secant pulse. The availability of a proper selective adiabatic $\pi/2$ pulse would also undo this advantage.

T2 Methods [177–188]

Differences in T_2 between solutes and solvent may provide a suitable means of suppression for some specific applications [177–183]. T_2 differences can be exploited when using spin echo sequences (Fig. 17, [184–187]). In general, acquisition is started on top of the echo at $T_E = 2\tau n$, and $n = 1$ is used. Echo sequences have two major disadvantages. First, compounds of interest with short T_2s may be lost and, second, intensity modulation due to scalar coupling evolution (J-modulation) may occur during TE. The first problem is inherent to the use of spin echoes, but the second can be solved by assuring that n is large and $\tau \ll 1/J$. In this situation, the effect of a pulse train is similar to that of a spin lock for the coupled spins [187] and no modulation occurs. For chemical shifts (δ in Hz) the requirement is $\tau \ll 1/\delta$. Of course relaxation ($T_1\rho$ for water on resonance, T_2 for water far enough off resonance) is still effective, and when large n-values are used, water suppression is attained. A disadvantage of high pulse repetition rate is that, like in traditional spin-locking experiments, chemical exchange effects depend on the repetition rate. Other disadvantages are heat deposition and problems with implementation in localization schemes. The Carr–Purcell–Meiboom–Gill sequence (CPMG) for large n and short τ has been applied successfully to erythrocyte solutions [177] and also to an in vivo

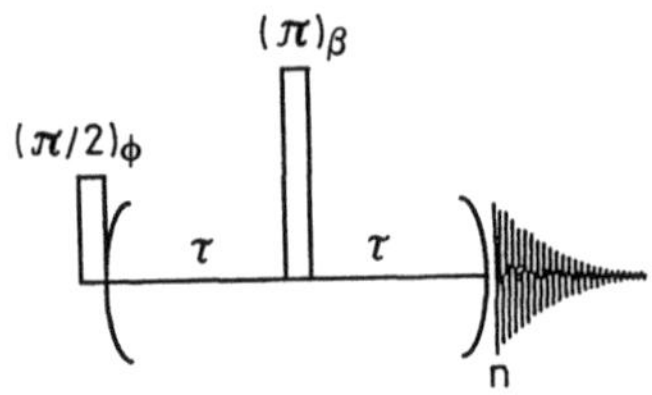

Fig. 17. Carr–Purcell–Meiboom–Gill (CPMG) echo sequence. The phase of the π pulse is chosen to avoid incremental phase errors ($\beta = \phi \pm 90$ without changing as a function of n, or $\beta = +\phi$ or $-\phi$, for odd and even n, respectively). The final TE $= 2\tau$n. For τ short with respect to evolution of J, an effective spinlock is attained, avoiding J-modulation [177, 185–187]

brain study [188]. An impressive example of its possibilities is demonstrated in Fig. 18, where the proton spectrum of a suspension of red blood cells is shown at $2\tau n = 270$ ms. No J-modulation is apparent.

The main problem with nonselective T_2 suppression is the achievement of sufficient T_2 differences between solute and solvents. Although natural differences can sometimes be exploited, e.g. in vivo when studying methyl groups of metabolites like lactate, cholines, creatine, and N-acetyl compounds, it is often necessary to artificially decrease the solvent T_2. This can be accomplished by addition of relaxation agents like paramagnetic ions (or complexes) [177, 181] or exchange reagents like ammonium chloride [177, 182]. The general name chosen for the exchange approach is WATR (Water Suppression by T_2 Relaxation [182]). It should be noticed that the paramagnetic approach also depends on exchange, namely that between solvent molecules (protons) in the bulk and in the coordination sphere of the ion [189]. For fast exchange, the relaxation efficiency depends on the square of the chemical shift difference between bulk water and water in the complex (Sect. 3.5, [189–193]) and thus becomes higher at higher field. The major disadvantage of adding artificial relaxation media is the possible induction of physiological and biochemical changes.

The strong decrease in T_2 of water upon going to higher magnetic field strengths [177, 180] benefits suppression at these fields. Although a T_2 decrease is expected in vivo due to exchange between bound water (e.g. in hydration layers of membranes) and bulk water [192], it has also been found in solutions. This may be a consequence of radiation damping (T_2^*) [1], or, less likely, of diffusion losses due to evolution through inhomogeneous magnetic fields (Sect. 3.4, Eq. (13)). The latter is not a real relaxation effect, but accelerates signal decay as a function of T_E and may be confused for one. It can be avoided by using a train of π pulses. Since both effects depend on the experimental situation (rf coil loading, shimming), they may be hard to reproduce. Radiation damping effects, if unwanted (e.g., when proper relaxation times are to be measured for water), may be suppressed by the use of gradients.

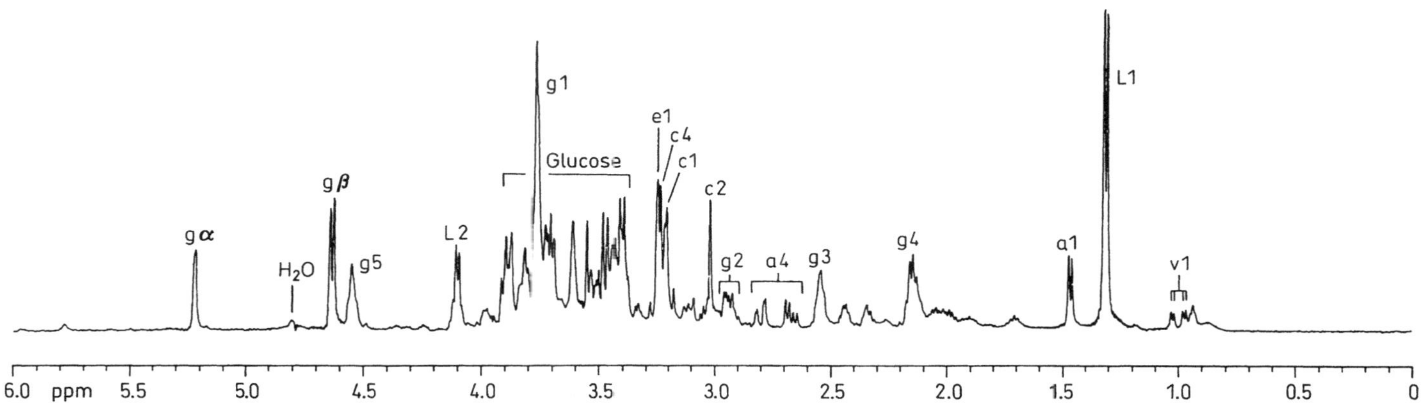

Fig. 18. 500 MHz CPMG ^{1}H NMR spectrum of intact red blood cells (not washed after separation from plasma), measured using the sequence in Fig. 17 with $\tau = 320\,\mu s$ and $TE = 2\tau n = 270$ ms. Assignments are: gα, gβ: α, β-protons of glucose anomers; g1–g5: glutathione resonances; g6: glycine; c1, c4: cholines; c2: creatine; L1, 2: lactate; e: ergothionine; a1: alanine; a4: aspartic acid; v1: valine. Reproduced, with permission, from Rabenstein DL, Millis KK, Strauss EJ (1988) Anal Chem 60: 1380A. Copyright (1988) American Chemical Society

3.4 Diffusion

A nonmagnetic nuclear property that can be used for suppression purposes is diffusion [194, 195]. Pulse sequences can be sensitized to diffusion by application of magnetic field gradients (B_0 [184, 185, 194, 196–198] or B_1 [195, 199]). For a spin-echo sequence, the signal reduction S/S_0 due to free diffusion of a single compound is [196–198]:

$$S/S_0 = e^{-(1/12)\gamma^2 G_0{}^2 D(TE)^3} + e^{-\gamma^2 G^2 \delta^2 (\Delta - \delta/3) D} \tag{13}$$

G and G_0 denote the strength of pulsed (Fig. 19) and static field gradients, resp. D is the particular diffusion constant and δ and Δ are the length and time between the pulsed gradients, resp; $t_{dif} = (\Delta - \delta/3)$ is the diffusion time. Influence of exchange and restriction increases when t_{dif} increases. Cross-terms in G and G_0 are neglected. For multi-compartment systems, this formula has to be extended to include relaxation time differences and exchange [200, 201]. When diffusion is restricted, the formula is not valid anymore. In general, terms in G_0 are negligible in vitro, but they may contribute for large samples in vivo at higher fields. We will concentrate on the pulsed gradients, which can be manipulated to increase diffusion weighting and thus suppression of rapidly-diffusing components. Figure 19 shows three possible echo sequences that can be used for diffusion suppression. Which one to use will mainly depend on

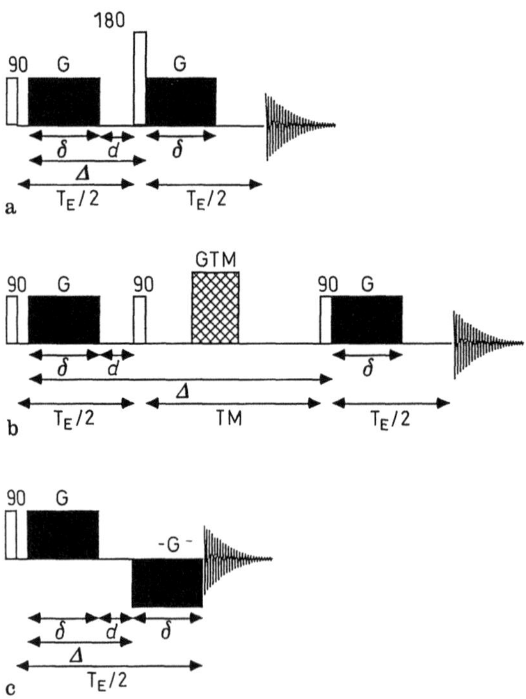

Fig. 19 a–c. Pulse sequences used to attain diffusion weighting: **a)** spin echo (SE), **b)** stimulated echo (STE), **c)** gradient-recalled echo (GRE). Gradient pairs are in black, rf pulses in white. δ is the length of diffusion gradient G; Δ is the time between the starts of the gradients. Delay d may be necessary for gradient settling. The length of rf pulses is neglected. Gradient GTM in the STE destroys nonstimulated echoes. Reproduced, with permission, from van Zijl PCM, Moonen CTW (1990) J Magn Reson 87: 18

experimental demands. When not much gradient strength is available, but when short T_Es are required, the stimulated echo (STE) is the sequence of choice. Otherwise a spin echo (SE) or the simple gradient-recalled echo (GRE) may be more convenient to use. The GRE can be implemented as a diffusion filter anywhere in the sequence where the magnetization to be dispersed is transverse.

General advantages of diffusion suppression are 1) independence on lineshape or shimming; 2) protons resonating at the water frequency can be studied; 3) no high-power rf irradiation is required; 4) no phase cycling is necessary; 5) suppression efficiency does not depend on accurate calibration of pulse lengths; 6) it is easy to implement in any sequence, especially when strong gradients are available. 7) When t_{dif} can be kept short and when the diffusion filter (e.g. a GRE) is applied just before acquisition, exchangeable protons can be studied. 8) It can be applied to any nucleus, when enough gradient strength is available. The major disadvantage is that solutes are also suppressed if the difference in D with the solvent is not large enough. Thus the method works better for larger solutes, which may be very important multidimensional NMR studies of solutions of macromolecules. For instance, suppression factors of 10^5 can be attained for solutions of molecules with $M > 4000$, while the solute is only suppressed by a factor of 1.5–2.0 [194]. Regretfully, most metabolites studied in in vivo NMR are small and diffusion suppression is generally not the method of choice. However, in specific cases, it may be very useful. Diffusion-weighted spectroscopy has been used to detect glycogen in solution [202, 203] and for the suppression of extracellular water and metabolites in

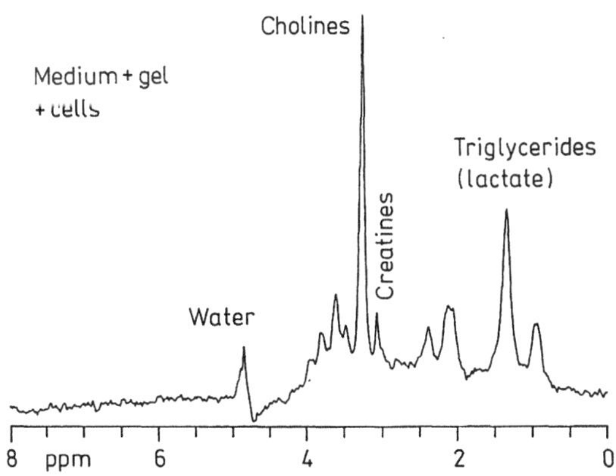

Fig. 20. 200 MHz ^{1}H-NMR spectrum of intracellular metabolites in perfused adriamycin-resistant MCF-7 human breast cancer cells. The cells occupy less than 1% of the total intra- plus extracellular water volume, but all water signal is completely suppressed using a combination of diffusion weighting ($\gamma^2 G^2 \delta^2 (\Delta - \delta/3) = 0.77 \times 10^{10}$ s/m^2 to suppress extracellular water and small metabolites) and CHESS pulses (3,2-DRYSTEAM, 1-lobe sinc, 8 ms) to suppress intracellular water and other less mobile water components. The total suppression factor is about 10^6. Reproduced, with permission, from van Zijl PCM, Moonen CTW, Faustino P, Pekar J, Kaplan O, Cohen JS (1991) Proc Natl Acad Sci USA 88: 3228

perfused cell cultures [204]. In the latter study, slower-diffusing water (e.g. intracellular) was suppressed using CHESS pulses (Sect. 3.1), and the total suppression factor for water was about 10^6 (Fig. 20). This method will also be important for the study of perfused organs.

A general disadvantage of diffusion techniques is that measurements are purposely sensitized to any motion, which is not desirable for in vivo studies. At present, additional disadvantages of B_0 methods are the need for enough gradient strength and the occurrence of residual gradient effects (Sect. 2.2). However, shielded gradients of up to 2 T/m are already commercially available at fields as high as 400–500 MHz. Very stable power supplies are also available. The use of B_1 gradients has also been suggested [195, 199] and has been applied to remove the 2,3-DPG signals in blood [205] to facilitate the study of intracellular phosphate in the hart muscle.

3.5 *Exchange*

There are several ways in which exchange properties of protons can be exploited to reduce the water signal. The principles of most of these have already been discussed in previous paragraphs. All are based on transfer of saturation from a more rapidly saturated medium to the solvent. Saturation can be accomplished using rf irradiation, gradient dephasing or T_2 relaxation. The efficiency of the effect depends on the lifetime of the protons in each pool, the proton concentration in each pool and the saturation accomplished in each pool. The effect is counteracted by T_1 relaxation. In practice, only the T_2 method (WATR, Sect. 3.3) has been applied for suppression purposes. In the case of rapid exchange between two pools, the formula for the effective T_2 of water is [189]:

$$(1/T_2)_{\mathrm{eff}} = x_\mathrm{a}(1/T_2)_\mathrm{a} + x_\mathrm{b}(1/T_2)_\mathrm{b} + x_\mathrm{a}x_\mathrm{b}\tau(2\pi\nu)^2 \tag{14}$$

in which a, b are the two pools (fractions $x_\mathrm{a}, x_\mathrm{b}$), τ is the lifetime of the protons in the rapidly relaxing pool and ν is the chemical shift difference in Hz between the pools. Thus this effect will be more advantageous for suppression at higher fields (Sect. 3.3, [177, 192]).

4 Conclusions

It will be clear from the previous section that the modern NMR spectroscopist has a variety of efficient water suppression techniques available to his/her disposal. Preference for a certain method will of course depend on the specific application and on the availability of certain hardware. In this concluding section we evaluate the possibilities of present suppression technology. The reader is referred to the specific sections for the detailed literature citations.

4.1 In Vivo Suppression

There has been a significant increase in reports of in vivo proton NMR studies over the past few years. Some excellent results have been obtained with conventional techniques using rf presaturation, binomial pulses (combined with phase cycling) and relaxation differences. However, the future of in vivo water suppression appears to be in gradient technology, which provides a set of single scan experiments. Thus the interference of motion with suppression efficiency, which is a potential in vivo problem when using phase cycling, has been strongly reduced. The magnetic field inhomogeneity problem of course remains, but is reduced due to the availability of high quality gradient localization methods. Present technology therefore allows efficient chemical-shift-selective suppression and spatial localization to be performed in a single scan. This has allowed the recording to high resolution brain spectra in humans and animals with echo times as short as 6–20 ms.

There are several reasons for the popularity of chemical shift methods (Sect. 3.1) over other techniques that have more inherent advantages. The most important is that the difference in relaxation (T_1, T_2) and diffusion properties between water and the majority of NMR-visible metabolites is not large enough. Although T_2 relaxation differences will increase at higher field, T_2 methods (Sect. 3.3) will not be able to achieve the excellent performance at short T_Es of the chemical shift methods. Diffusion (Sect. 3.4) and nonselective T_1 methods (Sect. 3.3) presently seem mainly suitable for the study of macromolecules and may only be important for specific applications. Homonuclear scalar coupling methods (Sect. 3.2) have the disadvantage of long preparation times (about 68 ms), but may be preferable over methods that use echo times in this order of magnitude or longer, due to the advantage of frequency independent suppression. When selective pulses are added, an additional advantage is found in lipid suppression. Proton-detected heteronuclear scalar coupling methods (Sect. 3.2) may suffer from insensitivity for natural abundance studies. However, when enriched compounds are used, they allow the study of metabolites at T_Es shorter than 10 ms with efficient frequency-independent water and lipid suppression and with proton sensitivity.

Important disadvantages of chemical shift methods are the often inherent variation of intensity and phase as a function of frequency and the dependence on B_1 homogeneity. This is especially annoying when concentration ratios have to be evaluated. In this respect, the availability of shaped pulses has substantially improved frequency profiles, while advances in the design of adiabatic pulses have alleviated problems with B_1 inhomogeneity. The adiabatic inversion pulse has drastically improved the performance of selective WEFT (Sect. 3.1). The development of a selective adiabatic excitation pulse with a rectangular frequency profile should allow effective water suppression with a single CHESS combination (Sect. 3.1). This will be important for the use of echo sequences instead of stimulated echo sequences for short-T_E spectroscopy and spectroscopic imaging. Since these methods are effective in a single scan, do not require high rf power, and have an almost-rectangular frequency profile they will eventually replace the conventional techniques in the future.

4.2 Ex Vivo Suppression

Suppression difficulties encountered in the study of cell suspensions and of perfused cells and organs are similar to those in vivo, with the additional problem of water and metabolites in the perfusion medium. Depending on the ratio of intracellular and extracellular water volumes, this may be a very serious obstacle. When this ratio is low, an additional problem that has to be solved is the suppression of extracellular metabolites. Due to a difference of a factor of about 5–10 in effective diffusion constant between intra- and extracellular compounds, this problem can be solved with diffusion methods (Sect. 3.4).

4.3 In Vitro Suppression

An important class of in vitro experiments deals with the study of macromolecules. Effective suppression of large diagonal peaks is of importance in multidimensional studies to avoid t_1 noise. One of the major obstacles in study of large molecules is the homogeneous broadening of the water line, that reduces the utility of chemical shift suppression methods. Again, the most simple solution for this problem is the use of diffusion methods (Sect. 3.4), which provide frequency independent suppression. Contrary to other methods, the suppression improves when the molecular weight increases. When strong (shielded) gradients are available, a short diffusion filter allows efficient suppression without loss of exchangeable protons (Sect. 3.4).

Another feature of many gradient methods is the possibility to select for certain coherence orders in a single scan. Water protons in solution can only generate single quantum coherence. Homonuclear multiple-quantum methods (for orders two or higher) and heteronuclear proton-detected coherence-transfer methods (Sect. 3.2) therefore allow efficient frequency-independent water suppression. When gradients are available this can be achieved without phase cycling. These methods are effective for any molecular weight. The availability of these single scan methods also allows the rapid accumulation of multidimensional spectra with inherent quadrature detection in the t_1, t_2, etc.-domains.

Diffusion and gradient-enhanced scalar coupling methods are so effective for water suppression that lyophilization procedures are no longer necessary. Compounds can be measured in neat water. Another advantage of using gradients is the possibility to avoid radiation damping effects by strategically placing dephasing-rephasing gradient pairs in the sequence. Present disadvantages of gradient techniques are that spinning is not possible when using x- and y-gradients and that the lock may have to be off during gradient pulses. However, these problems are not really disturbing and the considerations discussed here should present strong arguments for the implementation of high-performance gradients in high-resolution NMR equipment. As a matter of fact, our recent experience shows that the lock can recover after gradient pulses, allowing long-term experiments.

4.4 General Remarks

The sensitive proton nucleus is a prime candidate for application in clinical spectroscopy. Recent technical advances provide the possibility of effective suppression of unwanted resonances in in vivo proton spectra for echo times as short as a few milliseconds [13, 63–67]. This presently allows spectroscopic imaging with a spatial resolution comparable to or better than positron emission tomography (ca. 1 ml) in a clinically feasible time resolution of about 30 min [77, 206, 207]. Gradient-enhanced proton-detected heteronuclear techniques should allow imaging of fluxes for enriched compounds [163], without interference of water and lipid resonances.

Conventional suppression methods presently prevail in both medical and solution NMR. However, gradient methods seem to be the choice for the future. In this respect it is important to notice that the development of high performance gradients is a result of experimental demands for in vivo imaging and spectroscopy. Together with the many advances in the field of shaped and adiabatic pulses, these developments indicate that, in addition to providing new image technology, the biomedical NMR field is starting to provide useful spin-offs for NMR spectroscopy and that many more advances are expected to come from this most-rapidly growing section of the NMR community.

Acknowledgements: The authors are grateful to Drs. D van Ormondt, R Hurd, C Sotak, M Garwood, and A Bax for providing (p)reprints of their work. Drs. DI Hoult, PR Luyten, JA den Hollander, M Décorps, P Blondet, JR Alger, C Sotak, and DL Rabenstein kindly provided photographic copies for some of figures in this paper. Drs. A Knüttel, DI Hoult, R Hurd, B John, M von Kienlin, and JS Cohen are acknowledged for stimulating discussions. We are grateful to Dr. G Matson for helpful comments with the manuscripts.

5 References

1. Meier JE, Marshall AG (1990) In: Berliner LJ, Reuben J (eds) Biological magnetic resonance 9: 199. Plenum Press, New York
2. Hore PJ (1989) Methods in Enzym 176: 64; (1983) J Magn Reson 55: 283; (1983) J Magn Reson 54: 539
3. Lindon JC, Ferrige AG (1980) Progress NMR Spectr 14: 27
4. Redfield AG, Kunz SD (1979) In: Opella SJ, Lu P (eds) NMR in biochemistry. Dekker, New York, p 225; Redfield AG (1978) Methods in Enzym 49: 253

5. Aue WP (1986) Rev Magn Reson Med 1: 21
6. Bottomley PA (1989) Radiology 170: 1; Bottomley PA (1987) Ann NY Acad Sci 508: 333
7. Ordidge RJ, Mansfield P, Lohman JAB, Prime SB (1987) Ann NY Acad Sci 508: 376
8. Aue WP, Muller S, Cross TA, Seelig J (1984) J Magn Reson 56: 350
9. Bottomley PA, Foster TH, Darrow RD (1984) J Magn Reson 59: 338
10. Ordidge RJ, Connelly A, Lohman J (1986) J Magn Reson 66: 283
11. Luyten PR, Marien AJH, Sijtsma B, den Hollander JA (1986) J Magn Reson 67: 148
12. Frahm J, Merboldt KD, Hänicke W (1987) J Magn Reson 72: 502; Granot J (1986) J Magn Reson 70: 488; McKinnon G (1986) 5th Ann Mtg Soc Magn Reson Med (Works in Progress) p 168; Kimmich R, Hoepfel D (1987) J Magn Reson 72: 379
13. Frahm J, Bruhn H, Gyngell ML, Merboldt KD, Hänicke W, Sauter R (1989) Magn Reson Med 9: 79; (1989) Magn Reson Med 11: 47; Bruhn H, Frahm J, Gyngell ML, Merboldt KD, Hänicke W, Sauter R (1989) Magn Reson Med 9: 126; Frahm J, Merboldt KD, Michaelis T, Bruhn H, Gyngell ML, Hänicke W (1990) 9th Ann Mtg Soc Magn Reson Med p 133
14. Allen L, Eberly JH (1975) In: Optical resonance and two-level atoms. Wiley, New York
15. Hioe FT (1984) Phys Rev A30: 2100
16. Baum J, Tycko R, Pines A (1983) J Chem Phys 79: 4643; (1985) Phys Rev A32: 3435
17. Silver MS, Joseph RI, Hoult DI (1984) J Magn Reson 59: 347; (1985) Phys Rev A31: 2753
18. Bendall MR, Pegg DT (1986) J Magn Reson 67: 376
19. Ugurbil K, Garwood M, Rath AR (1988) J Magn Reson 80: 448
20. Warren WS (1988) Science 242: 878; Warren WS, Silver MS (1988) Adv Magn Reson 12: 247
21. Morris PG, McIntyre DJO, Rourke DE, Ngo JT (1989) NMR in Biomed 2: 257; Ngo JT, Morris PG (1987) Magn Reson Med 5: 217; (1986) Biochem Soc Trans 14: 1271
22. Conolly S, Macovski A (1985) 4th Ann Mtg Soc Magn Reson Med p 958
23. Murdoch J, Lent AH, Kritzer MR (1987) J Magn Reson 74: 226. Lent AH, Kritzer MR (1985) 4th Ann Mtg Soc Magn Reson Med, p 1015
24. Kirkpatrick S, Gelatt Jr CD, Vecchi MP (1983) Science 22: 671
25. Hardy CJ, Bottomley PA, O'Donnel M, Roemer P (1988) 77: 233
26. Tomlinson BL, Hill HDW (1973) J Chem Phys 59: 1775
27. Hsu AT, Hunter WW, Smallbrock P, Marshall AG (1987) J Magn Reson 72: 75
28. Brandes R, Kearns DR (1985) J Magn Reson 52: 120
29. Levitt MH (1986) Progr NMR Spectrosc 18: 61; Freeman R, Kempsell SP, Levitt MH (1980) J Magn Reson 38: 453
30. Tycko R (1984) Chem Phys Letters 111: 462
31. Vold RL, Waugh JS, Klein MP, Phelps DE (1968) J Chem Phys 48: 3831
32. Mansfield P, Chapman B (1986) J Phys E 19: 540
33. Turner R, Bowley RM (1986) J Phys E 19: 876
34. Roemer PB, Edelstein WA, Hickey JS (1986) 5th Ann Mtg Soc Magn Reson Med p 1067
35. Jensen DJ, Brey WW, Delayre JL, Narayana PA (1987) Med Phys 14: 859
36. von Kienlin M, Hedges LK, Olson AW (1990) 9th Ann Mtg Soc Magn Reson Med p 538
37. Marshall AG, Marcus T, Sallos J (1979) J Magn Reson 35: 227
38. Davies S, Bauer C, Barker P, Freeman R (1985) J Magn Reson 64: 155
39. Delsuc MA, Lallemand JY (1986) J Magn Reson 69: 504
40. Cooper JW (1976) In: Levy GC (ed) Topics in ^{13}C NMR Spectroscopy 2: 391. Wiley, New York; Cooper JW (1976) J Magn Reson 22: 345; Cooper JW, Machay IS, Pawle GB (1977) J Magn Reson 28: 405
41. Roth K, Kimber BJ, Feeney J (1980) J Magn Reson 41: 302
42. Haasnoot CA, Hilbers CW (1983) Biopolymers 22: 1259
43. Bielecki A, Levitt MH (1989) J Magn Reson 82: 562
44. Marion D, Ikura M, Bax A (1989) J Magn Reson 84: 425
45. Kuroda Y, Wada A, Yamazaki T, Nagayama K (1989) J Magn Reson 84: 604; (1990) J Magn Reson 88: 141
46. van Ormondt D et al. this book
47. Tsang P, Wright PE, Rance M (1990) J Magn Reson 88: 210
48. Plateau P, Dumas C, Gueron M (1983) J Magn Reson 54: 46
49. Callaghan PT, MacKay AL, Pauls KP, Soderman O, Bloom M (1984) J Magn Reson 56: 101
50. Loaiza F, McCoy M, Hammes S, Warren WS (1988) J Magn Reson 76: 504
51. Hoult D (1979) J Magn Reson 35: 69
52. Loaiza F, McCoy M, Hammes S, Warren WS (1988) J Magn Reson 77: 175
53. Friedrich J, Davies S, Freeman R (1987) J Magn Reson 75: 390

54. Emsley L, Bodenhausen G (1989) Magn Reson Med 10: 273
55. Barkhuysen H, de Beer R, van Ormondt D (1987) J Magn Reson 73: 553; Barkhuysen H, de Beer R, Bovee WMMJ, van Ormondt D (1985) J Magn Reson 61: 465
56. Kung SY, Rao DVB, Arun KS (1983) J Opt Soc Am 73: 1799
57. Kumaresan R, Tufts DW (1982) IEEE Trans Acoust Speech Signal Proc 30: 833
58. Abragam A (1961) In: The principles of nuclear magnetism. Oxford Univ Press, New York, p 73
59. Kaiser R, Bartholdi E, Ernst RR (1974) J Chem Phys 60, 2966
60. Haase A, Frahm J, Hänicke W, Matthaei D (1985) Phys Med Biol 30: 341
61. Doddrell DM, Galloway GJ, Brooks WM, Field J, Bulsing JM, Irving MG, Baddeley H (1986) J Magn Reson 70: 176
62. Brereton IM, Galloway GJ, Field J, Marshman MF, Doddrell DM (1989) J Magn Reson 81: 411; Brereton IM, Field J, Moxon LN, Irving MG, Doddrell DM (1989) Magn Reson Med 9: 118
63. Moonen CTW, van Zijl PCM (1990) J Magn Reson 88: 28
64. Griffey RH, Flamig DP (1990) J Magn Reson 88: 161
65. van Zijl PCM, Moonen CTW, Alger JR, Cohen JS, Chesnick AS (1989) Magn Reson Med 10: 256
66. Moonen CTW, van Zijl PCM, Gillen J, Daly P, von Kienlin M, Cohen JS, Wolf G (1989) NMR in Biomed 2: 201
67. van Zijl PCM, Moonen CTW (1989) 8th Ann Mtg Soc Magn Reson Med, p 488; (1990) 9th Ann Mg Soc Magn Reson Med, p 67
68. Barker GJ, Mareci TH (1989) J Magn Reson 83: 11
69. Patt SL, Sykes BD (1972) J Chem Phys 56: 3182
70. Benz FW, Feeney J, Roberts GCK (1972) J Magn Reson 8: 114
71. Krugh TR, Schaefer WC (1975) J Magn Reson 19: 99; Mooberry ES, Krugh TR (1975) J Magn Reson 17: 128
72. Inubushi T, Becker ED (1983) J Magn Reson 51: 128
73. Gupta RK (1976) J Magn Reson 24: 461; Gupta RK, Gupta P (1979) J Magn Reson 34: 657
74. Morris GA, Freeman R (1978) J Magn Reson 29: 433
75. Haasnoot CAG (1983) J Magn Reson 52: 153
76. Kao LF, Hruby VJ (1986) J Magn Reson 70: 394
77. Segebarth CM, Baleriaux DF, Luyten PR, van Rijen PC, Tulleken CAF, den Hollander JA (1990) Magn Reson Med 13: 62
78. Schaefer J (1972) J Magn Reson 6: 670
79. Jesson JP, Maekin P, Kniessel G (1973) J Amer Chem Soc 95: 618
80. Campbell ID, Dobson CM, Ratcliffe RG (1977) J Magn Reson 27: 455; Campbell ID, Dobson CM, Jeminet J, Williams RJP (1974) FEBS Letters 49: 115
81. Hoult DI (1976) J Magn Reson 21: 337
82. Rabenstein DL, Isab AA, Brown DW (1980) J Magn Reson 41: 361
83. Cutnell JD, Dallas J, La Mar GN (1980) J Magn Reson 41: 213
84. Krishna NR (1976) J Magn Reson 22: 555
85. Blondet P, Décorps M, Albrand JP (1986) J Magn Reson 69: 403
86. Piveteau D, Delsuc MA, Guittet E, Lallemand JY (1987) J Magn Reson 71: 347
87. Canet D, Boudot D, Brondeau J (1988) J Magn Reson 79; 377
88. Starcuk Z, Pucek L, Fiala R, Starzuk Jr Z (1988) J Magn Reson 80: 344
89. Behar KL, den Hollander JA, Stromski ME, Ogino T, Shulman RG, Petroff OAC, Prichard JW (1983) Proc Natl Acad Sci USA 80: 4945
90. Plateau P, Gueron M (1982) J Amer Chem Soc 104: 7310
91. Turner DL (1983) J Magn Reson 54: 146
92. Sklenar V, Starcuk Z (1982) J Magn Reson 50: 495
93. Bodenhausen G, Freeman R, Turner DL (1977) J Magn Reson 27: 511
94. Starcuk Z, Sklenar V (1985) J Magn Reson 61: 567
95. Morris GA, Smith KI, Waterton JC (1986) J Magn Reson 68: 526
96. Wang C, Pardi A (1987) J Magn Reson 71: 154
97. Starcuk Z, Sklenar V (1986) J Magn Reson 66: 391
98. Hall MP, Hore PJ (1986) J Magn Reson 70: 350
99. Galloway GJ, Haseler LJ, Marshman MF, Wiliams DH, Doddrell DM (1987) J Magn Reson 74: 184
100. Sklenar V, Bax A (1987) J Magn Reson 74: 469; Bax A, Sklenar V, Clore GM, Gronenborn AM (1987) J Amer Chem Soc 109: 6511

101. Blondet P, Décorps M, Confort S, Albrand JP (1987) J Magn Reson 75: 434
102. von Kienlin M, Décorps M, Albrand JP, Foray MF, Blondet P (1988) J Magn Reson 76: 169
103. Brindle KM, Porteous R, Campbell ID (1984) J Magn Reson 56: 543
104. Hetherington HP, Avison MJ, Shulman RG (1985) Proc Natl Acad Sci USA 82: 3115
105. Williams SR, Gadian DG, Proctor E (1986) J Magn Reson 66: 562; Gadian DG, Proctor E, Williams SR, Cady EB, Gadiner RM (1985) Magn Reson Med 3: 150
106. Levitt MH, Roberts MF (1987) J Magn Reson 71: 576
107. Levitt MH (1988) J Chem Phys 88: 3481
108. Levitt MH, Sudmeier JL, Bachovchin WW (1987) J Amer Chem Soc 109: 6540
109. Morris GA, Silveston ACT, Waterton JC (1989) J Magn Reson 81: 641
110. Hull WE (1987) In: Croasmun WR, Carlson RMK (eds) Two-dimensional NMR spectroscopy, VCH Publ, New York, p 166
111. Bendall MR, Gordon RE (1983) J Magn Reson 53: 365
112. Garwood M, Ross BD, Mitchell SL, Hendrich K, Merkle H (1988) 7th Ann Mtg Soc Magn Reson Med, p 754; Ross BD, Merkle H, Hendrich K, Staewen RS, Garwood M (1991) Magn Reson Med, in press
113. Alger JR, Brunetti A, Nagashima G, Hossman KA (1989) Magn Reson Med 11: 73
114. Redfield AG, Kunz SD, Ralph EK (1975) J Magn Reson 19: 114
115. Alexander S (1961) Rev Sci Instr 32: 1066
116. Redfield AG, Gupta RK (1971) J Chem Phys 54: 1418
117. Hall LD, Sukumar S, Talagala SL (1984) J Magn Reson 56: 275
118. Yao C, Simplaceanu V, Lin AKLC, Ho C (1986) J Magn Reson 66: 43
119. Sklenar V, Tschudin R, Bax A (1987) J Magn Reson 75: 352; Sklenar V, Bax A (1987) J Magn Reson 75: 378
120. Dadok J, Sprecher RF (1974) J Magn Reson 13: 243
121. Gupta RK, Ferretti JA, Becker ED (1974) J Magn Reson 13: 275
122. Ogino T, Arata Y, Fujiwara S, Shaun H, Beppu T (1978) J Magn Reson 31: 523
123. Ernst RR, Bodenhausen G, Wokaun A (1987) In: Principles of nuclear magnetic resonance in one and two dimensions. Clarendon Press, Oxford
124. Dumoulin CL (1985) J Magn Reson 64: 38; Dumoulin CL, Vatis D (1986) Magn Reson Med 3: 282
125. Nosel W, Trimble LA, Shen JF, Allen PS (1989) Magn Reson Med 11: 398
126. Bax A, de Jong PG, Mehlkopf AF, Smidt J (1980) Chem Phys Letters 69: 567
127. Sotak CH, Freeman DM (1988) J Magn Reson 77: 382; Sotak CH, Freeman DM, Hurd RE (1988) J Magn Reson 78: 355
128. Hurd RE, Freeman DM (1989) Proc Natl Acad Sci USA 86: 4402; Freeman RM, Sotak CH, Muller HH, Young SW, Hurd RE (1990) Magn Reson Med 14: 321
129. Knüttel A, Kimmich R (1989) Magn Reson Med 10: 404
130. Doddrell DM, Brereton IM, Moxon LN, Galloway GJ (1989) Magn Reson Med 9: 132
131. Sotak CH (1990) J Magn Reson 90: 198
132. de Graaf AA, Luyten PR, Marien AJH, den Hollander JA, Bovee WMMJ (1990) 9th Ann Mtg Soc Magn Reson Med p 138
133. van Dijk JE, Bovee WMMJ (1990) 9th Ann Mtg Soc Magn Reson Med, p 24
134. Trimble LA, Shen JF, Wilman AH, Allen PS (1990) J Magn Reson 86: 191
135. Shen JF, Allen PS (1990) 9th Ann Mtg Soc Magn Reson Med, p 27
136. Knüttel A, Kimmich R, Spohn K-H (1990) J Magn Reson 86: 526; Knüttel A, Spohn K-H, Kimmich R (1990) J Magn Reson 86: 542
137. Swanson S, Yeung HN (1990) 31st Exp NMR Conf, p 197
138. Hurd R, John B (1991) J Magn Reson 91: 648; (1991) J Magn Reson 92: 658
139. van Zijl PCM, John B, Hurd RE (1990) 32nd Exp NMR Conf, p 76; Vuister GW, Boelens R, Kaptein R, Hurd RE, John B, van Zijl PCM, J Am Chem Soc, submitted; van Zijl PCM, Moonen CTW (1990) 32nd Exp NMR Conf, p 39
140. Sillerud LO, Hulsteyn DB, Fink NH (1990) 9th Ann Mtg Soc Magn Reson Med, p 25; Bendall MR, Pegg DT, Doddrell DM, Field J (1983) J Magn Reson 51: 520
141. Brooks WM, Irving MG, Simpson SJ, Doddrell DM (1984) J Magn Reson 56: 521
142. Freeman R, Mareci TH, Morris GA (1981) J Magn Reson 42: 341
143. Bendall MR, Pegg DT, Doddrell DM (1981) J Magn Reson 45: 8
144. Bulsing JM, Brooks WM, Field J, Doddrell DM (1984) Chem Phys Letters 104: 229
145. Doddrell DM, Nicholls KM, Sanders JKM (1984) FEBS Letters 170: 73
146. Bulsing JM, Doddrell DM (1986) J Magn Reson 68: 52

147. Dumoulin CL, Williams EA (1986) J Magn Reson 66: 86; Dumoulin CL (1986) Magn Reson Med 3: 90; Hardy CJ, Dumoulin CL (1987) Magn Reson Med 5: 58
148. Sørensen OW (1989) Progr NMR Spectr 21: 503; Sørensen OW, Ernst RR (1983) J Magn Reson 51: 477
149. von Kienlin M, Albrand JP, Authier B, Blondet P, Lotito S, Décorps M (1987) 75: 371
150. von Kienlin M, Moonen CTW, van der Toorn A, van Zijl PCM (1991) J Magn Reson 93: 423
151. Ziegler A, Albrand JP, Remy C, Décorps M (1990) 9th Ann Mtg Soc Magn Reson Med, p 1234
152. Brereton IM, Crozier S, Field J, Doddrell D (1991) J Magn Reson 93: 54
153. Barker P, Freeman R (1985) J Magn Reson 64: 334
154. Knüttel A, Spohn KH, Kimmich R (1991) Magn Reson Med 17: 470
155. Knüttel A, Kimmich R (1989) Magn Reson Med 9: 254; (1989) J Magn Reson 83: 335
156. Campbell ID, Dobson CM (1979) Meth Biochem Anal 25: 1
157. Freeman R (1965) J Chem Phys 43: 3087
158. Marchal JP, Canet D (1975) J Amer Chem Soc 97: 6581
159. Bendall MR, Pegg DT, Doddrell DM, Field J (1981) J Amer Chem Soc 103: 934
160. Sillerud LO, Alger JR, Shulman RG (1981) J Magn Reson 45: 142
161. Foxhall DL, Cohen JS, Tschudin RG, Bax A (1983) J Magn Reson 51: 330; Cohen JS, Chen CW, Bax A (1984) J Magn Reson 59: 181
162. Rothman DL, Behar KL, Hetherington HP, Shulman RG (1984) Proc Natl Acad Sci USA 81: 6330; Rothman DL, Arias-Mendoza F, Shulman GI, Shulman RG (1984) J Magn Reson 60: 430
163. Rothman DL, Behar KL, Hetherington HP, den Hollander JA, Bendall MR, Petroff OAC, Shulman RG (1985) Proc Natl Acad Sci USA 82: 1633; Fitzpatrick SM, Hetherington HP, Behar KL, Shulman RG (1990) J Cerebr Blood Flow Metab 10: 170; Rothman DL, Novotny EJ, Shulman GI, Howseman AM, Mason GF, Petroff OAC, Nixon T, Hanstock CC, Prichard JW, Shulman RG (1990) 9th Ann Mtg Soc Magn Reson Med, p 986
164. Knüttel A, Kimmich R, Spohn KH (1989) J Magn Reson 81: 570
165. Knüttel A, Kimmich R (1989) Magn Reson Med 10: 411
166. Counsell CJR, Levitt MH, Ernst RR (1985) J Magn Reson 64: 470
167. Maudsley AA, Wokaun A, Ernst RR (1978) Chem Phys Letters 55: 9
168. Garbow JR, Weitekamp DP, Pines A (1983) J Chem Phys 79: 5301
169. Weitekamp DP (1983) In: Waugh JS (ed) Adv in magnetic resonance 11: 111. Academic Press, New York
170. Sørensen OW, Eich GW, Levitt MH, Bodenhausen G, Ernst RR (1983) Progr NMR Spectr 16: 163
171. Bodenhausen G, Kogler H, Ernst RR (1984) J Magn Reson 58: 370; Müller N, Bodenhausen G, Ernst RR (1987) J Magn Reson 75: 297
172. Bain AD (1984) J Magn Reson 56: 418
173. Garbow JR, Weitekamp DP, Pines A (1982) Chem Phys Letters 93: 504
174. Becker ED, Ferretti JA, Farrar TC (1969) J Amer Chem Soc 91: 7784
175. Shoup RR, Becker ED (1972) J Magn Reson 8: 298
176. Hochmann J, Kellerhals H (1980) J Magn Reson 38: 23
177. Rabenstein DL, Millis KK, Strauss EJ (1989) Anal Chem 60: 1380A; Rabenstein DL (1984) J Biophys Biochem Methods 9: 277
178. Daniels A, Williams RJP, Wright PE (1976) Nature 261: 321
179. Brown FF, Campbell ID, Kuchel PW, Rabenstein DL (1977) FEBS Letters 82: 12
180. Rabenstein DL (1978) Anal Chem 50: 1265A; Rabenstein DL, Isab AA (1979) J Magn Reson 36: 281
181. Bryant RG, Eads TM (1985) J Magn Reson 64: 312; Eads TM, Kennedy SD, Bryant RG (1986) Anal Chem 58: 1752
182. Rabenstein DL, Fan S, Nakashima TT (1985) J Magn Reson 64: 541
183. Bales JR, Bell JD, Nicholson JK, Sadler PJ, Timbrell JA, Hughes HD, Bennett PN, Williams R (1988) Magn Reson Med 6: 300, and refs therein
184. Hahn EL (1950) Phys Rev 80: 580
185. Carr HE, Purcell EM (1954) Phys Rev 94: 630
186. Meiboom S, Gill D (1958) Rev Sci Instrum 29: 688
187. Freeman R, Hill HDW (1975) In: Jackman LM, Cotton FA (eds) Dynamic nuclear magnetic resonance spectroscopy. Academic Press, New York, p 131
188. Williams SR, Gadian DG, Proctor E, Sprague DB, Talbot DF, Young IR, Brown FF (1985) J. Magn Reson 63: 406

189. Inagaki F, Miyazawa T (1981) Progr NMR Spectr 14: 67
190. Woessner DE (1961) J Chem Phys 35: 41
191. Luz Z, Meiboom S (1963) J Chem Phys 39: 366
192. Winkler H, Michel D (1985) Adv Colloid Interface Sci 23: 149
193. Zhong J, Gore JC, Armitage IM (1989) Magn Reson Med 11: 295
194. van Zijl PCM, Moonen CTW (1990) J Magn Reson 87: 18
195. Karczmar GS, Twieg DB, Lawry TJ, Matson GB, Weiner MW (1988) Magn Reson Med 7: 111
196. Stejskal EO, Tanner JE (1965) J Chem Phys 42: 288; Tanner JE, Stejskal EO (1968) J Chem Phys 49: 1768
197. Tanner JE (1970) J Chem Phys 52: 2523
198. Kärger J, Pfeifer H, Heink W (1988) In Waugh JS (ed) Adv in magnetic resonance, 12: 1. Academic Press, New York
199. Canet D, Piter B, Belmajdoub A, Brondeau J, Boubel JC, Elbayed K (1989) J Magn Reson 81: 1
200. Andrasko J (1976) J Magn Reson 21: 479; Andrasko J (1976) Biochim Biophys Acta 428: 304
201. Hills BP, Belton PS (1989) In: Webb GA (ed) Ann reports on NMR spectroscopy 21: 99. Academic Press, London
202. Wang D-J, Ikai I, Okuda M, Wang Z, Zhao P, Leigh JS (1990) 9th Annual Meeting Soc Magn Reson Med p 1100
203. Avison MJ, Shulman RG, private communication
204. van Zijl PCM, Moonen CTW, Faustino P, Pekar J, Kaplan O, Cohen JS (1991) Proc Natl Acad Sci USA, 88: 3228
205. Gober JR, Schwartz GG, Schaefer S, Massie B, Matson GB, Weiner MW, Karczmar GS (1989) 8th Ann Mtg Soc Magn Reson Med, p 36
206. Moonen CTW, Sobering G, van Zijl PCM, Gillen J, Daly P (1990) 9th Ann Mtg Soc Magn Reson Med, p 139
207. Moonen CTW, van Zijl PCM, Frank JA, Le Bihan D, Becker ED (1990) Science 250: 53

B_1 Insensitive Adiabatic RF Pulses

Michael Garwood and Kâmil Uğurbil

Center for Magnetic Resonance Research and Departments of Radiology, Biochemistry & Medicine, University of Minnesota Medical School, 385 East River Road, Minneapolis, Minnesota 55455, USA

Table of Contents

NMR Basic Principles and Progress, Vol. 26

1 Introduction

Inhomogeneities in the B_1 field pattern of RF transmitters are often the cause of major problems in applications of the NMR phenomenon. This simply follows from the fact that NMR experiments in general rely on rotation of magnetization vectors about the B_1 field through an angle $\gamma \int B_1(t)dt$, where γ is the gyromagnetic ratio. Consequently, even very simple tasks such as spin-excitation becomes non-uniform over the sample and this deleterious effect gets compounded in more complex and demanding pulse sequences employed in contemporary NMR applications ranging from high resolution studies of molecular structure to imaging and localized spectroscopy in vivo. This problem is particularly pronounced in the human and intact animal investigations in vivo where geometrical constraints frequently require the use of transmitters and detectors, such as surface coils, which generate extremely non-uniform B_1 field patterns.

Motivated by the importance of this problem, significant efforts have been made to design pulses, such as composite pulses, that try to compensate for the B_1 non-uniformities, or to develop new B_1 insensitive schemes for inducing the desired spin transformations. The latter approach, which is the subject of this review, relies on the well known adiabatic principles [1]. While B_1-insensitive composite pulses have been developed (for a review, see Ref. 2), they generally provide B_1 compensation at the expense of substantially diminished off-resonance performance (i.e. bandwidth). The recently developed "adiabatic pulses" can execute most of the desired rotations over large variations in B_1 (e.g. 50 to 100 fold) and over a wide off-resonance range.

Initial efforts with adiabatic pulses focussed on variations on the classical adiabatic passage and half-passage experiments which are capable of spin inversion and excitation [3–6] but cannot be used for the multitude of other rotations desired in contemporary NMR experiments. The advent of **plane** rotation adiabatic pulses based on the inversion of the effective field ($\mathbf{B}^e$) [7, 8] initiated the development of numerous pulses or pulse trains so that most of what can be done with conventional pulses using homogeneous coils can now be performed with adiabatic pulses with unprecedented insensitivity to B_1 inhomogeneity. To list a few accomplishments, adiabatic pulses can now be employed for variable angle plane rotations, solvent suppression, editing based on J-modulation, heteronuclear polarization transfer, and slice and volume selection. The fundamental principles involved in these pulses as well as the tasks they can currently accomplish will be briefly reviewed in this chapter.

2 Reference Frames Necessary for the Analysis of the Problem

All adiabatic transformations, including the classical adiabatic fast-passage or half-passage are best visualized and examined in a rotating reference frame that precesses at the instantaneous frequency of the RF field, ω_{rf}. In this reference frame (which we call the **frequency** frame and use x', y', and z' to designate its three orthogonal axes), the $\mathbf{B}_1$ vector has a constant orientation unless a discontinuous phase shift is executed. If the frequency of the pulse deviates from the Larmor frequency (ω_L) of the spins, the longitudinal magnetic field encountered in the **frequency** frame will equal $\Delta\omega/\gamma$, where $\Delta\omega = (\omega_L - \omega_{rf})$ and γ is the gyromagnetic ratio. The effective field $\mathbf{B}^e$ is the vector sum of $\mathbf{B}_1$ and $\Delta\boldsymbol{\omega}/\gamma$, and is tilted away from the transverse plane by the angle α as depicted in Fig. 1a. The orientation of the $\mathbf{B}_1$ vector in the transverse plane is arbitrary (but constant in the absence of phase jumps) and will *always* be taken to be along the x' axis at the *beginning* of the pulse. In adiabatic half or full passage, B_1 phase in the frequency frame is constant and B_1 remains along x' throughout the pulse; this, however, is not necessarily the case for the recently developed "*plane rotation*" pulses.

In all adiabatic transformations, the orientation and the magnitude of $\mathbf{B}^e$ vector is modulated. The analysis of the motion imparted on the magnetization vectors by this time dependent $\mathbf{B}^e$ field requires that the problem is examined in a *second rotating* frame where the orientation of $\mathbf{B}^e$ is constant (except for discontinuous inversions as utilized in the "*plane rotation*" pulses). Relative to the *frequency* frame, this *second rotating* frame (designated by the orthogonal axes x'', y'', and z'' in Fig. 1) changes its orientation together with $\mathbf{B}^e$ at the instantaneous angular velocity of $d\alpha/dt$. As the B^e vector rotates about the y' axis due to B_1 and $\Delta\omega$ modulation, so do the x'' and y'' axes of the second rotating frame as illustrated in Fig. 1a; however, the y' and y'' axes of the two reference frames remain collinear in this particular simple case (Fig. 1a). Consequently, the problem as viewed in this *second rotating frame* involves a new effective field $\mathbf{b}^e$ (Fig. 1b) which is the vector sum of $B^e[\mathbf{z}'']$ and $\gamma^{-1}(-d\alpha/dt)[\mathbf{y}'']$, where $[\mathbf{z}'']$ and $[\mathbf{y}'']$ are unit vectors that identify the axes along which the vectors lie.

The reference frames described above are also utilized in the analysis of the more complex adiabatic **plane rotation** pulses. For the simple case of adiabatic full- or half-passage, the relative orientations and the rotations of the vectors as well as the two reference frames depicted in Fig. 1 are exact and complete. However, with the plane rotation pulses, both the orientations and the rotations become more complicated. Therefore, with respect to these transformations several conventions must be established.

In adiabatic **plane rotation** pulses, $\mathbf{B}^e$ is suddenly inverted in the frequency frame during the pulse; convenient analysis of the problem dictates that this "$\mathbf{B}^e$-flip" also appears as an inversion of $\mathbf{B}^e$ in the *second rotating* frame so that

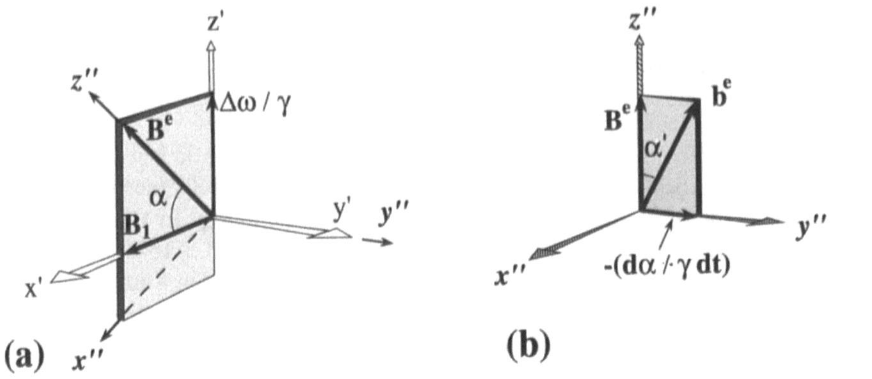

Fig. 1 a, b. The $\mathbf{B}_1$, $\Delta\omega/\gamma$ and effective field $\mathbf{B}^e$ in the rotating "frequency" frame (designated by x', y', z' coordinates) which precesses at the instantaneous *frequency* of the RF. The orientation of the "*second rotating*" frame designated by axes x'', y'', z'' relative to the frequency frame is also shown. **b.** The effective magnetic fields seen in the "second rotating" frame

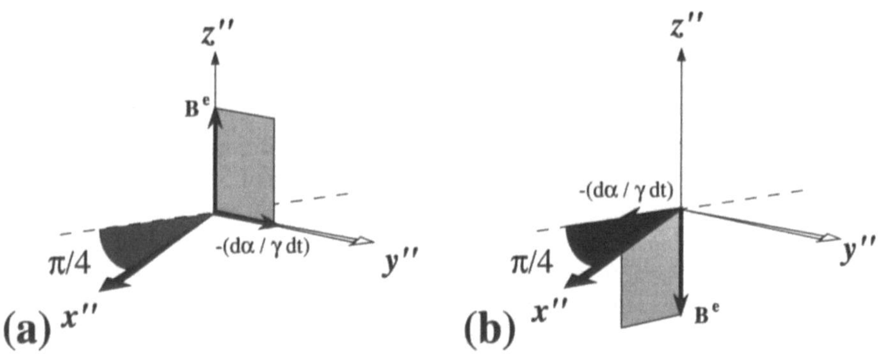

Fig. 2. The consequences of the discontinuous $\mathbf{B}^e$ inversion, the $\mathbf{B}_1$ phase shift by $5\pi/4$ to alter the axis about which $\mathbf{B}^e$ rotates, and changing the direction of $\mathbf{B}^e$ rotation during the BIR-4 pulse, as viewed in the second rotating frame. $\mathbf{B}^e$ rotates in the plane defined by the $\mathbf{B}_1$ and $\Delta\omega/\gamma$ vectors and the $d\alpha/\gamma\, dt$ vector is always perpendicular to this plane

this frame does not change its orientation relative to the *frequency* frame during these sudden alterations in $\mathbf{B}^e$ direction. Thus, the $\mathbf{B}^e$ vector is always collinear with the z'' axis of the second rotating frame but can acquire a positive or negative value. In addition to the $\mathbf{B}^e$ inversion, some pulses, such as variable angle adiabatic pulse BIR-4, utilize discontinuous shifts of the axis about which $\mathbf{B}^e$ rotates in the frequency frame. This action translates into changing the phase of the $\gamma^{-1}(-d\alpha/dt)$ vector in both the *frequency* and the *second rotating* frames. Finally, the direction of $\mathbf{B}^e$ precession can be, and often is, reversed in plane rotation pulses. This inverts the $-\gamma^{-1}(d\alpha/dt)$ vector in both frames, which is equivalent to changing the phase of the $-\gamma^{-1}(d\alpha/dt)$ vector by π. For example, $\pi/2$ plane rotation by BIR-4 requires that at $T/4$, where T is pulse duration, $\mathbf{B}^e$ is inverted, B_1 phase is shifted by $5\pi/4$ in order to shift the axis about which $\mathbf{B}^e$ rotates by $\pi/4$, and the direction of $\mathbf{B}^e$ precession is altered from clockwise to counterclockwise. The consequence of these discontinuous shifts as viewed in the second rotating frame is illustrated in Fig. 2. In the event that $\gamma^{-1}(d\alpha/dt)$ is no longer along one of the primary orthogonal axes of these reference frames, the orientation of the second rotating frame is altered relative the frequency frame with respect to *all* axes.

3 Adiabatic Fast-Passage

3.1 The Classical Adiabatic Passage Experiment

In a classical adiabatic fast passage, RF frequency is swept at a constant rate while B_1 remains invariant. At the onset of the pulse, the RF frequency deviates substantially from the spin Larmor frequency so that $|\Delta\omega/\gamma| \gg |B_1|$ and consequently α is approximately equal to $\pi/2$ or $-\pi/2$ corresponding to positive or negative $\Delta\omega$, respectively. As $|\Delta\omega|$ decreases progressively, it goes through zero and continues to change linearly to reestablish $|\Delta\omega/\gamma| \gg |B_1|$ but with a final value which is of opposite sign compared to the initial value. Thus, taking $\Delta\omega > 0$ at the beginning, $\mathbf{B}^e$ starts out collinear with the z' axis (i.e. $\alpha \sim \pi/2$ and $\mathbf{B}^e \sim \Delta\omega/\gamma[\mathbf{z}'']$), then rotates to be aligned with $\mathbf{B}_1$ (i.e. $\alpha = 0$ and $\mathbf{B}^e = \mathbf{B}_1$) and finally rotates further to lie along the negative z-axis (i.e. $\alpha \sim -\pi/2$, and $\mathbf{B}^e \sim -\Delta\omega/\gamma[\mathbf{z}''])$. The consequences of this RF pattern on the magnetization vectors can be easily deduced in the second rotating frame[1] (Fig. 1b); an initially longitudinal magnetization vector $\mathbf{M}$ will precess in the second rotating frame about $\mathbf{b}^e$ so that $\mathbf{M}$ will always remain within an angle $2\alpha'$ of $\mathbf{B}^e$. The adiabatic condition is satisfied when $|d\alpha/dt| \ll |\gamma B^e|$ so that α' is small and $\mathbf{b}^e \approx \mathbf{B}^e$. Thus, if $\mathbf{M}$ is collinear with $\mathbf{B}^e$ initially, it will remain so throughout the period when the RF is on provided that the adiabatic condition is satisfied. This translates into inversion of $\mathbf{M}$ because $\mathbf{B}^e$ rotates by π about the y' axis of the frequency frame.

Adiabatic half-passage is simply half of the full inversion so that the RF is terminated when $\Delta\omega = 0$. In this case, $\mathbf{B}^e$ rotates by $\pi/2$ only, changing its orientation from z' to x' axes in the frequency frame. Consequently, $\mathbf{M}$ which is initially along the z' axis, will rotate onto the transverse plane within an angle of $2\alpha'$ of the x'-axis.

The most important point to note about adiabatic half- or full-passage is that provided the adiabatic condition is fulfilled, only the **inversion** (i.e. $\mathbf{m}_z \rightarrow -\mathbf{m}_z$) or **excitation** ($\mathbf{m}_z \rightarrow \mathbf{m}_{xy}$) transformations are well defined without requiring detailed knowledge of the $\mathbf{B}_1$ and $\mathbf{B}^e$ fields. For these transformations, there exists a single rotation axis (in our case y' axis) and the transformation induced is either a π or $\pi/2$ rotation about this axis. For a magnetization vector

[1] The use of the second rotating frame suffices only when $\mathbf{b}^e$ is not changing its orientation rapidly relative to its magnitude (i.e. $|d\mathbf{b}^e/dt| \ll |\gamma\mathbf{b}^e|$)

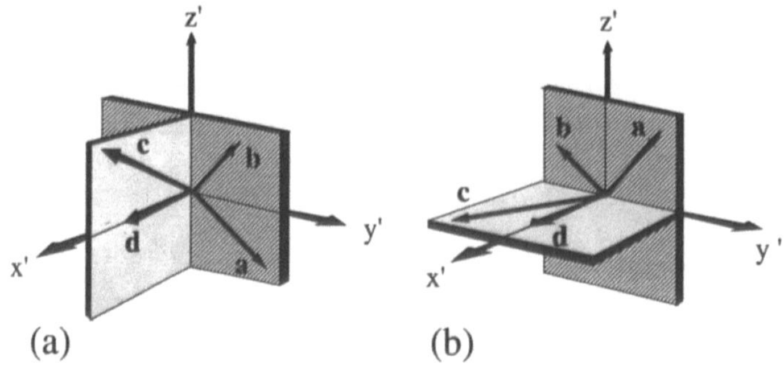

Fig. 3. The effect of a $+\pi/2$ plane rotation about the x' axis on vectors contained in the $x'z'$ plane (vectors **c** and **d**) and the $z'y'$ plane (vectors **a** and **b**)

M with both longitudinal and transverse components at the **onset** of the RF pulse, however, the transformation induced by adiabatic half or full passage is much more complex and cannot be described by a *single* rotation axis and rotation angle. In this case, the longitudinal component of **M** will still rotate by π or $\pi/2$ about the y' axis irrespective of the B_1 magnitude within the confines of the adiabatic condition; in contrast, the trajectory and final orientation of the transverse component of **M** will depend on the details of the pulse, and will require a second rotation axis to describe it. Irrespective of the specific nature of this motion, however, it can be stated that adiabatic full- or half passage cannot execute "plane" or "sphere" rotations. A "plane" rotation pulse will rotate **all** vectors with arbitrary orientations about a single rotation axis and through the same angle. All vectors contained in *any* plane perpendicular to or along the rotation axis will rotate about the same axis and through the same angle. For example, a $+\pi/2$ plane rotation about the x' axis alters the two planes and the vectors contained in them as illustrated in Fig. 3. Thus, $+\pi/2$ and π "plane" rotations about the x' axis will convert the vector (m_x, m_y, m_z) to $(m_x, -m_z, m_y)$ and $(m_x, -m_y, -m_z)$, respectively. Conventional "square" pulses induce "**plane**" rotations about the B_1 axis. Such rotations are essential in many applications. For example, all spin echo sequences and T_2 measurements require π plane rotation pulses, whereas stimulated echo sequences require $\pi/2$ plane rotation pulses.

It can be easily surmised that the *inversion* and *excitation* transformations induced by adiabatic passage and half-passage, respectively, can be relatively independent of B_1 magnitude since the requirement is that $|d\alpha/dt| \ll |\gamma B^e|$. In the "classical" adiabatic passage where B_1 is invariant and RF frequency is linearly swept, the adiabatic condition reduces to satisfying the inequality

$$|d\Delta\omega/dt| \ll \gamma^2 (B^e)^3 (B_1)^{-1} \tag{1}$$

where $d\Delta\omega/dt$ is a constant because the frequency sweep is linear in time. Since $B^e \geqq B_1$ always, the condition dictated by Eq. (1) is satisfied at a sufficiently large B_1 for a given $d\Delta\omega/dt$; additional increases in B_1 magnitude and hence power results in further improvements in fulfilling the adiabatic requirement.

It is, therefore, accurate to say, as is often claimed carelessly and indiscriminately, that for this case adiabatic transformations work above a certain power level and their performance gets better with increasing power. However, it should be mentioned that this is not universally correct. Most often used B_1 and $\Delta\omega$ modulation functions in fact also impart an *upper limit* on the operational B_1 or pulse power above which adiabatic half of full passage will not work.

3.2 Adiabatic Passage Based on Hyperbolic or Trigonometric Modulation Functions

Adiabatic pulses can be specified by defining their amplitude, frequency, and/or phase modulation functions. Continuing to restrict the discussion to a sample of spins with a single Larmor frequency, the magnetic fields encountered in the **frequency** frame can be written as

$$\mathbf{B_1}(t) = B_1^0 f_{\mathrm{B}}(t)[\mathbf{x}'] \quad \text{and} \quad \gamma^{-1}\boldsymbol{\Delta\omega} = \gamma^{-1}\Delta\omega^0 f_\omega(t)[\mathbf{z}'] \tag{2}$$

where $f_{\mathrm{B}}(t)$ and $f_\omega(t)$ are unitless normalized modulation functions, and B_1^0 and $\Delta\omega^0$ are the modulation amplitudes of B_1 and $\Delta\omega$, respectively. In a reference frame rotating at a fixed frequency, for example at the Larmor frequency of the spins, the B_1 vector can be described as a time dependent complex number

$$\mathbf{B_1}(t) = B_1^0 f_{\mathrm{B}}(t)\exp\left(\mathrm{i}\Delta\omega^0 \int_0^t f_\omega(t')\mathrm{d}t'\right) \tag{3}$$

Where the real and imaginary parts of $\mathbf{B_1}(t)$ are the x and y components, respectively, in the rotating frame, and the exponent of the exponential is the phase of B_1. We will restrict all future discussion to the formalism employed in Eq. (2) which is appropriate for the frequency frame.

Adiabatic half or full passage can be attained by a variety of $\Delta\omega$ and B_1 modulation schemes. Besides satisfying the adiabatic condition, it is only required that the $\mathbf{B}^\mathrm{e}$ is initially along the z' axis (i.e. $\mathbf{B}^\mathrm{e} \approx \gamma^{-1}\Delta\omega[\mathbf{z}'])$ and rotates onto the transverse plane to be approximately equal to $\mathbf{B_1}$ at the end to achieve spin excitation (adiabatic half-passage) or rotates further to become $-\gamma^{-1}\Delta\omega[\mathbf{z}']$ to execute an inversion (adiabatic passage). These requirements can be fulfilled using both B_1 and $\Delta\omega$ modulations. In this case, adiabatic *half*-passage necessitates that $B_1 \sim 0$ and $\Delta\omega \neq 0$ at the beginning of the pulse, and $B^\mathrm{e} \sim B_1$ and $\Delta\omega \sim 0$ at the end. Similarly, adiabatic *full* passage would have the boundary conditions $B_1 = 0$, $\Delta\omega \neq 0$ initially, $B_1 \neq 0$, $\Delta\omega = 0$ in the middle, and at the end $B_1 = 0$, $\Delta\omega \neq 0$ and of opposite sign relative to $\Delta\omega$ at the beginning of the pulse.

The adiabatic excitation and inversion can be achieved, for example, using

$$f_{\mathrm{B}}(t) = \operatorname{sech}\beta t \quad f_\omega(t) = \tanh\beta t \tag{4}$$

with t varying from $-T$ to 0 for adiabatic excitation and $-T$ to $+T$ for inversion; β defines the cut-off points for the sech and tanh functions and is typically chosen to set sech $(\beta T) = 0.01$ or $\beta T = 5.3$. This pair of modulation functions defines the well known sech/tanh pulse [3,4].

Similarly, the boundary conditions for B_1 and $\Delta\omega$ can be fulfilled using $\sin\theta(t)$ and $\cos\theta(t)$ functions for f_B and f_ω, respectively, with θ varying with time from 0 to π for adiabatic inversion and from 0 to $\pi/2$ for adiabatic half-passage [6].

The ability to satisfy the adiabatic condition with these modulation functions will dictate the B_1 range where these pulses will execute excitation or inversion transformations. This can be evaluated by examining the inequality $|\gamma B^e(d\alpha/dt)^{-1}| \gg 1$. For the sin/cos pair it can be shown that

$$|\gamma B^e(d\alpha/dt)^{-1}| = (4T_{90}\Delta\omega^0/2\pi)v^{-1}[\cos^2\theta + v^2\sin^2\theta]^{3/2} \tag{5}$$

or

$$|\gamma B^e(d\alpha/dt)^{-1}| = (4T_{90}\gamma B_1^0/2\pi)(v)[v^{-2}\cos^2\theta + \sin^2\theta]^{3/2} \tag{6}$$

where $v = \gamma B_1^0/\Delta\omega^0$ and T_{90} is the pulse duration required to rotate $\mathbf{B}^e$ by $\pi/2$ (i.e. $\theta(t) = \pi t/2T_{90}$); thus, T_{90} is the pulse duration for adiabatic *half-passage* and $2T_{90}$ would be the pulse duration for adiabatic *inversion*. The use of T_{90} rather than the pulse duration T is often preferred because all adiabatic pulses are composed of segments that rotate $\mathbf{B}^e$ by $\pi/2$ (i.e. segments that execute adiabatic half-passage).

In Eqs. (5) and (6), the parameters $(4T_{90}\Delta\omega^0/2\pi)$ and $(4T_{90}\gamma B_1^0/2\pi)$, which we will denote with R and Q, respectively, and v are dimensionless. It is of interest to note that $4T_{90}$ is the time it takes for $\mathbf{B}^e$ to rotate by 2π, $(2\pi/4T_{90})$ is the *average* frequency of $\mathbf{B}^e$ rotation, and $(4T_{90}\gamma B_1^0/2\pi) = 1$ for a constant amplitude and frequency square pulse whose B_1 magnitude is simply B_1^0.

It can be deduced from Eqs. (5) and (6) that two parameters are adequate to define the adiabaticity of the pulse when we are restricted to on-resonance performance; one of these two parameters is $(\gamma B_1^0/\Delta\omega^0)$ or v, and the other can be either $\gamma B_1^0 T$, $\Delta\omega^0 T$, $4T_{90}\Delta\omega^0/2\pi$ or $4T_{90}\gamma B_1^0/2\pi$. Of course, it is also possible to use other parameters that are directly proportional to these; for example following the convention employed in composite pulses, it is possible to use $2\gamma B_1^0 T/\pi$ as one of the parameters. This particular parameter is used by Garwood and Ke and is called $\omega_{1\max}/\omega_1^0$ [9]. The experimental definition of the pulse also needs the parameter R or $\Delta\omega^0 T$; if the pulse is executed as a phase modulated pulse, the phase pattern, ϕ, is given by the equation

$$\phi(t) = \Delta\omega^0 T \int_0^1 f_\omega(\tau)d\tau. \tag{7}$$

Thus, a constant that is proportional to $\Delta\omega^0 T$ or R must be specified; running the pulse for different T values defines $\Delta\omega^0$ and hence $\gamma B_1^0/\Delta\omega^0$ for a given B_1 field generated by a coil. If the pulse is executed using hardware capable of direct frequency modulation, then the frequency pattern generated requires the specification of $\Delta\omega^0$, and running the pulse for a given period T defines $\Delta\omega^0 T$.

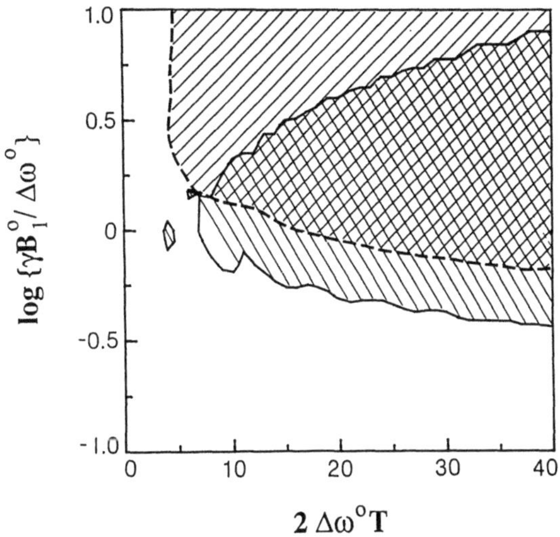

Fig. 4. Contour plot depicting the on-resonance performance of the adiabatic refocusing pulse BIREF-1 as a function of $\gamma B_1^0/\Delta\omega^0$ vs $2\Delta\omega^0 T$ when based on sin/cos or sech/tanh modulation functions, where modulation functions are specififed as f_B/f_ω. See Figs. 8 and 9 for details of BIREF-1; $2\Delta\omega^0 T$ is equal to R for this pulse. The 95% contour level is shown with a *solid line* for the sin/cos based pulse and a *dashed line* for the sech/tanh-based pulse. In the area bounded by the 95% contour line and identified with the *diagonal stripes*, the pulses achieve refocusing that preserves $\geqq 95\%$ of the initial, prepulse magnetization

Figure 4 displays contour plots, obtained using Bloch Equation calculations, depicting the R and $\gamma B_1^0/\Delta\omega^0$ region where sech/tanh[2] and sin/cos refocusing pulse BIREF-1 works with $\geqq 95\%$ efficiency [10]; R is equal to $2\Delta\omega^0 T$ for this pulse. For a given R, the sin/cos pulse has a lower and an upper limit for B_1^0 values over which it will work. The same is true for the sech/tanh pulse; however, the upper limit is much higher and does not appear in Fig. 4. Nevertheless, the sech/tanh pulse will also fail at some high γB_1^0 value when pulse duration and $\Delta\omega^0$ are fixed. The performance of these pulses simply follow from the ability to fulfill the adiabatic condition and can be predicted from Eqs. (5) and (6). In fact, as previously shown [10], it suffices to consider the initial and the end of the pulse for a qualitative analysis since the most severe demand on the adiabatic condition occurs at these time points for these modulation functions. For the sin/cos pair

$$\begin{aligned} |\gamma B^e (d\alpha/dt)^{-1}| &\sim R v^{-1} \cos^3\theta \quad \text{at} \quad \theta \sim 0 \\ &\sim R v^2 \sin^3\theta \quad \text{at} \quad \theta \sim \pi/2. \end{aligned} \tag{8}$$

If $v \ll 1$, the adiabatic condition becomes difficult to satisfy at $\theta \sim \pi/2$ and conversely the same problem arises for $v \gg 1$ at $\theta \sim 0$. Thus, at a given R and $\Delta\omega^0$ value, satisfactory performance has upper and lower bounds in B_1^0 and hence power. For the sech/tanh pulse

$$\begin{aligned} |\gamma B^0 (d\alpha/dt)^{-1}| &\sim R v^{-1} (\beta T \operatorname{sech} \beta t)^{-1} \quad \text{at} \quad t \sim -T \\ &\sim R v (\beta T)^{-1} \operatorname{sech}^2 \beta t \quad \text{at} \quad t \sim 0 \end{aligned} \tag{9}$$

Typically $\beta T = 5.3$ and $\operatorname{sech} \beta T = 0.01$. Thus, adiabaticity is proportional to $R/(0.053\,v)$ at the beginning ($t \sim -T$), and to $Rv/5.3$ at $t \sim 0$, respectively.

[2] For simplicity, the modulation functions are often specified using the format f_B/f_ω

Therefore, with these modulation functions, high v (i.e. $\gamma B_1^0/\Delta\omega^0$) values are tolerated much better relative to the sin/cos pulse. If Rv^{-1} (in other words $R(\gamma B_1^0/\Delta\omega^0)$) is not much larger than 0.053, however, the adiabatic condition will still fail at the beginning of the pulse. Hence, these modulation functions have a *higher* upper limit in power compared to sin/cos functions, but they do none-the-less have an upper limit.

3.3 Off-Resonance Performance and Slice Selectivity of Adiabatic Passage

When the pulses are applied so that $\Delta\omega$ is nonvanishing when $f_\omega(t) = 0$, then the general equations for the magnetic fields present in the *frequency* frame become

$$\mathbf{B}_1(t) = B_1^0 f_B(t)[\mathbf{x}'] \quad \text{and} \quad \gamma^{-1}\boldsymbol{\Delta\omega}(t) = \gamma^{-1}[\Delta\omega^0 f_\omega(t) + \Omega][\mathbf{z}'] \tag{10}$$

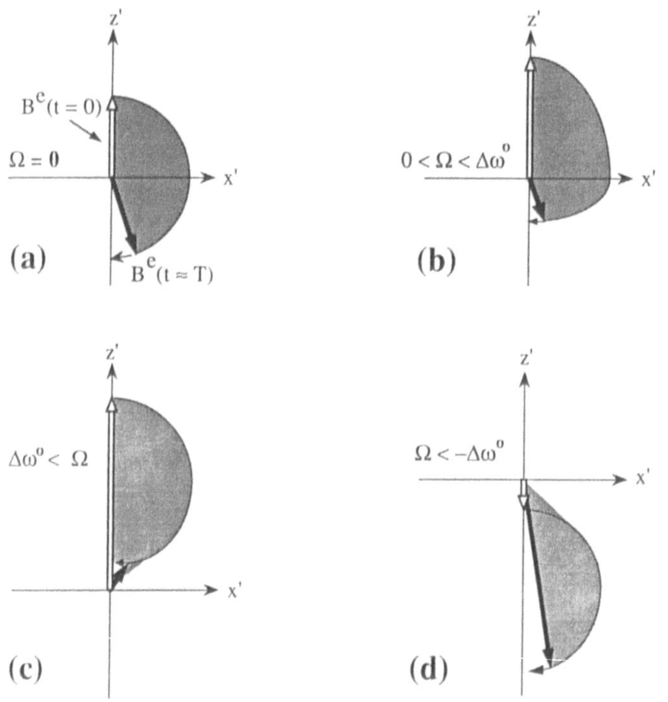

Fig. 5. $\mathbf{B}^e$ trajectory during adiabatic inversion at different values of resonance offset, Ω. The vector shown along the *z' axis is* $\mathbf{B}^e$ at the beginning of the pulse ($t = 0$), and the solid black vector is $\mathbf{B}^e$ towards the end of the pulse ($t \sim T$). The *gray area* is the area swept by the $\mathbf{B}^e$ vector as it rotates during the pulse; the *solid line* at the boundary of the gray area is the trajectory traced by the tip of the $\mathbf{B}^e$ vector. An initially longitudinal magnetization vector will remain collinear with $\mathbf{B}^e$ and follow it provided the adiabatic condition is satisfied

where Ω is the "resonance offset" in rad/sec. By necessity, Ω must be present when the sample contains spins with more than one Larmor frequency because the condition $\Delta\omega = 0$ when $f_\omega(t) = 0$ cannot be simultaneously satisfied for all spins.

For adiabatic *inversion* and *excitation*, we are concerned only with magnetization vectors that are initially parallel to $\mathbf{B}^e$ and remain so throughout the pulse and follow $\mathbf{B}^e$. Therefore, in order to visualize the transition that occurs, all we need to do is to define the $\mathbf{B}^e$ trajectory. Let us consider the case when $\Delta\omega^0 > 0$ and $f_\omega(t)$ goes from $+1$ to -1 during the inversion pulse; at the beginning and the end of the pulse, $\gamma\mathbf{B}^e$ equals $+\Delta\omega[z']$ and $-\Delta\omega[z']$, respectively. When $\Omega = 0$, $\mathbf{B}^e$ trajectory is a simple rotation about the y' axis of the *frequency* frame (Fig. 5a). When Ω is non-vanishing, $\gamma B^e = (\Delta\omega^0 + \Omega)$ at the beginning of the pulse and $(\Omega - \Delta\omega^0)$ at the end. Thus, the $\mathbf{B}^e$ trajectory still defines an inversion provided that $|\Omega| < |\Delta\omega^0|$ (Fig. 5b). However, if $|\Omega| > |\Delta\omega^0|$, then $\mathbf{B}^e$ is no longer inverted during the pulse (Fig. 5c and 5d). For example, if $\Omega > \Delta\omega^0$, $\mathbf{B}^e$ moves away from z'-axis as $\mathbf{B}_1(t)$ magnitude increases and $\Delta\omega$ decreases; however, at the end of the pulse, $\mathbf{B}^e$ again goes

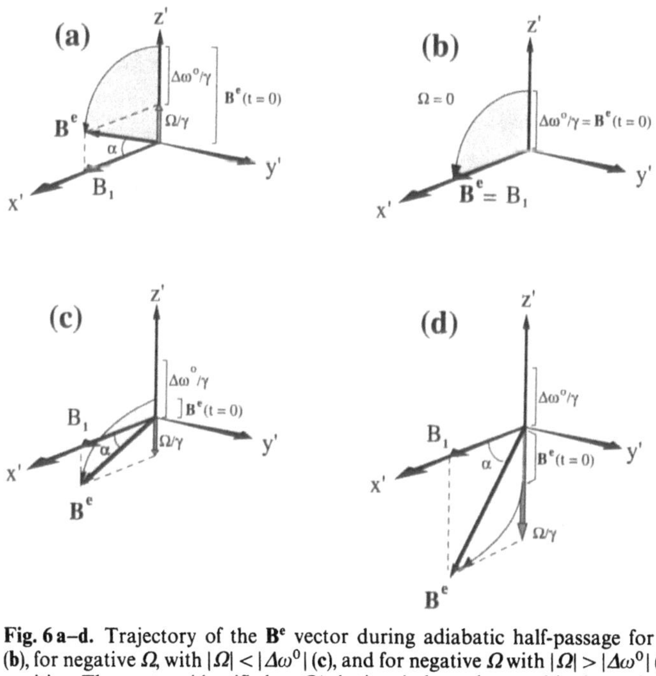

Fig. 6a–d. Trajectory of the $\mathbf{B}^e$ vector during adiabatic half-passage for $0 < \Omega < \Delta\omega^0$ (**a**), $\Omega = 0$ (**b**), for negative Ω, with $|\Omega| < |\Delta\omega^0|$ (**c**), and for negative Ω with $|\Omega| > |\Delta\omega^0|$ (**d**). $\Delta\omega^0$ is assumed to be positive. The vectors identified as Ω/γ is time independent and is shown in each figure. The $\mathbf{B}_1$ and $\mathbf{B}^e$ vectors shown depict these vectors at the end of the pulse. At the onset of the pulse ($t = 0$), $\mathbf{B}_1 = 0$ and $\mathbf{B}^e = \gamma^{-1}(\Delta\omega^0 + \Omega)[\mathbf{z}']$

back to be alligned along the $+z'$ direction because $\Delta\omega$ remains positive (Fig. 5c). The boundary between the ability to invert and not to invert is defined by the condition $|\Omega| = |\Delta\omega^0|$; for $\Omega > |\Delta\omega^0|$ or $\Omega < -|\Delta\omega^0|$, the magnetization vector is returned to its initial orientation, and it is inverted if and only if $|\Omega| < |\Delta\omega^0|$. Consequently, an inversion "slice" is defined. At or near $|\Omega| = |\Delta\omega^0|$, the pulse fails to satisfy the adiabatic condition completely; these regions are the two "edges" of the slice where signal excitation occurs leading to non-vanishing transverse magnetization as a result of the non-adiabaticity.

The same type of considerations easily lead to the conclusion that adiabatic **excitation** cannot be slice selective. Again let us consider the case where $\Delta\omega^0 > 0$, B_1 is along $+x'$ axis and $f_\omega(t)$ goes from $+1$ to 0. If $\Omega > 0$, $\mathbf{B}^e$ rotates counter-clock-wise about the y' axis and is left above the transverse plane making an angle α with the x'-axis (Fig. 6a) where $\alpha = \arctan(\Omega/\gamma B_1)$. A magnetization vector $\mathbf{M}$ initially along $+z'$, will follow B^e and remain parallel to it. Signal detected is equal to $|M|\cos\alpha$. If $-\Delta\omega^0 < \Omega < 0$, $\mathbf{B}^e$ rotates in the same direction but through the transverse plane as shown in Fig. 6c. However, when $\Omega < -\Delta\omega^0$, $\mathbf{B}^e$ is along $-z'$ initially and is rotated **clockwise** about y'-axis; in this case, a magnetization vector $\mathbf{M}$ that is initially along $+z'$ direction will remain *antiparallel* to $\mathbf{B}^e$ and will also rotate *clockwise* about to y' axis, generating a signal along the negative x'-axis equal to $-|M|\cos\alpha$ (Fig. 6d). In particular, a point of singularity occurs when $\Omega = -\Delta\omega^0$ in which case $\Delta\omega = 0$ at the beginning of the pulse and the adiabatic condition fails; in this case signal excitation becomes B_1 dependent because the magnetization vector precesses about the B_1 vector at the beginning of the pulse. Thus, adiabatic half-passage has a very asymmetric off-resonance performance.

3.4 Optimized Modulation Functions

3.4.1 On Resonance

It is in principle possible to start with the adiabatic condition and construct optimized functions which work over a predetermined $\gamma B_1^0/\Delta\omega^0$ and R or Q range. One possible approach to this problem was recently described [10]. This particular solution is based on reducing the unknowns in the problem by choosing the modulation functions to be one of the commonly used functions such as sin/cos or sech/tanh, and driving them not linearly with the time, but with an arbitrary function of time, $\theta(t)$. Letting

$$B_1(t) = B_1^0 f_B(\theta(t)) \quad \text{and} \quad \Delta\omega(t) = \Delta\omega^0 f_\omega(\theta(t)), \tag{11}$$

we obtain

$$\begin{aligned}\gamma B_1^e(d\alpha/dt)^{-1} = {} & (T\Delta\omega^0)(v)^{-1}(v^2 f_B{}^2(\theta) + f_\omega{}^2(\theta))^{3/2} \\ & \cdot\{(f_B(df_\omega/d\theta) - f_\omega(df_B/d\theta)\}^{-1}(d\theta/d\tau)^{-1},\end{aligned} \tag{12}$$

where $v = \gamma B_1^0/\Delta\omega^0$ as given before, and τ is *normalized time* defined as t/T. In this case, for any initial choice of f_B and f_ω, it is possible to construct a function $f_{min}(\theta)$ so that in a pre-specified $(\gamma B_1^0/\Delta\omega^0)$ range

$$|v^{-1}[(v^2 f_B{}^2(\theta) + f_\omega{}^2(\theta)]^{3/2} \cdot \{(f_B(df_\omega/d\theta) - f_\omega(df_B/d\theta)\}^{-1}| \geqq |f_{min}(\theta)| \tag{13}$$

The desired $\theta(\tau)$ is then constructed by numerical integration and interpolation by equating

$$(d\theta/d\tau) = q f_{min}(\theta) \tag{14}$$

where q is given by the specific integral

$$q = \int_{\theta_1}^{\theta_2} [f_{min}(\theta)]^{-1} d\theta \tag{15}$$

which is performed between θ_1 and θ_2, the maximum and minimum values acquired by θ as τ goes from 0 to 1. Combining Eqs. (13) through (15), the adiabatic condition can be expressed as

$$|\gamma B^e (d\alpha/dt)^{-1}| \geqq T\Delta\omega^0 q^{-1}. \tag{16}$$

In this approach, $f_{min}(\theta)$ and consequently q are determined by the initial choice of f_ω and f_B function pairs and the $(\gamma B_1^0/\Delta\omega^0)$ range over which satisfactory performance is desired. For a given q then, there will exist a $T\Delta\omega^0$ or R value above which the pulse will work essentially uniformly well over the desired $\gamma B_1^0/\Delta\omega^0$ range. The modulation function thus obtained are described by the nomenclature NOM (f_B/f_ω; v: a–b), where a and b specify optimization limits in $\gamma B_1^0/\Delta\omega^0$ or v. It should be emphasized that in this approach numerical integration is used to solve Eq. (14) to obtain $\theta(\tau)$; the desired result is not based on a computerized "search" algorithm. It has also been shown that analytical solutions are possible under somewhat more restricted conditions [11] using the approach taken in the NOM procedure [10].

Experimentally, a given range in B_1 inhomogeneity generated by a coil can be operationally covered at different values of $\gamma B_1^0/\Delta\omega^0$. For example, if the coil has 50 fold variation in B_1, NOM pulses can be constructed to operate over $\gamma B_1^0/\Delta\omega^0$ range of 1 to 50 or 0.1 to 5. There may exist many different experimental considerations which dictate this choice. For example, it is generally desirable to operate at low Q (i.e. low $\gamma T B_1^0$) and low $\gamma B_1^0/\Delta\omega^0$ values for spectroscopy applications where minimum possible power deposition, reduced peak power, and a large off-resonance range is desired. Based on Eq. (16), then, the most desirable choice is the one that gives the minimum $q\gamma B_1^0/\Delta\omega^0$ value. A systematic study to find the "optimal" range was not performed in this approach; never-the-less, Bloch equation simulations and experimental data demonstrated that in $\gamma B_1^0/\Delta\omega^0$ ranges that can be attained with small surface coils and approximately 100 W amplifiers, NOM functions were superior to their parent functions f_B and f_ω in three ways: 1) the range of B_1 insensitivity can be *predetermined*, 2) they can operate over a much larger range of variation in B_1

(insensitivity to factor 10 to 100 variation in B_1 is easily obtainable) and 3) they can be constructed to achieve the larger B_1 insensitivity at less $\gamma B_1^0 T$ values (i.e. *peak* power required is less).

3.4.2 On- and Off-Resonance

Optimization of modulation functions for on-resonance performance does not guarantee similar improvements off-resonance. However, the procedure described for on-resonance can be extended to cover off-resonance behavior as well [12]. For the more general case, we must start the analysis of the problem with the formalism introduced in Eq. (10) for $\mathbf{B_1}$ and $\Delta\omega$. Furthermore, the appropriate dimensionless parameter to introduce in evaluating off-resonance behavior is $\Omega/\gamma B_1^0$; this logically follows from the fact that off-resonance performance of all pulses, adiabatic or not, deteriorate with diminishing B_1^0. The reasons for this B_1 dependence can be easily visualized for normal and time-reversed[3] adiabatic half-passage. During adiabatic half-passage, $\mathbf{B^e}$ will be rotated so that at the end of the pulse the angle between the $\mathbf{B^e}$ vector and the transverse plane is equal to arctan $(\Omega/\gamma B_1^0)$ (Fig. 6a–c); provided the adiabatic condition is fulfilled, an initially longitudinal magnetization vector $\mathbf{M}$ will follow $\mathbf{B^e}$ and the signal detected will be $\mathrm{M}\cos\{\arctan(\Omega/\gamma B_1^0)\}$. If we are interested in detecting 95% or more of the initial magnetization, the cosine of this final angle must be $\geqq 0.95$; this dictates that $\arctan(\Omega/\gamma B_1^0) \sim 18°$ or less, which in turn requires that, $\Omega/\gamma B_1^0$ must be 0.33 or less. Thus, in constructing modulation function, the most that should be required with respect to off-resonance performance is that the adiabatic condition is fulfilled for $-0.33 < \Omega/\gamma B_1^0 < 0.33$; forcing the modulation functions to satisfy the adiabatic condition over a larger $\Omega/\gamma B_1^0$ would not improve the magnitude of the signal detected.

In the context of off-resonance performance, a second point of concern exists with adiabatic half-passage where $f_B \approx 0$ and $f_\omega \neq 0$ should define the initial conditions. Letting $\xi = \Omega/\gamma B_1^0$, Eq. (10) can be rewritten as

$$\mathbf{B_1}(t) = B_1^0 f_B(t)[\mathbf{x}'] \quad \text{and} \quad \gamma^{-1}\mathbf{\Delta\omega}(t) = \gamma^{-1}\Delta\omega^0[f_\omega(t) + v\xi][\mathbf{z}'] \tag{17}$$

As can be easily surmized from Eq. (17) and as previously mentioned, a point of singularity arises when $v\xi$ equals $+1$ or -1 for negative and positive frequency modulation functions, respectively; this leads to $\mathbf{\Delta\omega} = 0$ at the beginning of adiabatic half-passage and the applied RF behaves like a conventional pulse, inducing a precession of the magnetization vector *about* the $\mathbf{B_1}$ vector. Avoiding this point of singularity means that the operational v range should be much less than 3 when $-0.33 < \xi < 0.33$. In full passage inversion pulses, these points of singularity defines the "slice" boundaries where the pulse behaves non-adiabatically.

[3] Time-reversed adiabatic half-passage starts out with $\mathbf{B^e} \sim \mathbf{B_1}$ and $\Delta\omega \sim 0$, and ends with $\mathbf{B_1} \sim 0$ and $\mathbf{B^e} \sim \gamma^{-1}\Delta\omega[z']$

The NOM procedure can be expanded to incorporate off-resonance performance within the constraints of the two points discussed above. The off-resonance analog of Eq. (12) is

$$|\gamma B_1^e(d\alpha/dt)^{-1}| = |(T\Delta\omega^0)(v)^{-1}\{v^2 f_B{}^2(\theta) + (v\xi + f_\omega(\theta))^2\}^{3/2} \cdot\{(f_B(df_\omega/d\theta) - (v\xi + f_\omega(\theta))(df_B/d\theta)\}^{-1}(d\theta/d\tau)^{-1}|, \tag{18}$$

The rest of the procedure is similar to the on-resonance calculation. First, initial functions for f_ω and f_B are chosen, and a v range is specified over which 95% or better performance is desired. Next, for each θ, the magnitude of the function

$$(T\Delta\omega^0)(v)^{-1}\{v^2 f_B{}^2(\theta) + (v\xi + f_\omega(\theta))^2\}^{3/2} \cdot\{(f_B(df_\omega/d\theta) - (v\xi + f_\omega(\theta))(df_B/d\theta)\}^{-1}(d\theta/d\tau)^{-1}, \tag{19}$$

is examined over the predetermined v and ξ ranges and the minimum value is extracted numerically or analytically; this minimum value for the given θ is equated to $f_{min}(\theta)$. The modulation functions are then constructed from the $f_{min}(\theta)$ function exactly as in on-resonance NOM procedure.

For reasons already outlined, sech/tanh modulation pair works at relatively higher $\gamma B_1^0/\Delta\omega^0$ values. As such they do not represent a good choice for adiabatic half-passage applications where off-resonance performance is of concern. However, interchanging the modulation functions to tanh/sech results in a scheme that fulfills the adiabatic condition at lower $\gamma B_1^0/\Delta\omega^0$ values. This pair was chosen as the starting point and pulses were constructed for $-0.33 < \xi < 0.33$ and over a predetermined 50 fold range in $\gamma B_1^0/\Delta\omega^0$ starting at different $\gamma B_1^0/\Delta\omega^0$ values. Figure 7a illustrates the calculated contour plots depicting the operational $\gamma B_1^0/\Delta\omega^0$ *versus* ξ range covered by adiabatic half-passage based on one such NOM function pair; this is NOM (tanh/sech; $v = 0.02 - 1$; $|\xi| < 0.33$). Similar plots are illustrated for sech/tach and tanh/tan modulation function; the latter pair were chosen because they approximate the NOM functions. In these calculations, $\Delta\omega^0 T/2\pi$ was 60 for all functions and the vertical axis can be converted to $\gamma B_1^0 T/2\pi$ simply by multiplying $\gamma B_1^0/\Delta\omega^0$ with 60. In this type of plot, what is desirable is that the contour lines cover approximately equal area at positive and negative ξ for $-0.3 < \xi < 0.3$, and reach as low as possible in $\gamma B_1^0/\Delta\omega^0$ (i.e. v). The NOM pulse covers a 50 fold range (and more (not shown in figure)) in $\gamma B_1^0/\Delta\omega^0$, starting at 0.017 at the 95% performance level. Please note that the vertical scale for the sech/tanh plot is different. The sech/tanh modulation functions achieve a minimum operational $\gamma B_1^0/\Delta\omega^0$ of 0.63 (at $\xi = 0$), and can only cover 10 fold range before the off-resonance performance becomes *asymmetric*. This asymmetry is due to the fact that at these high $\gamma B_1^0/\Delta\omega^0$ values, maximum operational ξ must decrease in order to avoid the singularity that exist when $v\xi = 1$. Thus, despite its past popularity, sech/tanh modulation functions are suboptimal with half-passage even though they are excellent for slice selection with adiabatic inversion pulses.

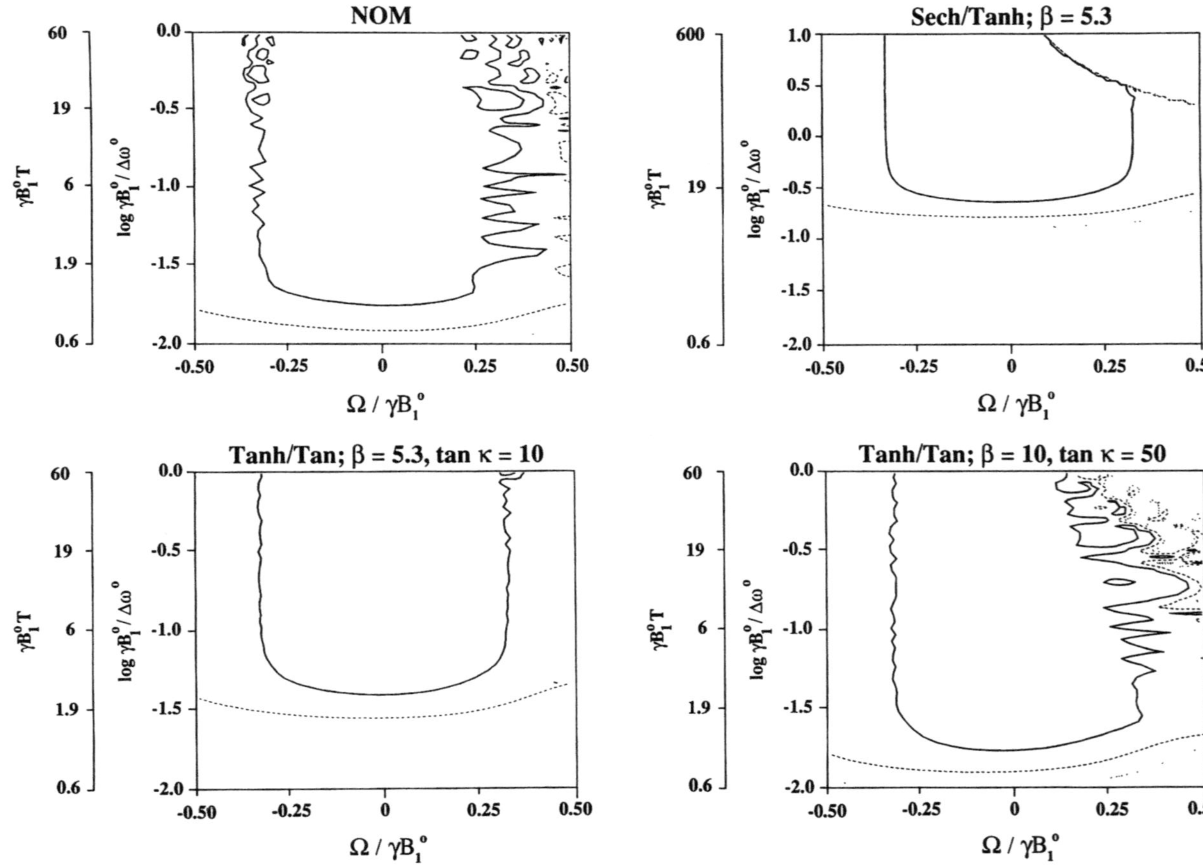

Fig. 7 a–d. Contour plots of the M_x component of the $M_z \rightarrow M_x$ transformation calculated for adiabatic excitation using different modulation functions. **a** On and off-resonance optimized pulse, NOM (tanh/sech; $v = 0.02–1.0$; 95%). **b** sech/tanh ($\beta T = 5.3$), **c** tanh/tan with $\beta T = 5.3$ and 10 as the maximum value of tan, and **d** tanh/tan with $\beta T = 10$, and 50 as the maximum value of tan Modulation functions are specified as f_B/f_ω. The *solid contour line* marks the "95%" contour; in the area bounded by this contour line, > 95% of the initial longitudinal magnetization is placed along x'. The *dashed* and *dotted lines* identify the 85% and 75% contours, respectively. All pulses were calculated for $\Delta\omega^0 T/2\pi$ of 60. The $\gamma B_1^0 T/2\pi$ scale is non-linear and is obtained by multiplying $\gamma B_1^0/\Delta\omega^0$ with 60

The lower two plots in Fig. 7 are based on tanh/tan modulation with two different cut-offs for the maximum value of the tan function which was normalized to vary between 0 and 1; thus $f_{\omega}(t) = [\tan \kappa]^{-1}[\tan \kappa t/T]$. The tanh/tan modulation was chosen because it approximates NOM. While these analytical pairs do not work as well as NOM, they represent an excellent compromise and out perform sech/tanh.

A simple calculation can illustrate the desirability of NOM over sech/tanh. If these adiabatic half-passage pulses, calculated for $\Delta\omega^0 T/2\pi$ of 60, are applied with a pulse length of 1 msec; then $\Delta\omega^0/2\pi = 60\,000$ Hz. This implies that the NOM pulse requires a minimum γB_1^0 of 1020 Hz (i.e. 60 000 Hz times the minimum operational $\gamma B_1^0/\Delta\omega^0$ of 0.017) to achieve 95% performance efficiency; in contrast, sech/tanh pulse will need a minimum γB_1^0 of 14 228 Hz.

4 Adiabatic Plane Rotation Pulses

4.1 Principles of Adiabatic Plane Rotation

The construction of **plane** rotating adiabatic pulses are based on two principles: 1) in the *second rotating* frame, the transformation induced on *any* magnetization vector must be an identity transformation, and 2) the second rotating frame must undergo the desired rotation relative to the *frequency* frame during the pulse. With respect to the latter principle, it must be kept in mind that the two reference frames are superimposable at the beginning of the RF train and that the second rotating frame follows $\mathbf{B}^e$ except during *discontinuous* jumps in the orientation of $\mathbf{B}^e$.

During adiabatic half-passage or half-passage run in reverse, the *orientation* of $\mathbf{b}^e$ in the second rotating frame will not change significantly as the magnitudes of B^e and $d\alpha/dt$ are modulated during the pulse provided that the adiabatic condition remains fulfilled and consequently $\mathbf{b}^e \approx \mathbf{B}^e$. In the case, the motion of a magnetization vector **M** of fixed magnitude but of *arbitrary initial* orientation will describe a simple precession about $\mathbf{B}^e$ in the second rotating frame. Thus, the component of **M** initially collinear with $\mathbf{B}^e$ will remain so, and the component of **M** perpendicular to $\mathbf{B}^e$ will remain on the plane perpendicular to $\mathbf{B}^e$ but will precess through an angle $\boldsymbol{\beta}$ given by $\gamma \int \mathbf{M} \times \mathbf{B}^e(t)dt$. If we *invert* the $\mathbf{B}^e$ vector and apply an RF pattern so that $\gamma \int \mathbf{M} \times \mathbf{B}^e(t)dt$ is now exactly $-\boldsymbol{\beta}$ following the $\mathbf{B}^e$ inversion, we will achieve an identity transformation on **M** in the *second rotating frame*. In other words, the adiabatic plane rotation pulses must execute a spin-echo experiment in the second rotating frame by one or more refocussing actions achieved by inverting the $\mathbf{B}^e$. When the adiabatic condition is fulfilled throughout the pulse, then the angle between **M** and $\mathbf{B}^e$ remains virtually unaltered and consequently, it becomes sufficient to require that $\int \mathbf{B}^e(t)dt = 0$ to assure that **M** is left at or near its initial position in the *second rotating frame* at the end of the pulse.

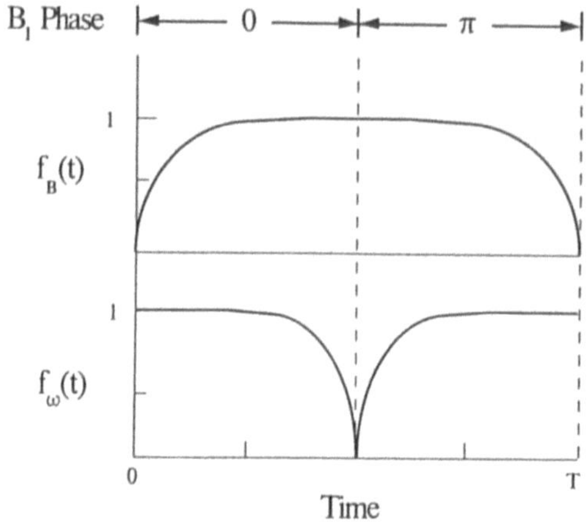

Fig. 8. f_B, f_ω and B_1 phase for the refocusing pulse BIREF-1. f_B and f_ω were drawn to illustrate salient points and do not represent unique or commonly used functions. Important aspects are the values acquired by these functions at $t = 0$, $T/2$ and T. Many different modulation functions will be suitable provided they fulfill the boundary conditions for $t = 0$, $T/2$ and T. However, tolerance to B_1 homogeneity and efficiency of the pulse will depend on the exact nature of the modulation functions

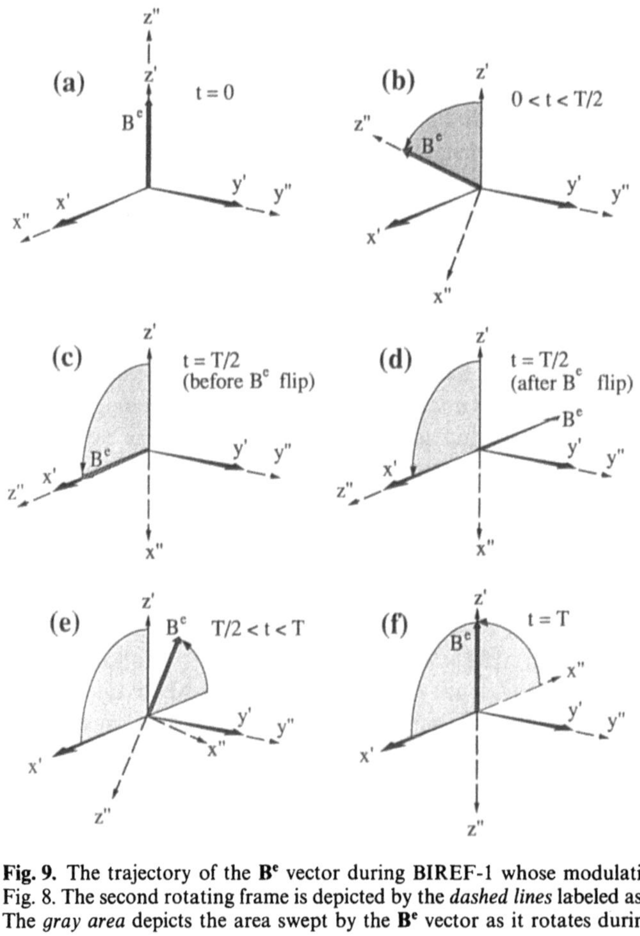

Fig. 9. The trajectory of the $\mathbf{B}^e$ vector during BIREF-1 whose modulation functions are given in Fig. 8. The second rotating frame is depicted by the *dashed lines* labeled as x'', y'' and z'' coordinates. The *gray area* depicts the area swept by the $\mathbf{B}^e$ vector as it rotates during the pulse

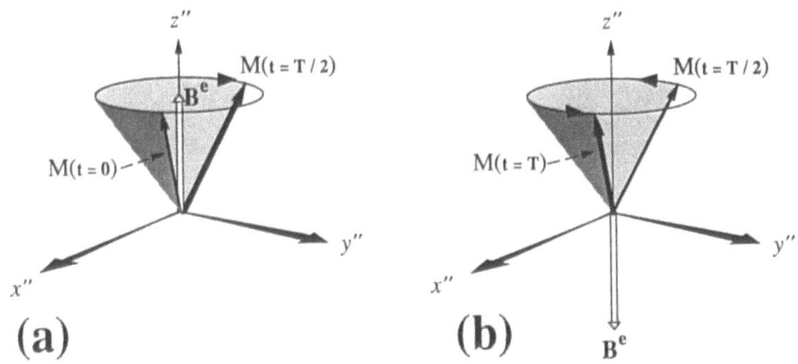

Fig. 10 a, b. The motion of an arbitrary magnetization vector **M** in the second rotating frame before (**a**) and after (**b**) the $\mathbf{B}^e$ inversion in the middle of the pulse ($t = T/2$). The $\mathbf{B}^e$ inversion returns **M** to its initial position at the end of the pulse

Consider for example the pulse given by f_B and f_ω patterns shown in Fig. 8. This is BIREF-1, a plane rotation refocusing pulse that executes a π rotation about the y' [8]. The $\mathbf{B}^e$ inversion is achieved in the middle of the pulse when $\mathbf{B}^e = \mathbf{B}_1$. The time evolution of the $\mathbf{B}^e$ vector and the second rotating frame relative to the frequency frame is illustrated in Fig. 9 for the case where $\Omega = 0$. Initially, the two pulse frames are superimposed and $\mathbf{B}^e = \gamma^{-1}\Delta\omega^0[\mathbf{z}']$. In the first half of the pulse, $\mathbf{B}^e$ and the second rotating frame rotate counterclockwise about the y' axis (Fig. 9a–c). The shaded region in Fig. 9 depicts the area swept by the $\mathbf{B}^e$ vector during its rotation. In the middle of the pulse $\mathbf{B}^e = \mathbf{B}_1$ (provided $\Omega = 0$) and a π phase shift in $\mathbf{B}_1$ inverts $\mathbf{B}_1$ and consequently $\mathbf{B}^e$ (Fig. 9c and 9d). The second half of the pulse is equivalent to the first half except for time reversal and the B_1 phase. Because of symmetry, the magnitude of $\mathbf{B}_1$ and consequently $\int|\mathbf{B}^e(t)|dt$ is the same for the two halves of the pulse, and $\int\mathbf{B}^e(t)\,dt$ for the whole pulse is zero in the second rotating frame. Thus, in the *second rotating frame*, an arbitrary magnetization vector **M** presesses first in one direction and then in the opposite direction so that it is finally returned to its original position (Fig. 10); however, the second rotating frame has executed a π rotation about the y' axis of the frequency frame as depicted in Fig. 9.

In general, the $\mathbf{B}^e$ inversion can also be executed when $\mathbf{B}^e = \gamma^{-1}\boldsymbol{\Delta\omega}$; in this case $\Delta\omega$ changes sign to achieve the $\mathbf{B}^e$ inversion. Examples of this are the refocusing pulse BIREF-2(8) and the variable angle pulse BIR-4 described further on.

Off-resonance, the problem gets somewhat more complicated. Proper execution of the desired plane rotation dictates that $\int|\mathbf{B}^e(t)|dt$ must still be zero when $\Omega \neq 0$. Even when this criteria is fulfilled, however, off-resonance performance will still fail with increasing $\Omega/\gamma\mathbf{B}_1^0$ ratio for other reasons. In case of BIREF-1, for example, $\int|\mathbf{B}^e(t)|dt = 0$ for the whole pulse even for non-vanishing Ω; however, $\int\mathbf{M} \times \mathbf{B}^e(t)\,dt$ deviates from zero with non-vanishing Ω because $\mathbf{B}^e$ is not truly inverted by the π phase shift in $\mathbf{B}_1$; instead, $\mathbf{B}^e$ is

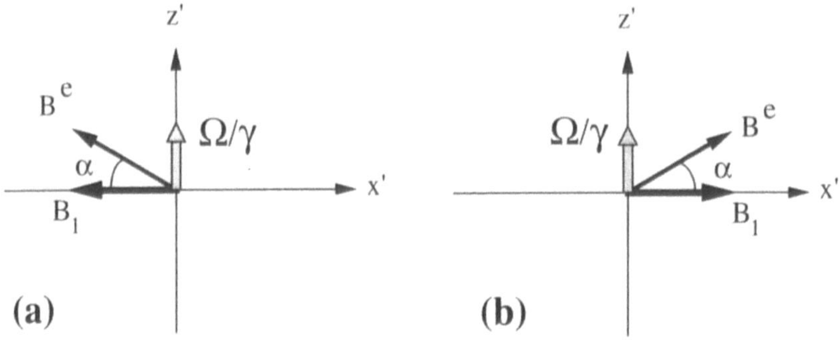

Fig. 11 a, b. $\mathbf{B}^e$, $\mathbf{B}_1$ and Ω/γ vectors immediately before (**a**) and after (**b**) the π shift in $\mathbf{B}_1$ phase during BIREF-1

reoriented so that it makes an angle of [π-(2 arctan $\Omega/\gamma B_1^0$)] with respect to its orientation prior to the π phase jump in B_1 (Fig. 11).

Based on these principles, $\pi/2$ plane rotation pulses were described first [7] followed by refocusing π [8, 13–15] pulses and variable angle pulses [9, 16, 17]. The most useful general purpose pulse is the variable angle BIR-4 pulse [9, 16].

4.2 General Purpose Variable Angle Adiabatic Plane Rotation Pulse BIR-4

BIR-4 is composed of four sections where B^e rotates by $\pi/2$; each of these sections are effectively adiabatic half-passage or adiabatic half-passage *in reverse*. The section between $T/4$ and $3T/4$ is simply an inversion pulse; thus, the pulse can also be viewed as composed of three sections, an adiabatic half passage in reverse, adiabatic inversion, and adiabatic half passage. This pulse has two $\mathbf{B}^e$ inversions executed when $\mathbf{B}^e = \gamma^{-1}\mathbf{\Delta\omega}$ at $T/4$ and $3T/4$, where T is the pulse length. $\mathbf{B}^e$ inversions are simply changes in the sign of $\Delta\omega$. The f_B, $\Delta\omega$, and B_1 phase functions in the frequency frame are depicted for BIR-4 in Fig. 12. The modulation functions used in this figure approximate tanh and tan for f_B and f_ω, respectively. As previously discussed, with respect to the operational B_1 range and off-resonance performance, the specific choice of functions is important. However, the desired transformations can be attained by a variety of modulation functions provided they are continuous for each half-passage section, and they fulfill the boundary conditions for f_B and f_ω at $t = 0, T/4, 3T/4$ and T. The discontinuous change in the sign of f_ω at $T/4$ and $3T/4$ and the B_1 phase are critical for achieving variable angle plane rotations. The B_1 phase jumps $\Delta\phi_1$ and $\Delta\phi_2$ are defined so that relative to an *initial* and *arbitrary* phase of ϕ, the phase of the B_1 in the *frequency* frame is $\phi + \Delta\phi_1$ between $T/4$ and $3T/4$ and $[\phi + \Delta\phi_1 + \Delta\phi_2]$ after $3T/4$; experimentally, if the pulse is executed using amplitude and phase modulation, the instantaneous pulse phase must discontinuously change by $\Delta\phi_1$ and an additional amount of $\Delta\phi_2$ at $T/4$ and $3T/4$, respectively.

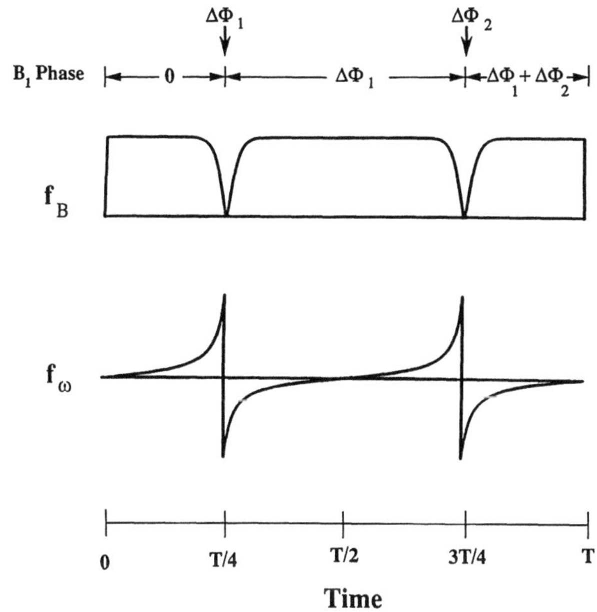

Fig. 12. Variable angle, plane rotation BIR-4 pulse. Modulation functions shown are based on the tanh/tan function pair

The BIR-4 pulse can achieve a **plane** rotation through an arbitrary angle, the magnitude of which is determined by the phase jumps $\Delta\phi_1$ and $\Delta\phi_2$. In general, for $\Delta\phi_1 = \pi + \delta$ and $\Delta\phi_2 = -(\pi + \beta)$, BIR-4 induces a plane rotation through an angle $-(\delta + \beta)$ about the *initial* B_1 axis (i.e. a *clockwise* rotation by $(\delta + \beta)$ about the x' axis of the frequency frame) followed by a rotation about the z' axis of the frequency frame (i.e. a phase shift) by $(\delta - \beta)$ radians; the latter rotation occurs because B_1 phase at the end is shifted by $(\delta - \beta)$ relative to its initial value. When $\Delta\phi_1 = -\Delta\phi_2 = \pi + \delta/2$, the rotation axis for the whole pulse is the initial B_1 axis (which is the same as the final B_1 axis), and δ is the angle of rotation or the "flip angle".

The rotations undergone by the *second rotating* frame relative to the *frequency frame* are depicted in Fig. 13 for $\Delta\phi_1 = -\Delta\phi_2 = \pi + \delta/2$ and with $\Delta\omega^0 > 0$. In this figure, $R_q(\theta)$ describes an operator which executes a rotation through an angle of θ about an axis identified with q. During BIR-4, $\mathbf{B}^e$ is initially along the x' axis and rotates clockwise by $\pi/2$ about the y' axis of the *frequency frame*, going from the x' to the z' axis. This is described by the rotation operator $R_{y'}(-\pi/2)$. Since B^e is always collinear with the x'' axis of the *second rotating frame*, the coordinate system described by the x'', y'', and z'' axes also rotates by $-\pi/2$ about the y' axis of the *frequency frame*. At $t = T/4$ (Fig. 13b), $\mathbf{B}^e$ is inverted, the axis of $\mathbf{B}^e$ rotation is displaced away from y' axis by an angle of $\delta/2$ by shifting the $\mathbf{B}_1$ phase by $(\pi + \delta/2)$, and the direction of

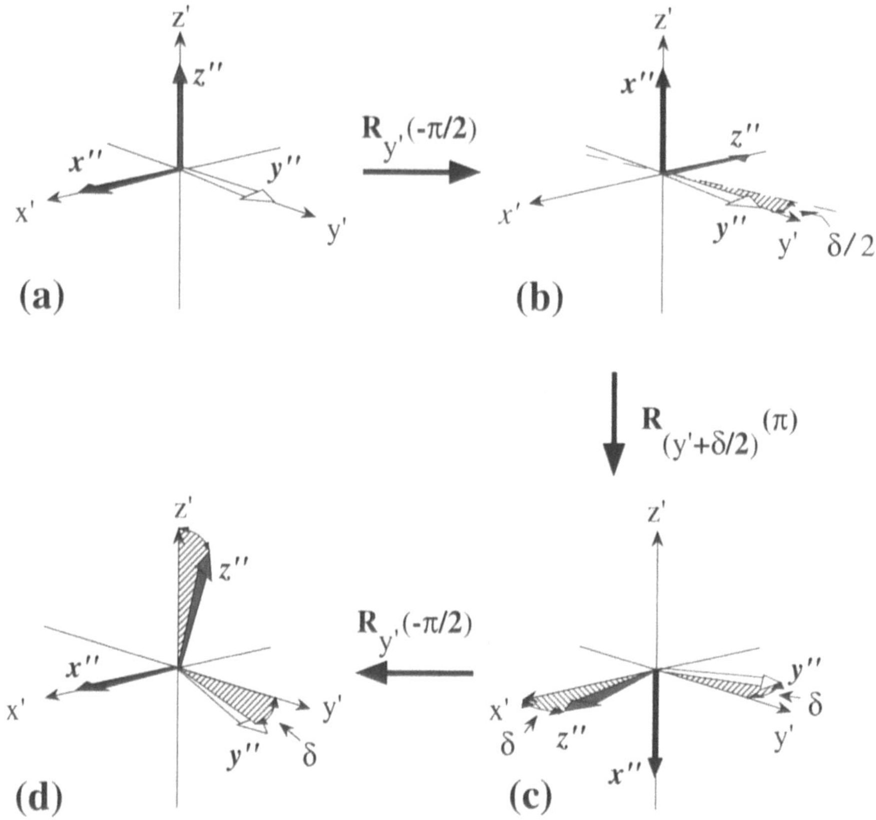

Fig. 13. The transformations of the second rotating frame relative to the frequency frame during a BIR-4 pulse that induces a net clockwise plane rotation of δ radians about the x' axis of the frequency frame

$\mathbf{B}^e$ precession in the plane defined by $\boldsymbol{\Delta\omega}$ and $\mathbf{B}_1$ vectors is reversed. During the subsequent "inversion" section of BIR-4 between $T/3$ and $3T/4$, x'', y'', z'' coordinates rotate by $+\pi$ about an axis with a phase of $(y' + \delta/2)$ (depicted by the dashed line in Fig. 13b). Finally, at $3T/4$ (Fig. 13c) the direction and the axis of $\mathbf{B}^e$ rotation reverts back to the initial conditions and $-\pi/2$ rotation about the y' axis is executed. The net effect is described by the product of the three rotation operators, $R_{y'}(-\pi/2)\ R_{(y'+\delta/2)}(\pi) R_{y'}(-\pi/2)$, and is equivalent to the *second rotating frame* experiencing a *net* clockwise rotation by δ about the x' axis of the *frequency frame.* During plane rotation adiabatic pulses, the magnetization vectors undergo an *identity* transformation in the *second rotating frame* (except in special cases discussed further on); therefore, the rotation of this reference frame in fact describes the net effect of the pulse on any magnetization vector. The corresponding transformations for the specific case of $\pi/2$ plane rotation (i.e. $\delta = \pi/2$) by BIR-4 is illustrated in Fig. 14; the only

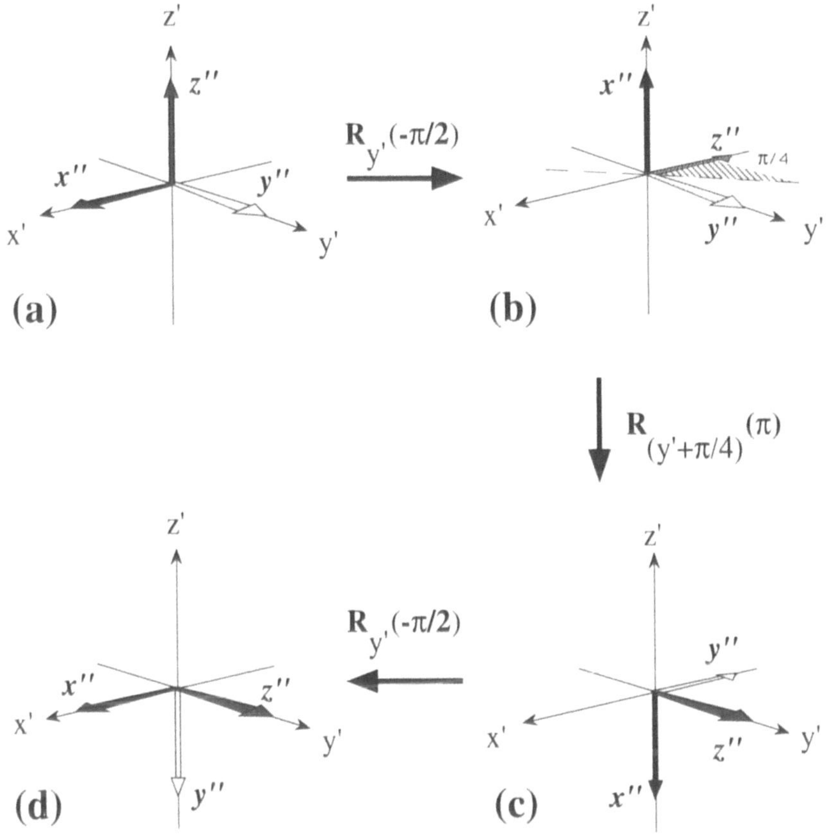

Fig. 14. Transformations of the second rotating frame during BIR-4 that induces a net $\pi/2$ rotation about the x' axis of the frequency frame

difference is that the inversion section of the pulse executes a π rotation described by $R_{(y'+\pi/4)}(\pi)$.

It should be noted that for $\delta = 0$, BIR-4 not only inverts $\mathbf{B}^e$ at $T/4$ and $3T/4$, but it inverts $\mathbf{b}^e$ as well. This can be surmized from Fig. 2. Therefore, to a large degree the adiabatic condition can be relaxed for $\delta \sim 0$; in fact, provided that $\mathbf{b}^e$ does not change its orientation rapidly compared to $\gamma \mathbf{b}^e$, the time symmetry of BIR-4 assures accurate execution of the identity transformation for small angle δ flips without demanding strict adherence to the adiabatic condition. Even for $\delta = \pi/2$, some compensation due to partial $\mathbf{b}^e$ inversion occurs, and the power requirements of BIR-4 consequently do not exceed that of adiabatic half-passage even though BIR-4 is composed of four adiabatic half-passage segments (Fig. 15).

Composite pulses of the type $90°_\alpha 180°_\beta 90°_\gamma$ provide a convenient tool to visualize the transformations induced by BIR-4. In this composite pulse notation, $90°_\alpha$ indicates a square pulse applied with a B_1 phase of α in the *frequency frame*, and of sufficient duration and amplitude to achieve a 90°

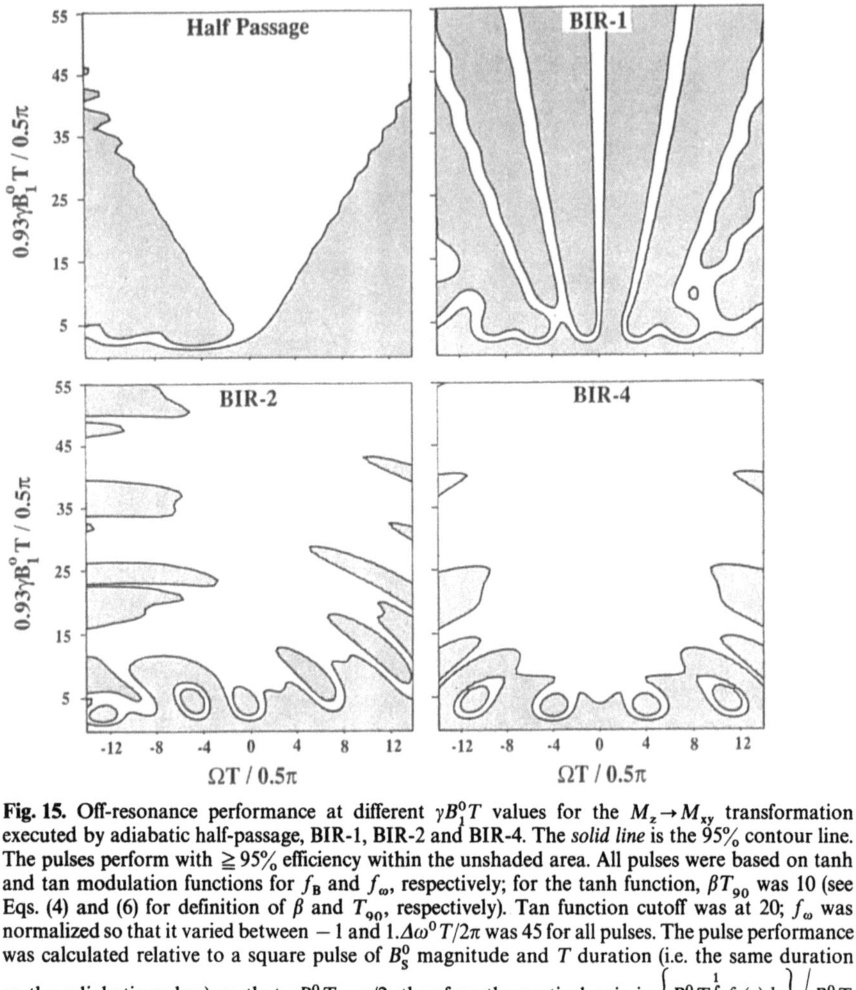

Fig. 15. Off-resonance performance at different $\gamma B_1^0 T$ values for the $M_z \rightarrow M_{xy}$ transformation executed by adiabatic half-passage, BIR-1, BIR-2 and BIR-4. The *solid line* is the 95% contour line. The pulses perform with $\geqq 95\%$ efficiency within the unshaded area. All pulses were based on tanh and tan modulation functions for f_B and f_ω, respectively; for the tanh function, βT_{90} was 10 (see Eqs. (4) and (6) for definition of β and T_{90}, respectively). Tan function cutoff was at 20; f_ω was normalized so that it varied between -1 and $1. \Delta\omega^0 T/2\pi$ was 45 for all pulses. The pulse performance was calculated relative to a square pulse of B_S^0 magnitude and T duration (i.e. the same duration as the adiabatic pulses) so that $\gamma B_S^0 T = \pi/2$; therefore the vertical axis is $\left\{ B_1^0 T \int_0^1 f_B(\tau) d\tau \right\} \Big/ B_S^0 T$. Since the integral is 0.93 for these modulation functions, the vertical axis becomes $0.93 B_1{}^0/B_S^0$ or $0.93\gamma B_1^0 T/0.5\pi$. Similarly, the horizontal axis is $\Omega/\gamma B_S^0$ or $\Omega T/0.5\pi$

clockwise rotation of a magnetization vector about the B_1 axis. In other words, $90°_\alpha$ is equivalent to a rotation depicted by the rotation operator $R_\alpha(-\pi/2)$. For the specific BIR-4 pulse implemented with $\Delta\phi_1 = -\Delta\phi_2 = \pi + \delta/2$ and with the $\mathbf{B_1}$ vector initially (and at the end) taken to be along the x'-axis of the *frequency* frame, the transformations undergone by the *second rotating frame* is described by the operator product $R_{y'}(-\pi/2) R_{y'+\delta/2}(\pi) R_{y'}(-\pi/2)$. Since $R_{y'+\delta/2}(\pi)$ is equal to $R_{y'+\pi+\delta/2}(-\pi)$, the analogous composite pulse to consider is $90°_{y'} 180°_{(y'+\pi+\delta/2)} 90°_{y'}$, where we are designating B_1 phases for the composite square pulse by specifying the axes in the frequency frame; thus,

$180°_{(y'+\pi+\delta/2)}$ means a 180° pulse applied with a B_1 phase of $(\pi+\delta/2)$ relative to the y' axis.

As previously emphasized, all adiabatic plane rotation pulses, including BIR-4 compensate for the B_1 induced dephasing. Thus, in the effort to conceptualize the transformation induced by these adiabatic pulses, the $\mathbf{B}_1$ inhomogeneities can be ignored and it suffices to consider only the transformation of the second rotating frame. For BIR-4 this can be accomplished simply by using the aforementioned composite pulse. Consider the vectors labeled x'', y'' and z'' in Figs. 13 and 14 to be magnetization vectors $\mathbf{M}_{x''}$, $\mathbf{M}_{y''}$, and $\mathbf{M}_{z''}$, initially parallel with the x', y' and z' axes, respectively, of the *frequency* frame. The effect of the composite pulse $90°_{y'}180_{y'+(\pi+\delta/2)}90°_{y'}$ (applied with homogeneous B_1) on these vectors is identical to the *net* transformation induced by BIR-4 (with $\Delta\phi_1 = -\Delta\phi_2 = \pi+\delta/2$) on the same vectors and consequently on the second rotating frame; however, the details of the path traversed by the magnetization vectors *during* the pulse differ for BIR-4 versus the composite pulse, and BIR-4 accomplishes the task with much larger B_1 insensitivity.

For BIR-4, $\int \mathbf{B}^e(t)dt$ and $\int \mathbf{M}\times\mathbf{B}^e(t)dt$ is zero for the whole pulse, even when Ω is non-zero provided that $|\Omega| < \Delta\omega^0$. Considering the on-resonance condition, $|\gamma\int \mathbf{M}\times\mathbf{B}^e(t)\,dt|$ for each of the four consecutive sections of the pulse are β, $-\beta$, $-\beta$, and β and add up to zero. Note that the sign of $\int \mathbf{M}\times\mathbf{B}^e(t)dt$ changes each time the $\mathbf{B}^e$ vector is inverted. With $\Omega\neq 0$, the corresponding rotation angles for the four sections become β_1, $-\beta_2$, $-\beta_1$ and β_2 respectively and still add up to zero even though they are no longer the same for all segments; this can be deduced from the fact that effective $\Delta\omega$ is altered, but B_1 is independent of Ω. When $|\Omega| > |\Delta\omega^0|$, the $\mathbf{B}^e$ inversions cannot be accomplished and the magnetization is simply dephased; in this sense, BIR-4 is slice selective; however, outside the slice it will dephase the magnetization rather than leave it unperturbed. Within the slice (i.e. when $|\Omega| < |\Delta\omega^0|$), BIR-4 will not work equally well for all Ω and will ultimately fail as $\Omega/\gamma B_1^0$ increases; the cause of this failure is that when $\Omega\neq 0$, $\mathbf{B}^e\neq\mathbf{B}_1$ initially but makes an angle equal to $\{\arctan\,\Omega/\gamma B_1^0\}$. Nevertheless, for a given B_1, off-resonance performance of BIR-4 exceeds that of adiabatic half-passage, previously described BIR-1 and BIR-2 pulses (Fig. 15). Most importantly unlike adiabatic half-passage, off-resonance behavior of BIR-4 is symmetric. It should also be noted that for a $\pi/2$ rotation, BIR-4 does not require any more power than adiabatic half-passage when driven by the same modulation functions. This is a consequence of the time symmetry of BIR-4, and can be understood **either** in terms of the vector diagrams utilized here or in terms of average Hamiltonian theory [9] or symmetric versions of recursive expansions [9] originally employed [18] in developing composite pulses. It is worthwhile to note that the symmetry principals involved in BIR-4 can be extended to construct composite pulses that display superior performance compared to previously described composite pulses [9]. For example $90°_0 180_{\pi+\delta/2} 90°_0$ is a composite pulse inspired by BIR-4 which achieves a general plane rotation through δ radians about an axis shifted in phase by $\pi/2$ relative to $\mathbf{B}_1$ (i.e. the rotation axis is perpendicular to the initial and final B_1 axis); the specific case of $\delta = \pi$, is the often used composite π pulse that partially compensates for B_1 inhomogeneities [19]. More detailed discussion of this is provided elsewhere [9].

5 Adiabatic Solvent Suppression, Editing and Polarization Transfer

One of the most unique features of BIR-4 is the fact that the rotation angle is dictated simply by the phase shifts $\Delta\phi_1$ and $\Delta\phi_2$. When $\Omega \neq 0$, these phase shifts can be achieved not only by shifting $\mathbf{B}_1$ phase but also by allowing magnetization vectors to precess on the transverse plane during finite intervals inserted within the BIR-4 pulse at the points where the phase jumps normally occur. During these intervals, the RF is turned off and the spin states evolve based on their resonance offset. Such intervals can be executed in the absence of or in addition to changes in $\mathbf{B}_1$ phase so that the net transformation is dictated by the sum of two events. The transformation in the *second rotating frame* is no longer an identity because of the evolution during the delays and the net transformation induced by the pulse on magnetization vectors is equivalent to the transformations of the second rotating frame plus the spin evolution. This general principle forms the basis of a whole series of **selective** pulses that discriminate according to spin evolution about the $\mathbf{B}_0$ axis. These include solvent selective adiabatic pulses (SSAP) [20], adiabatic editing based on spin–spin coupling (BISEP)[21], and adiabatic polarization transfer (BINEPT) [22]. These pulses or pulse trains accomplish their tasks with extremely high insensitivity to variations in B_1 magnitude (e.g. 50–100 fold variation in B_1 can be tolerated) and as such they permit the use of coils such as surface coils as if their B_1 profile is virtually homogeneous. These particular implementations should have utility not only in in vivo NMR spectroscopy but also in high resolution NMR where deletorious effects due to B_1 inhomogeneities are nonnegligible.

5.1 B_1 *Insensitive Solvent Suppression*

Consider BIR-4 with $\Delta\phi_1 = -\Delta\phi_2 = \pi$, so that the B_1 phase for this pulse is 0 initially, π between $T/4$ and $3T/4$ and 0 at the end. This pulse achieves an identity transformation. Now consider a delay of τ duration within this BIR-4 pulse exactly at $T/4$, at the end of the first segment. During this delay the pulse is turned-off and the magnetization vectors are allowed to precess on the

[4] Again, Ω is defined so that $\Delta\omega = \Omega$ when $f_\omega(t) = 0$

transverse plane of the frequency frame according to their resonance-offset[4], Ω. If $\Omega = 0$, the delay does not induce a precession and the magnetization vector encounters an *identity* transformation. As previously mentioned, this identity transformation results from $\mathbf{b}^e$, and not just $\mathbf{B}^e$ inversion and is accomplished admirably even with less than strict adherence to the adiabatic condition. If $\Omega \neq 0$, however, **M** will precess through an angle $\Omega\tau$ during the interpulse delay. This is equivalent to a B_1 phase jump of $-\Omega\tau$. We can now define an *effective* phase jump, $\Delta\phi^e$, where $\Delta\phi_1^e = \Delta\phi_1 - \Omega\tau$ and, in this case, $\Delta\phi_2^e = \Delta\phi_2$. With $\Delta\phi_1^e = \pi - \Omega\tau$ and $\Delta\phi_2^e = -\pi$, a magnetization vector with Ω resonance offset will undergo a plane rotation by $\Omega\tau$ about the $\mathbf{B}_1$ axis provided that $|\Omega| < |\Delta\omega^0|$; the $\mathbf{B}^e$ inversion does not occur outside of this range, as previously discussed, and the pulse fails. Note that even though $\Delta\phi_1^e \neq -\Delta\phi_2^e$, the B_1 phase is the same initially and at the end of the pulse because $\Delta\phi_1 = -\Delta\phi_2$; therefore the transformation is a simple rotation about the initial (and final) B_1 axis by $\Omega\tau$ radians and the signal detected is proportional to $\sin\Omega\tau$, with no phase roll across the spectrum.

This transformation and others that are discussed subsequent to this section can also be visualized by considering the problem in the **second rotating** frame; the general principle is that the overall rotation imparted on magnetization vectors is described by the transformation of this frame relative to the frequency frame, plus the evolution of the magnetization vector in the second rotating frame during the interpulse delay(s). In case of SSAP, the second rotating frame executes an identity transformation **relative** to the *frequency* frame during the pulse. On-resonance, magnetization vectors will also execute an identity transformation. Off-resonance, however, magnetization vectors precess about the x'' axis of the second rotating frame when $\gamma\mathbf{B}^e = \Omega[\mathbf{x}'']$ during the delay; because of the asymmetric delay, the rotation through $\Omega\tau$ is not eliminated during the remainder of the pulse. Consequently, at the end of this pulse magnetization vectors are left $\Omega\tau$ radians away from their initial orientation in the second rotating frame, having precessed through this angle about the x'' axis. Since the second rotating frame and the frequency frames are superimposable both at the beginning and at the end of this pulse, the net transformation observed in the frequency frame is a rotation about the B_1 axis by $\Omega\tau$.

As an example, consider an initially longitudinal magnetization vector **M** with $\Omega \neq 0$ and which is collinear with z' and z'' axes at the onset of the pulse. In addition to the rotation about $\mathbf{B}^e$ when the RF is on, **M** will precess through $\Omega\tau$ radians on the $z''y''$ plane during the delay; this additional rotation will not be refocused by the $\mathbf{B}^e$ inversions so that **M** will be left $\Omega\tau$ radians away from the z' axis of the *frequency* frame and relative to B_0 at the end of the pulse. Signal detected will be proportional to $\mathrm{M}\sin\Omega\tau$. Maximum excitation will occur for $\Omega\tau = \pi/2$. Similarly, refocusing and inversion transformations can be executed with full refocusing and inversion occurring at $\Omega\tau = \pi$.

In a spin-echo sequence, BIR-4 can be employed as solvent suppressive excitation and refocusing pulses. In order to achieve maximal signal detection

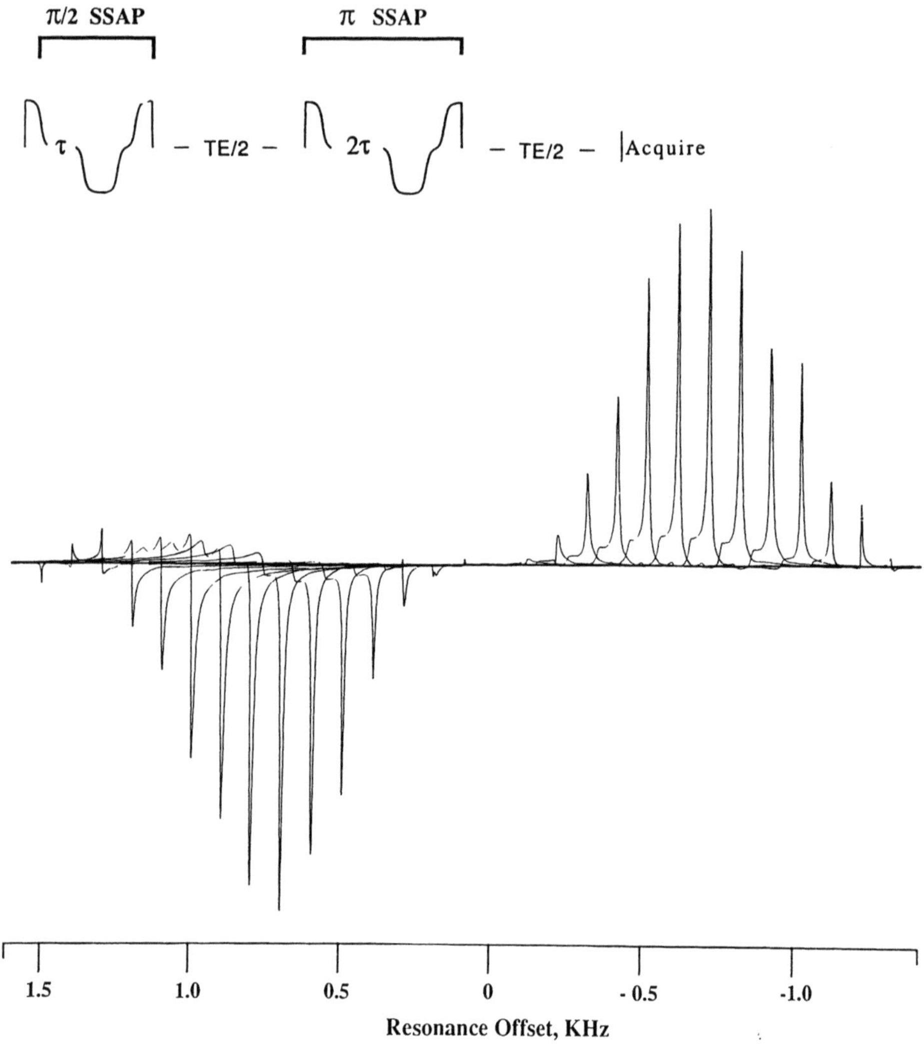

Fig. 16. The spin-echo, SSAP pulse sequence and its experimental performance as evaluated by a H_2O containing phantom and a surface coil, and shifting transmitter frequency relative to the H_2O resonance. The resulting signal detected is plotted as a function of offset. All peaks were processed with the same phase correction. The pulse is illustrated by its approximate B_1 shape; negative B_1 implies a π phase shift. The frequency (or phase) modulation of the adiabatic pulses are not shown

for a given Ω, the first BIR-4 pulse in the spin-echo sequence must be implemented with a delay of τ so that $\Omega\tau = \pi/2$; the second pulse (i.e. the refocusing pulse) must have a delay of 2τ so that $2\tau\Omega$ is equal to π. Figure 16 illustrates experimental data obtained by a spin echo SSAP sequence (shown in figure), by detecting the H_2O resonance as a function of transmitter offset (i.e. resonance offset). In the diagram illustrating the pulse sequence, the pulse shape is depicted by its approximate B_1 amplitude, and the interpulse delays are shown by τ and 2τ within each pulse; the negative B_1 amplitude signifies the π phase shift that is executed for the middle section of the pulse. A surface coil was employed with a container of H_2O placed against it; therefore, the B_1 was highly inhomogeneous over the sample. Note that unlike many solvent suppressive composite pulses such as 1331 and 2662 [23], SSAP pulses induce virtually no phase roll across the spectrum.

5.2 Editing Based on J Modulation (BISEP)

We will only consider heteronuclear editing when the spin that is not detected in the coupled spin system is a $S = 1/2$ nucleus. In fact, for simplicity of discussion we will refer to detection of ^1H nuclei coupled to ^{13}C spins, although the technique is general.

The B_1 amplitude and frequency profiles and the interpulse delays for the editing sequence (BISEP) are given in Fig. 17. The ^1H pulses are segmented BIR-4 pulses where B_1 phase is 0 initially, π between $T/4$ and $3T/4$, and 0 again at the end (i.e. $\Delta\phi_1 = -\Delta\phi_2 = \pi$ for the B_1 phase). This π phase shift for the middle section is depicted by the negative B_1 amplitude in the figure. The pulses have two delays at $T/4$ and $3T/4$. The ^{13}C pulses which must occur at the same time as the middle section of the ^1H BISEP pulses are simple adiabatic inversions. It is important to note that for BISEP, difference spectroscopy is not necessary to achieve editing. This sequence gives an identity transformation

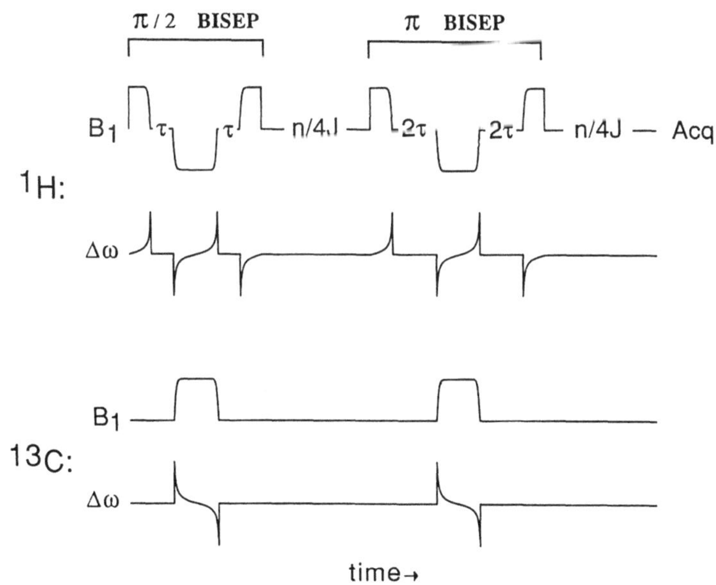

Fig. 17. Heteronuclear editing BISEP sequence. Negative B_1 implies a π phase shift

to the uncoupled spins and leaves the coupled spins on the transverse plane with π phase difference between the two ^1H magnetization vectors coupled, respectively, to "up" or "down" ^{13}C spin states (i.e. $|+>$ and $|->$ ^{13}C S_z states).

Let us first consider the effect of the $\pi/2$ BISEP pulse on the spin system. If the inversion pulse for the ^{13}C nucleus is not implemented, then the symmetric delays in BIR-4 applied at ^1H frequency assures that all pressessions of the ^1H spins due to chemical shift and coupling to other spins are refocussed and the pulse executes an *identity* transformation. In other words, $\Delta\phi_1^e = \pi + \Omega_e\tau$ and $\Delta\phi_2^e = -(\pi - \Omega_e\tau)$ where Ω_e is resonance offset due to chemical shift (Ω_{ch}) and J coupling; hence the transformation induced is an *identity*. If, however, an inversion pulse is applied to the ^{13}C spins, then $\Delta\phi_1^e = \pi + (\Omega_{ch} + 2\pi mJ)\tau$, and $\Delta\phi_2^e = -(\pi - (\Omega_{ch} - 2\pi mJ\tau))$ where J is in Hz and $m = \pm 1/2$ and represent $\langle S_z \rangle$ for the ^{13}C magnetization. The chemical shift term cancels and the net rotation induced by the pulse on the protons coupled to ^{13}C nuclei is a rotation about the B_1 axis by $4\pi mJ\tau$; thus, when $\tau = 1/4\,J$, the two components of the ^1H doublet will undergo $\pi/2$ and $-\pi/2$ rotations, respectively, about the B_1 axis. All other ^1H nuclei that are not coupled to ^{13}C spins will undergo an *identity* transformation. The antiphase state of the ^{13}C coupled ^1H spins can be detected as it is or can be made to evolve into an "in phase" coherent state prior to detection. The latter evolution requires a π pulse to refocus chemical shift and B_0 inhomogeneities. The requisite π pulse can also be an editing pulse so that it does not excite unwanted signals; this is an especially important consideration when there exists large ^1H resonances such as from H_2O, which require suppression. The π BISEP pulse is based on the same principle as the $\pi/2$ BISEP pulse; therefore, it is identical to the $\pi/2$ BISEP pulse except for the twice as long interpulse delays.

In terms of product operator formalism [24] the ^1H doublet spins at the end of the BISEP $\pi/2$ pulse have anti-phase coherence, $2I_xS_z$. By applying a second BISEP with $\tau = 1/2\ J$ (i.e. a π BISEP), it is possible to form in-phase coherence, I_y, at the top the resulting spin-echo when the appropriate echo delays, τ_e, are used. The effect of the τ_e delays and the π BISEP on the anti-phase coherence generated by the initial $\pi/2$ BISEP can be seen from product operator calculations:

$$\begin{aligned}
&2I_xS_z\\
&\downarrow \tau_e\\
&2I_xS_z\cos(\pi J\tau_e) + I_y\sin(\pi J\tau_e)\\
&\downarrow \pi\ \text{BISEP}\\
&-2I_xS_z\cos(\pi J\tau_e) - I_y\sin(\pi J\tau_e)\\
&\downarrow \tau_e\\
&2I_xS_z(\sin^2(\pi J\tau_e) - \cos^2(\pi J\tau_e)) - 2I_y\sin(\pi J\tau_e)\cos(\pi J\tau_e)\\
&= 2I_xS_z\cos(2\pi J\tau_e) - I_y\sin(2\pi J\tau_e).
\end{aligned} \tag{20}$$

The last expression above reduces to in-phase coherence when the echo delays preceeding and following the π BISEP are odd integer multiples of $1/4\,J$.

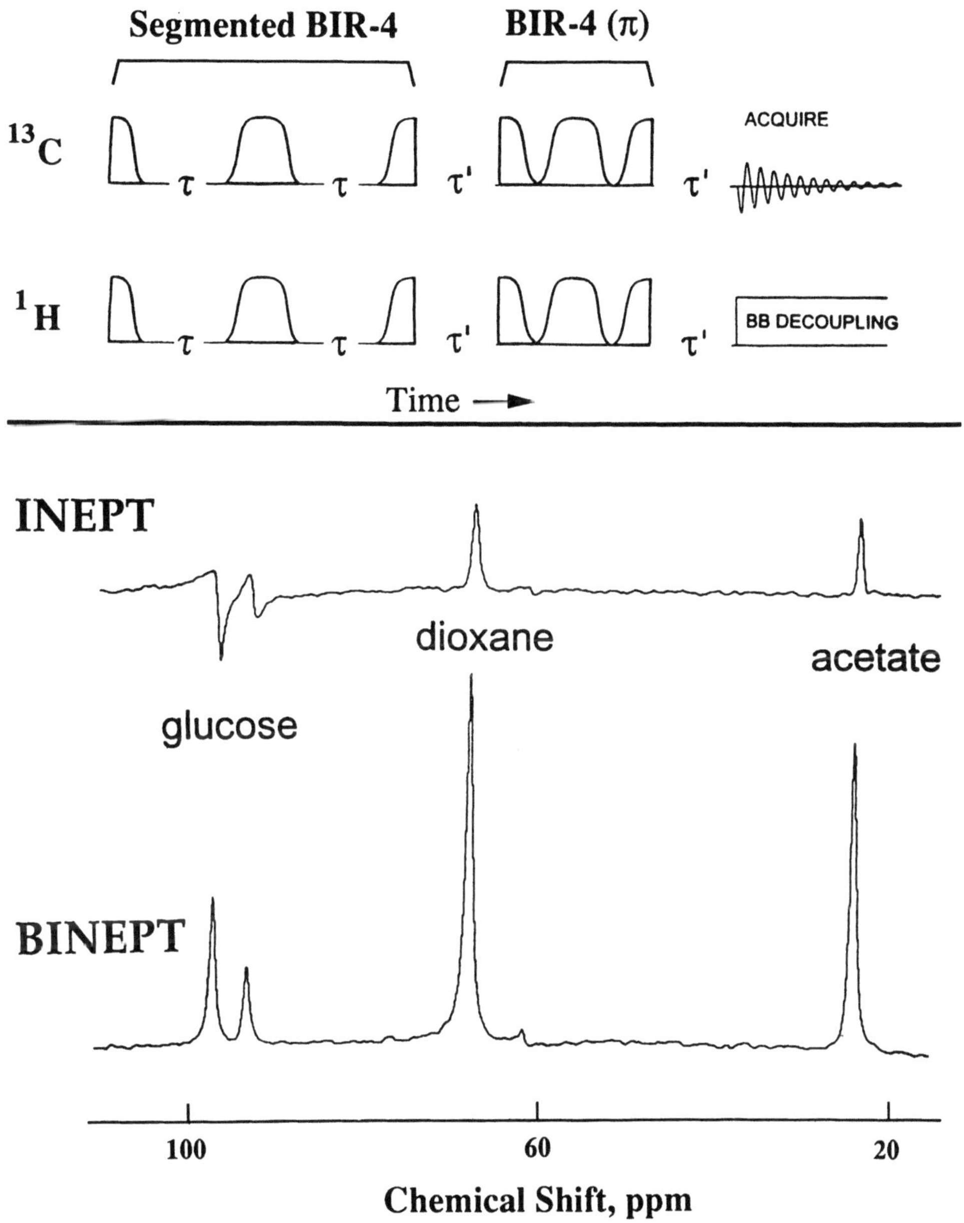

Fig. 18. Adiabatic Polarization Transfer Enhancement (BINEPT) of ^{13}C nuclei via a J-ordered state. The pulse sequence and experimental results obtained with INEPT and BINEPT, respectively, using a surface coil and a three chambered phantom with each chamber containing one of three compounds, [1-^{13}C]glucose, dioxane and [2-^{13}C] acetate. INEPT result was essentially the best obtained when the product of the square pulse length and pulse amplitude was varied over a range that covered the nominal $\pi/2$ and π pulse range for all of the three chambers

5.3 Adiabatic Polarization Transfer Enhancement (BINEPT)

Signal enhancement due to polarization transfer has been employed using sequences such as INEPT [25, 26] and DEPT [27]. The same can be accomplished using BIR-4 based sequences. Again we will use the specific case of polarization transfer from ^{1}H to ^{13}C nuclei for the discussion.

BINEPT sequence, which is the adiabatic analog of one version of INEPT is illustrated in Fig. 18a. The first three pulses applied both at the ^{1}H and ^{13}C frequencies simply constitute a segmented BIR-4 $\pi/2$ pulse with B_1 phase jumps of $\Delta\phi_1 = -\Delta\phi_2 = (\pi + \pi/4)$ and with equal and symmetric delays at $T/4$ and $3T/4$. The transformations induced by the segmented pulse at ^{1}H frequency can be understood in many ways. Because the delays are symmetric, this pulse will induce a $\pi/2$ rotation in all ^{1}H spins that are *not* coupled to ^{13}C irrespective of their resonance offset due to chemical shift, provided that $|\Omega| < |\Delta\omega^0|$; note that **all** ^{1}H spins will also execute $\pi/2$ rotations if the ^{13}C pulses are **not** applied. If a π pulse is applied to ^{13}C spins concurrently with the middle segment of ^{1}H BIR-4 pulse, then for the coupled ^{1}H spins $\Delta\phi_1^e = -\Delta\phi_2^e = (\pi + \pi/4 + 2\pi Jm\tau)$ where m represents $\langle S_z \rangle$ for ^{13}C nuclei; we have ignored the evolution due to chemical shift since it cancels out. Thus, if we are concerned with a CH moiety, for $\tau = 1/4J$, $\Delta\phi_1^e$ and $-\Delta\phi_2^e$ are the same and equal to either $(\pi + \pi/2)$ or simply π for $m = 1/2$ and $-1/2$, respectively, corresponding to overall rotations of π and 0 radians, respectively, for the two components of the ^{1}H doublet. This is the same transformation executed by the $(90°) - \tau - (180°) - \tau - (90°)$ pulse train applied at ^{1}H frequency together with the 180° ^{13}C pulse in the improved INEPT sequence [26]. The role of the second and unsegmented ^{1}H BIR-4 pulse in the BINEPT sequence is a simple inversion; as such, this pulse can in fact be replaced by any other adiabatic pulse capable of this particular spin transformation.

In the BINEPT sequence, a segmented BIR-4 pulse with $\Delta\phi_1 = -\Delta\phi_2 = \pi$ is applied to the ^{13}C nuclei with the two symmetric delays so that the middle segment of this pulse occurs at the same time as the middle segment of the ^{1}H BIR-4 pulse with the interpulse delays. For the ^{13}C magnetization induced by the polarization transfer, the only significant part of the ^{13}C segmented BIR-4 pulse is the middle and last parts, which are simply an adiabatic inversion and half-passage excitation respectively. The middle part of the segmented ^{13}C BIR-4 pulse is significant for the ^{1}H spin evolution because this segment by itself achieves an inversion transformation. The last section of the segmented ^{13}C pulse simply excites the enhanced ^{13}C polarization. Thus, only these last two segments are necessary for the enhanced polarization transfer; however, by using a complete but segmented BIR-4 with $\Delta\phi_1 = -\Delta\phi_2 = \pi$, we ensure that all uncoupled ^{13}C spins encounter an identity transformation. The second unsegmented BIR-4 applied at the ^{13}C frequency is important only as a *refocusing* pulse.

If we ignore B_1 dependent dephasing and rephasing that occurs, the B_1 insensitive adiabatic BIR-4 based BINEPT can also be understood using the composite pulse analog of BIR-4. As previously mentioned, the composite pulse $(90°)_{y'}\,180°_{(y'+\pi+\delta/2)}\,90°_{y'}$ is analogous to BIR-4 with $\Delta\phi_1 = -\Delta\phi_2 = \pi + \delta/2$, and both pulses execute a rotation of δ rads about the x' axis; the composite pulse, of course needs a relatively uniform B_1 field to achieve this transformation. Similarly the sequence

$$^{1}\text{H:}\ 90°_{y'} - \tau - 180°_{(y'+5\pi/4)} - \tau - 90°_{y'} - \tau' - \{90_{y'}180°_{(y'+3\pi/2)}90°_{y'}\} - \tau' - \text{Dec}$$

$$^{13}\text{C:}\ 90°_{y'} - \tau - 180°_{(y'+\pi)} - \tau' - 90°_{y'} - \tau' - \{90_{y'}180°_{(y'+3\pi/2)}90°_{y'}\} - \tau' - \text{Acq}$$

is exactly analogous to APTE in terms of the net rotations induced but of course does not have APTE's large B_1 insensitivity. Note that the above sequence is not simply INEPT in any of its previously described forms [22, 23]. The new BIR-4 inspired INEPT sequence given here in fact does possess some compensation for B_1 inhomogeneities which is absent in the previously described INEPT sequences; this B_1 insensitivity, however, is not and cannot be as large as that which can be achieved by BINEPT.

Figures 18b and c illustrate experimental data obtained with INEPT and BINEPT, respectively, using a surface coil which generates a highly *inhomogeneous* B_1 pattern. The sample was a phantom placed on one side of a surface coil. The phantom had three spatially distinct chambers parallel to the surface coil plane; the B_1 was different for each chamber due to the varying proximity to the coil; even within a given chamber, the B_1 was not uniform because of the finite volume of the chamber. Each chamber contained a different compound, [1-^{13}C] glucose, dioxane (natural abundance), and [2-^{13}C]acetate. INEPT is very sensitive to B_1 inhomogeneities and consequently failed completely at any power level; even when the "square" pulses were adjusted to yield "nominal" $\pi/2$ and π rotations in one of the three chambers, polarization transfer enhancement was not observed for that particular chamber because of B_1 inhomogeneities over the volume of the chamber. The best INEPT result is shown in Fig. 18b. With BINEPT, however, all chambers yielded the expected enhancement simultaneously (Fig. 18c) despite the large differences in B_1 among the chambers and the smaller but nonnegligible variations in B_1 within each chamber.

6 Adiabatic Slice and Volume Selection

6.1 *Slice Selection with Inversion Pulses*

As previously discussed, adiabatic inversion pulses invert only within a well-defined frequency range [3, 4]. This frequency selectivity can be employed in sequences such as ISIS [28] to select a slice (1D-ISIS) or 2 and 3 dimensional volumes. Minimum of two FID's obtained with and without the inversion pulses are needed to define the slice. Similarly, minimum of four and eight acquisitions are necessary to define the 2D and 3D volumes, respectively. It is also possible to select more than one distinct slice simultaneously by using somewhat more complex adiabatic inversion pulses [29, 30]. These "multislice" pulses in return can be used to obtain simultaneous multivoxel volume localization.

6.2 Slice Selection with Excitation Pulses

Gradient Modulated Adiabatic Excitation: Slice selection in this approach is based on the principle of exciting spins so that the phase of the transverse magnetization is shifted by π about a point in space. GMAX is simply adiabatic excitation implemented together with a gradient modulated in the same manner as the frequency modulation of the pulse [31]; consequently, the magnetic fields encountered in the *frequency* frame are

$$B_1(t) = B_1^0 f_B(t)[x'] \quad \text{and} \quad \gamma^{-1}\Delta\omega(t) = \gamma^{-1}(G_0(\mathbf{r}) + \Delta\omega^0) f_\omega(t) \qquad (21)$$

where $\mathbf{r}$ designates spatial coordinates, and $G_0(\mathbf{r}) f_\omega(t)$ is the gradient modulation function. A node exists at $G_0(\mathbf{r}) = -\Delta\omega^0$ where the spins precess about B_1 and are dephased in the $y'z'$ plane of the frequency frame due to the inhomogeneous B_1. In contrast, when $G_0(\mathbf{r}) \neq \Delta\omega^0$ adiabatic excitation occurs; however, the phase of the resultant transverse magnetization differs by π in regions where $G_0(\mathbf{r}) > \Delta\omega^0$ versus regions where $G_0(\mathbf{r}) < \Delta\omega^0$. In a second application, the node can be shifted by changing $\Delta\omega^0$ and/or $G_0(\mathbf{r})$ magnitude; if the B_1 phase in the second application is also π shifted relative to the first application, then the **sum** of the two FIDs will generate a slice. When implemented with optimized modulation function GMAX can define a slice over 100 fold or larger B_1 variation [31].

Similarly GMAX can be extended to 2D, and ultimately to 3D volume selection [32]. Figure 19 shows $B_1, \Delta\omega$, and gradient modulation functions of 3D-GMAX using $f_B(t) = \tanh(\beta t/T)$ and $f_\omega(t) = \operatorname{sech}(\beta t/T)$ for each adiabatic half-passage segment of length T; the first and last segments of the pulse are adiabatic half-passages, whereas the middle segment is a *time-reversed* adiabatic half-passage. In this implementation of 3D-GMAX, the gradient modulations remain the same in each application of the sequence, while the direction of the frequency sweeps are inverted. Between segments 2 and 3, a delay is inserted to allow G_y to decay and G_z to rise. Signal losses during this delay will not occur since the magnetization is collinear with the longitudinal axis at this time and the delay which is typically a few milliseconds is much shorter than the spin T_1s. In order to completely define the localized volume, eight applications of the pulse sequence are required. Table 1 gives the different combinations of the frequency sweeps along with the receiver phase inversion necessary to achieve full cancellation of signals outside the region of interest after eight acquisitions.

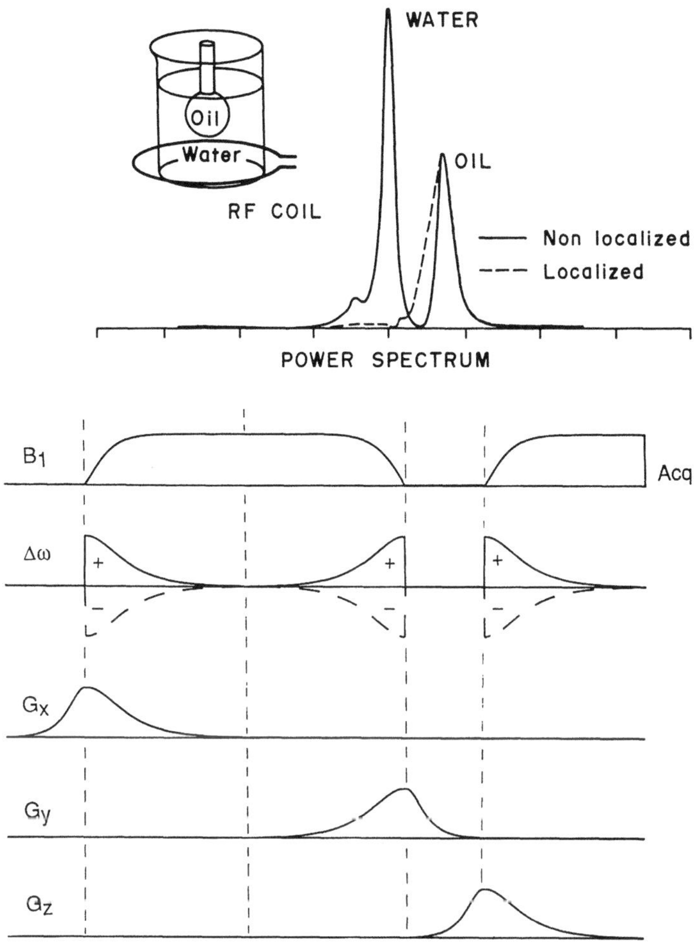

Fig. 19. 3D-GMAX sequence and experimental result

Table 1.

	$\Delta\omega$		Receiver
x	y	z	phase
+	+	+	+
+	+	−	−
+	−	−	+
+	−	+	−
−	+	−	+
−	+	+	−
−	−	+	+
−	−	−	−

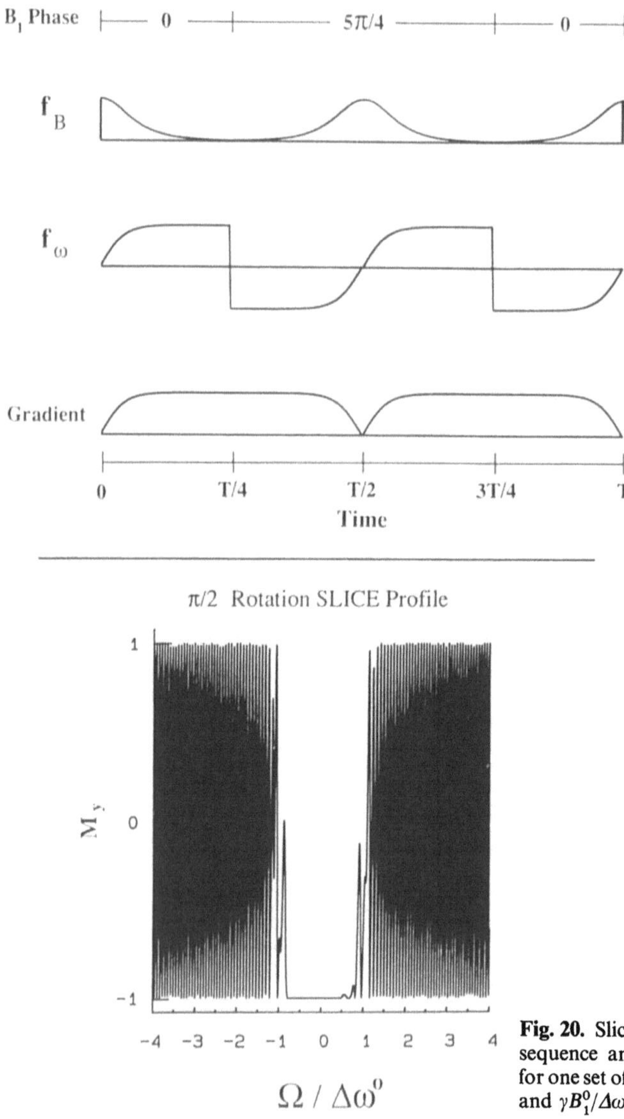

Fig. 20. Slice selective BIR-4 pulse sequence and calculated performance for one set of conditions ($\Delta\omega^0 T/2\pi = 20$ and $\gamma B_1^0/\Delta\omega^0 = 5$)

Figure 19 also illustrates the results of a phantom experiment using 3D-GMAX to select a volume within a spherical sample of oil which was immersed in a beaker of H_2O. Pulse transmission and signal detection was performed with a surface coil having a diameter which was approximately that of the beaker. Without the localization, both H_2O and oil resonances are detected (solid line in the figure). With localization (dashed line) the large H_2O signal is effectively suppressed. The slight broadening of the oil signal during the GMAX experiment resulted from eddy currents (the gradients were not shielded and eddy current compensation was not performed on this system).

6.3 Slice Selection with BIR-4

BIR-4 is a "slice-selective" pulse in the sense that it will accomplish the desired transformation only for $|\Omega| < |\Delta\omega^0|$; outside of this range it will dephase the magnetization as a function of Ω. In a spatially homogeneous sample or a sample where the dimensions of spatial inhomogeneities are large relative to this dephasing, signal cancellation will occur in the region where $|\Omega| > |\Delta\omega^0|$. Thus, BIR-4 can be applied in the presence of constant gradient to define a slice. Further suppression of the signal outside the slice can be achieved if $\Delta\omega^0$ and the gradient signs and B_1 phases are all reversed every other pulse; this reverses the direction of the dephasing outside the slice but does not effect signal within the slice. With a constant gradient, however, the profile within the slice will be somewhat rounded and B_1 dependent because the pulse performance will deteriorate when $\Omega/\gamma B_1$ exceeds ~ 0.4. This can be remedied and good slice profiles can be obtained if the gradients are modulated as shown in Fig. 20; the figure also illustrates the calculated slice profile and out-of-slice dephasing for $M_z \rightarrow M_y$ transformation under one set of conditions.

6.4 Slice Selection with Adiabatic Excitation (SSAX)

Described in terms of amplitude and frequency modulation, this slice selective adiabatic excitation or SSAX pulse [33] consists of two functionally distinct but contiguous segments. The initial segment is basically an adiabatic half-passage pulse with B_1 and frequency modulation; the subsequent segment couples decaying B_1 amplitude with zero frequency modulation. Taking B_1 to be along x' in a reference frame which rotates at the instantaneous frequency of the pulse (axes x', y', and z'), the effect of SSAX can be visualized as follows: SSAX first creates a distribution of spin orientations in the $x'z'$ plane as a function of frequency offset so that only spins with zero offset point along x'. Subsequently, spins not along x' are taken either back up to z' (positive offset) or down to $-z'$ (negative offset), whereas those along x' (zero offset) are allowed to remain in the transverse plane. For illustrative purposes, the following example is a simple version of a SSAX pulse constructed with tangent and sech functions.

$$B_1(t) = B_1^0 \operatorname{sech}[\beta t/T] \qquad (-T < t < 0)$$

$$\Delta\omega(t) = \Delta\omega^0 \tan[\kappa(1 - t/T)]/\tan[\kappa]$$

$$B_1(t) = B_1^0 \operatorname{sech}[\beta t/0.5\,T] \qquad (0 < t < 0.5\,T)$$

$$\Delta\omega(t) = 0$$

where T is pulse duration, and κ and β are chosen to set the modulation limits.

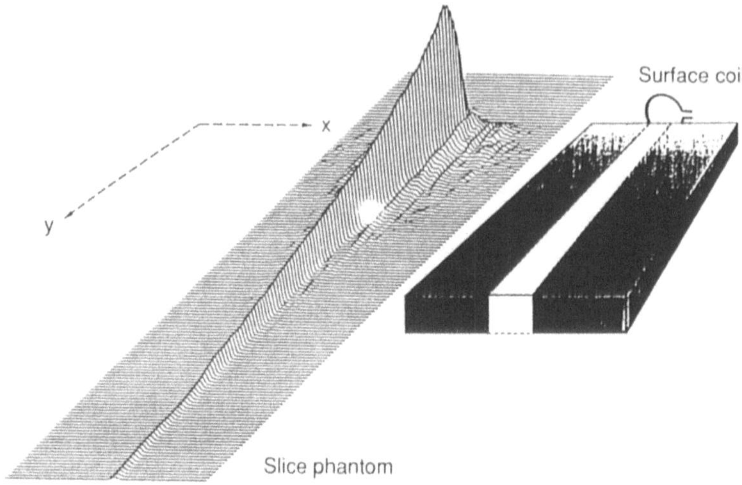

Fig. 21. Slice selection by adiabatic SSAX pulse using a surface coil

While the above example is written using tan and sech function, provided the boundary values remain the same, other modulation schemes can be used. Figure 21 illustrates the slice profile of SSAX obtained with a surface coil placed adjacent to the coil as illustrated. The phantom was already a plane of finite thickness that contained water; a slice was selected perpendicular to the plane of the coil. The slice position is indicated by the white region within the slice phantom (Fig. 21); signal intensity was obtained as a function of x and y coordinates using readout and one dimensional phase encode gradient. The signal intensity is decreasing because the B_1 magnitude hence coil sensitivity is diminishing with increasing distance from the coil.

7 Conclusion

In summary, it can be concluded that in the last few years, adiabatic pulses have evolved dramatically so that most of the experiments that can be performed by use of conventional pulses and homogeneous B_1's can be executed with adiabatic transformations, despite the presence of large B_1 inhomogeneities. This capability is of course indispensible in applications that are constrained to use coils such as surface coils which generate extremely inhomogeneous fields. In view of the fact that even high resolution NMR studies suffer from B_1 imperfections, such studies should also benefit from these adiabatic spin transformations.

8 References

1. Slichter CP (1979) Principles of magnetic resonance, 2nd edn, Springer, Berlin Heidelberg, New York, p. 24
2. Levitt MH (1968) Prog NMR Spectros 18: 61–122
3. Silver MS, Joseph RI, Hoult DI (1984) J Magn Reson 59: 347–351
4. Baum J, Tycko R, Pines A (1985) Phys Rev 32: 3435–3447
5. Hardy CJ, Edelstein WA, Vatis D (1986) J Magn Reson 66: 470–482
6. Bendall MR, Pegg DT (1986) J Magn Reson 67: 376–381
7. Ugurbil K, Garwood M, Bendall MR (1987) J Magn Reson 72: 177–185
8. Ugurbil K, Garwood M, Rath A, Bendall MR (1988) J Magn Reson 78: 472–497
9. Garwood M, Ke Y (1991) J Magn Reson 94: 511–525
10. Ugurbil K, Garwood M, Rath A (1988) J Magn Reson 80: 448–469
11. Town G, Rosenfeld D (1990) J Magn Reson 89: 170–175
12. Johnson AJ, Ugurbil K, Garwood M (1989) Soc Magn Reson Med Abstracts p. 24
13. Bendall MR, Garwood M, Ugurbil K, Pegg DT (1987) Magn Reson Med 4: 493–499
14. Connolly S, Nishimura D, Macovski A (1989) J Magn Reson 83: 324–334
15. Nuss M, Whittern D, Garwood M (1990) Soc Magn Reson Med Abstracts p. 128
16. Staewen RS, Johnson AJ, Ross B, Parrish T, Merkle H, Garwood M (1990) Invest Radiol 25: 559–567
17. Norris DG, Haase A (1989) Magn Reson Med 9: 435–440
18. Levitt MH, Ernst RR (1983) J Magn Reson 55: 247–259
19. Levitt MH, Freeman R (1979) J Magn Reson 33: 473–476
20. Ross BD, Merkle H, Hendrich K, Staewen RS, Garwood M (1990) Magn Reson Med. (in press)
21. Garwood M, Merkle H (1991) J Magn Reson 94: 180–185
22. Wei H, Merkle H, Garwood M, Kamil Ugurbil (1991) Soc Magn Reson Med Abstracts
23. Hore PJ (1983) J Magn Reson 55: 283–300
24. Sorensen OW, Eich GW, Levitt MH, Bodenhausen G, Ernst RR (1983) Prog NMR Spectros 16: 163
25. Morris GA, Freeman RJ (1979) J Am Chem Soc 101: 760
26. Burun DP, Ernst RR (1980) J Magn Reson 39: 163–168
27. Doddrell DM, Pegg DT, Bendall M (1982) J Magn Reson 48: 323–327
28. Ordridge RJ, Connelly A, Lohman JAB (1986) J Magn Reson 66: 283
29. Müller S, Sauter R, Weber H, Seeling J (1982) J Magn Reson 76: 155–161
30. Goelman G, Leigh JS (1991) J Magn Reson 91: 93–101
31. Johnson A, Garwood M, Ugurbil K (1989) J Magn Reson 81: 653–660
32. Tannus A, Garwood M, Panapucci H, Bonagamba TJ (1990) Soc Magn Reson Med Abstracts p. 1328
33. Johnson AJ, Ugurbil K, Garwood M (1988) Soc Magn Reson Med Abstracts p. 5660

Frequency Selective Excitation Using Phase-Compensated RF Pulses in One and Two Dimensions

Peter G. Morris

Department of Physics, University of Nottingham, University Park, Nottingham, NG7 2RD, UK

Table of Contents

Selective pulses are central to most MRI and many MRS applications, yet the versions used in most commercial systems are 'first generation' pulses, typically curtailed sinc pulses, derived from simple Fourier transform theory. Although such pulses work well enough for most straightforward applications, they are non-ideal in many respects, including quality of excitation profile, and their need for subsequent refocusing of the transverse magnetization. The difficulty lies in the non-linearity of the NMR spin system. Several approaches to the design of improved selective pulses are discussed, including a new analytical approach based on inverse scattering theory and numerical methods such as SPINCALC and simulated annealing. The application of k-space theory to pulse design is also discussed. Although only valid in the low tip angle approximation, this theory has proved remarkably successful even for the design of π pulses, and very importantly has clarified the design principles for phase-compensated and multidimensional pulses. A variety of 'clean' two-dimensional pulses have now been described, which have applications in 'zoom' imaging and localized spectroscopy. Some are difficult to implement on standard systems because of their requirement for rapid gradient reorientation. One solution is to divide the coverage of k-space into several steps. Such compromises offer the possibility of extending the principle to practical pulses of higher dimensionality for applications such as selection of an arbitrarily-shaped volumes in localized spectroscopy.

1 Introduction

1.1 Selective Pulses

Selective pulses were originally proposed by Garroway, Grannell and Mansfield [1] and have come to play a central role in the majority of magnetic resonance imaging (MRI) experiments, where they are used in conjunction with an applied linear magnetic field gradient to isolate a slice of controlled thickness and orientation from within an extended object [2, 3]. Selective pulses feature too in many of the more successful magnetic resonance spectroscopy (MRS) methods [3–5]. Although at present little exploited, they also hold great promise for conventional high resolution spectroscopy. Potential applications include solvent suppression without distortion of the frequency response, magnetization transfer and reduction of bandwidth in two- and three-dimensional high resolution experiments.

The guiding principle for the design of selective pulses has been Fourier transform theory. The logic behind this is illustrated in Fig. 1. In order to

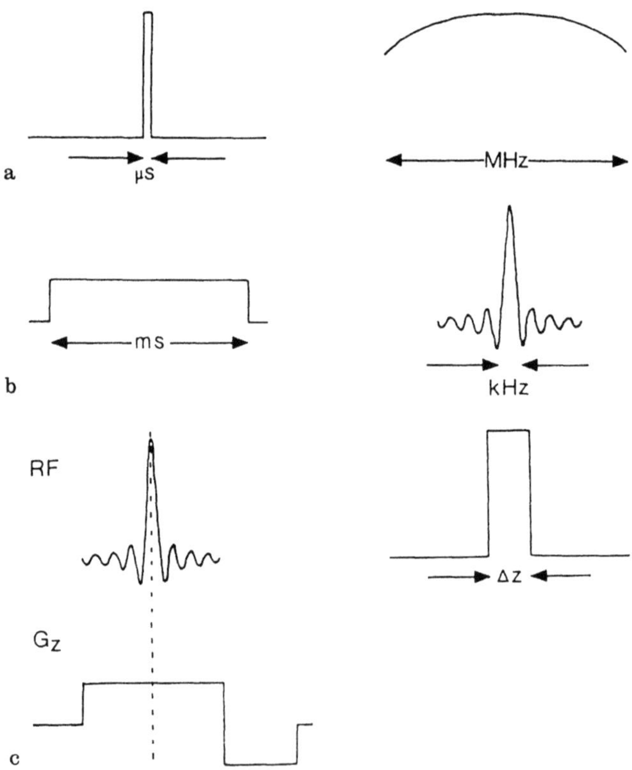

Fig. 1 a–c. Radiofrequency pulse envelopes and their Fourier transforms: **a** hard (non-selective) pulse; **b** soft pulse; **c** sinc pulse. (From [5])

achieve a slice with a square profile, the RF carrier is modulated by the Fourier transform of a square or top hat function, that is a sinc function. Unfortunately, the NMR spin system is non-linear in several respects and this approach only gives acceptable results in the case of small flip angles, for which the spin system is approximately linear. For example, $\pi/6$ sinc pulses give reasonably square profiles (Fig. 2(a)). Even for a $\pi/2$ flip angle, the response is tolerable (b), but the approximation breaks down badly for π pulses (c). The inappropriateness of the linear Fourier transform theory is manifest in the departure from perfectly square slice profiles. However, even if the theory were exact, there would still be practical difficulties associated with this approach. In particular, the sinc function decays rather slowly. Relaxation and other constraints preclude the application of a protracted RF pulse, so that it is, in practice, necessary to curtail it at some point. Typically this is done at a zero crossing and one talks of, for example, a five-lobe sinc function consisting of the central maximum with two lobes on each side (as in Fig. 2). This truncation gives rise to sidebands which correspond to the partial excitation of a sequence of planes parallel to the slice of interest. This can be alleviated by multiplying the sinc function by a second function such as a Gaussian, so that the pulse envelope decays smoothly to zero before the cut off [7]. Removal of the satellite excitations in this way is at the expense of the sharpness of the slice edges. A further problem, which has to be taken account of, is the first order phase shift generated by the pulse. Normally, this is taken care of by gradient reversal, but, as we shall see, it is possible to design pulses which have inherent phase-compensation.

How do we progress beyond the pulses generated by simple Fourier transform theory? Unfortunately, there appears to be no completely general method for inversion of the Bloch equations, that is, for deriving the RF pulse envelope required to achieve a target distribution of magnetization. Improvements in pulse performance have resulted from numerical optimization schemes such as optimal control theory. Such methods have been used by Conolly and Macovski [8] and Lent and Kritzer [9] to generate selective inversion pulses. However, optimal control theory based methods are prone to terminate prematurely since they work by minimizing a single number, the error functional, representing the difference between the actual and the desired responses. For most non-trivial problems, this must undergo transient increases in the approach to the global minimum. Recently, two optimization schemes, which are not prone to getting stuck in local minima, have enjoyed considerable success in the generation of novel pulses. These schemes, simulated annealing and a method based on a transformation to a switched reference frame and known as SPINCALC are discussed below and their application to the design of phase compensated $\pi/2$ pulses demonstrated.

The description of MRI methods based on k-space trajectories has proved to be an excellent method for analyzing and comparing the relative methods of different imaging methods [10]. The same k-space analysis has recently been applied to RF pulse design [11]. The method is an approximation which theoretically is only valid in the small tip-angle limit. In practice it works rather well

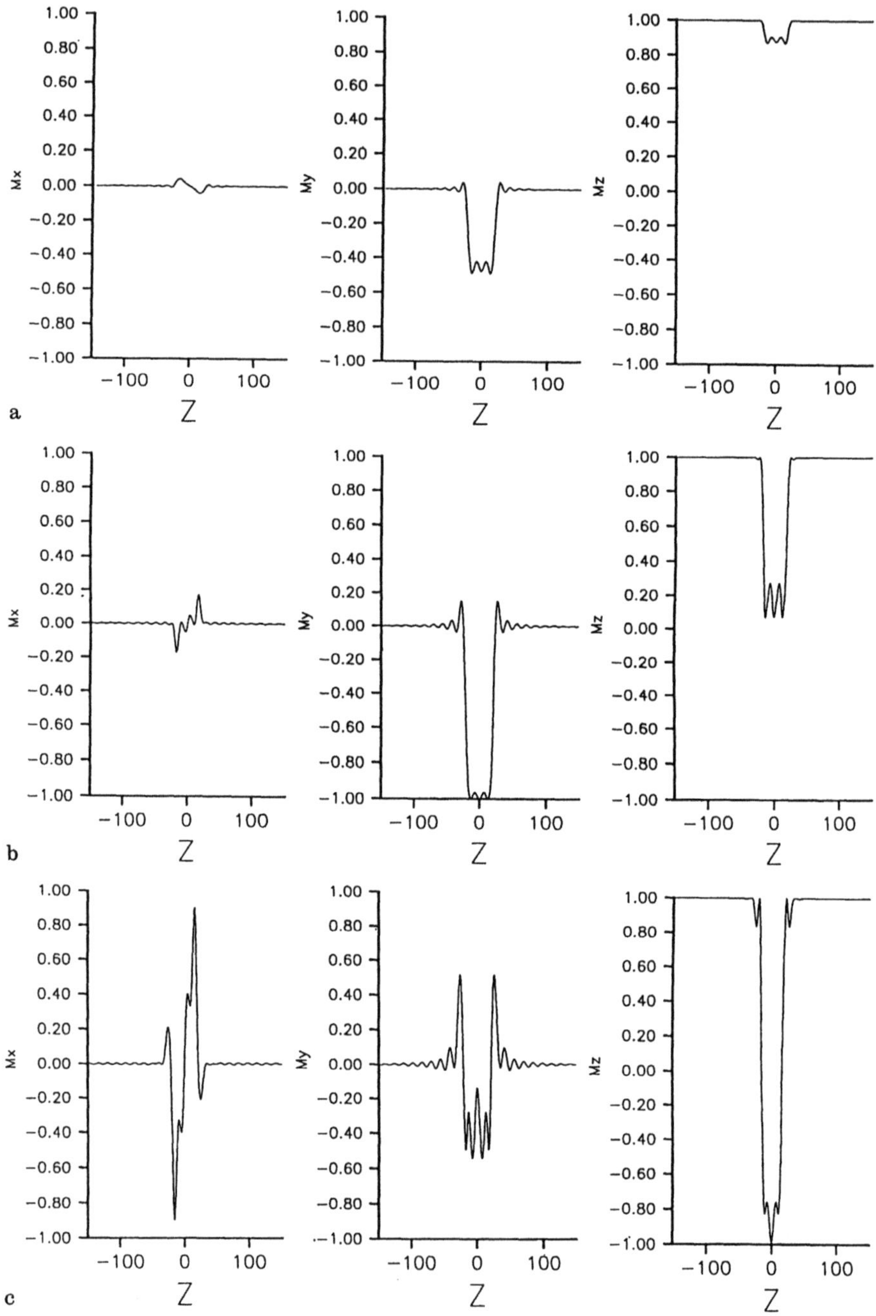

Fig. 2 a–c. The M_x, M_y and M_z magnetization profiles following application of $\pi/6$ **a**, $\pi/2$ **b** and π **c** five-lobe sinc selective pulses to a sample of uniform spin density in the presence of a linear field gradient G_z. In each case the magnetization was refocused by simulated gradient reversal to minimize the M_x component. (From [6])

even for π pulses. It has demonstrated very clearly that for multidimensional pulses it is neccessary for the gradient to map out the appropriate region of k-space. This leads to the choice of spiral rather than circular paths for the gradient vectors in the case of two-dimensional pulses (see below). The theory also suggests a choice for the RF envelope, namely, the appropriately weighted Fourier transform of the desired magnetization profile. The results are not perfect, but are surprisingly good and, at the very least, provide useful starting approximations for numerical optimization methods.

1.2 Phase-Compensated Pulses

A problem common to both MRI and MRS is the dephasing of the xy magnetization that occurs during selective pulses. (These are necessarily protracted in order to achieve their frequency selectivity.) For most pulses, the phase shift is approximately first order and corresponds to evolution at the relevant offset for half the pulse length [3]. If the RF pulses are spatially selective (i.e. applied in the presence of a magnetic field gradient), the transverse magnetization can be refocused using a negative gradient lobe, whose area is one half that of the positive lobe applied during the RF pulse. Refocusing is time consuming because it requires the gradients to be ramped down, ramped

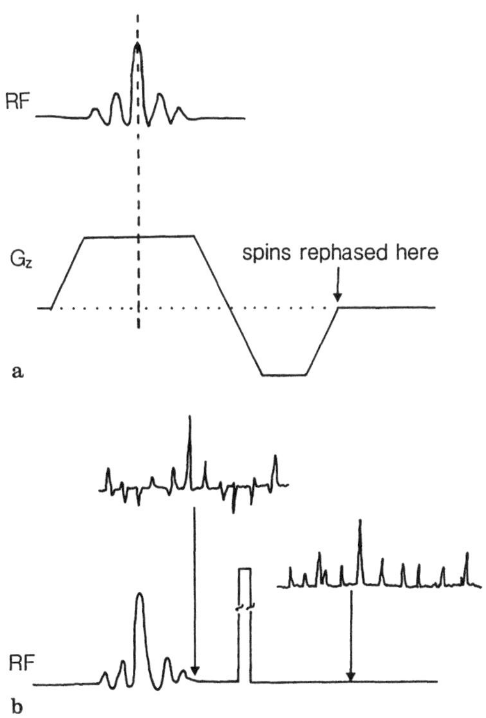

Fig. 3 a, b. Conventional selective pulses require the magnetization to be refocused, either, in the case of spatial localizaton, by gradient reversal **a**, or, in the case of high-resolution NMR, by using a hard π pulse **b**. (From [12])

up in the negative direction, left on for a time approaching half the pulse length and then ramped back to zero again (Fig. 3). Gradients cannot be switched too fast because of the rise times associated with the inductance of the gradient coils themselves, the rise times of the gradient drivers and the effects of eddy currents in the magnet bore, if properly shielded gradient sets are not employed. A similar problem exists in the case of high-resolution applications where the phase accumulation appears as a first order shift across the spectrum. Although no gradients are involved, refocusing is necessary and must be achieved using a hard π pulse. A delay corresponding to at least half the width of the selective pulse is incurred, that could prove disastrous in the case of short T_2 species. This could well be the case for two- or three-dimensional high-resolution NMR experiments on larger (> 10 kDa molecular weight) macromolecules or for selective saturation or inversion of the ^{31}P resonances of nucleotides.

1.3 Multidimensional Selective Pulses

Pulses which localize in two or three dimensions will be useful in many imaging and spectroscopic applications, particularly when there is motion present. In order to achieve excitation in more than one spatial dimension, it is necessary to shape and reorient the gradient during the modulation of the RF pulse. Field gradients with response times and linearities greatly improved over those offered on early commercial MRI systems are now widely available and it is possible to entertain the routine use of such pulses.

It is often desirable to image with high-resolution a specific region of interest lying deep within the body. A uniform increase in image resolution over the whole body would require a proportional increase in the number of gradient increments in Fourier based schemes in order to avoid aliasing artefacts. This would result in unacceptably long imaging times. Echo-planar imaging [13] does not suffer from this minimum performance constraint but the number of voxels in the switched gradient direction is limited on practical grounds to 128 or possibly 256. Techniques to avoid aliasing have been developed [14] for EPI, but a pulse, or combination of pulses to enable zoomed imaging or regional microscopy is highly attractive. Replacement of the initial conventional $\pi/2$ selective pulse with a two-dimensional $\pi/2$ pulse with selectivity normal to that of an ensuing π refocusing pulse will define a plane of finite extent.

Most volume selection sequences involve the successive application of three mutually orthogonal slice selective pulses. In addition, differencing methods may be required to eliminate the signal from the surrounding volume. For example, in the case of three-dimensional ISIS [15], eight separate experiments are required to define the ISIS cube. Such methods are prone to motional artefacts. Where motion is a problem, procedures which achieve localization in a single FID offer several advantages. Two-dimensional pulses can be combined with a standard slice-selective pulse to define a volume. It will also be possible in the near future to apply three-dimensional pulses which achieve the requisite localization in a single stage. Although such pulses reduce motional artefacts, they require high gradient slew rates and the necessarily incomplete coverage of k-space leads to aliasing artefacts which the user must ensure are located beyond the maximum extent of the sample. Nevertheless, we may speculate that techniques for pulse optimization may become sufficiently well developed and the cost performance ratio of computers sufficiently low that it will become possible to tailor quite sophisticated pulses on-line. Thus, one can envisage looking at a three-dimensional proton image (or a series of contiguous slices), defining from it a region of interest of arbitrary shape and calculating the appropriate three-dimensional selective pulse to excite it.

2 Analytical Solutions of the Bloch Equations

Specific solutions of the Bloch equations are rare. The classic, and, until recently, the only example was the hyperbolic secant pulse [16, 17], which gives perfect inversion over a slice. Unfortunately, it is not a refocusing ('pancake flip') pulse and so does not solve the problem of providing a good π pulse for multi-slice spin-echo imaging experiments. However, its use has been crucial to the success achieved by the ISIS method of spatial localization. It is an adiabatic pulse which has a large B_1 window (though not as commonly believed, an infinite one!).

A new analytic method for inverting the Bloch equations for non-interacting spins has been described [18], based on the inverse scattering transform (IST) described by Zakharov and Shabat [19]. The IST was originally developed to determine the potential giving rise to the measured scattering coefficients of a one-dimensional wave function. The problem of the inversion of the Bloch equations can be recast in this form and solved provided the desired magnetisation profile is described by a rational function. An application of this method is shown in Fig. 4 in which a top-hat magnetization profile is approximated by a 20th order rational function. The calculated pulse, which is the analytical solution to the problem, is shown in (a) and leads to magnetization profiles (b–d) which precisely match the target profile. Note that the pulse is phase-compensated and, when used for spatial selection, does not therefore require any gradient reversal. In principle the IST method can generate any desired magnetization profile.

3 Optimization Methods

3.1 Simulated Annealing

Simulated annealing [20, 21] is a relatively new optimization technique which has attracted widespread interest, due in part to its simplicity of concept, but

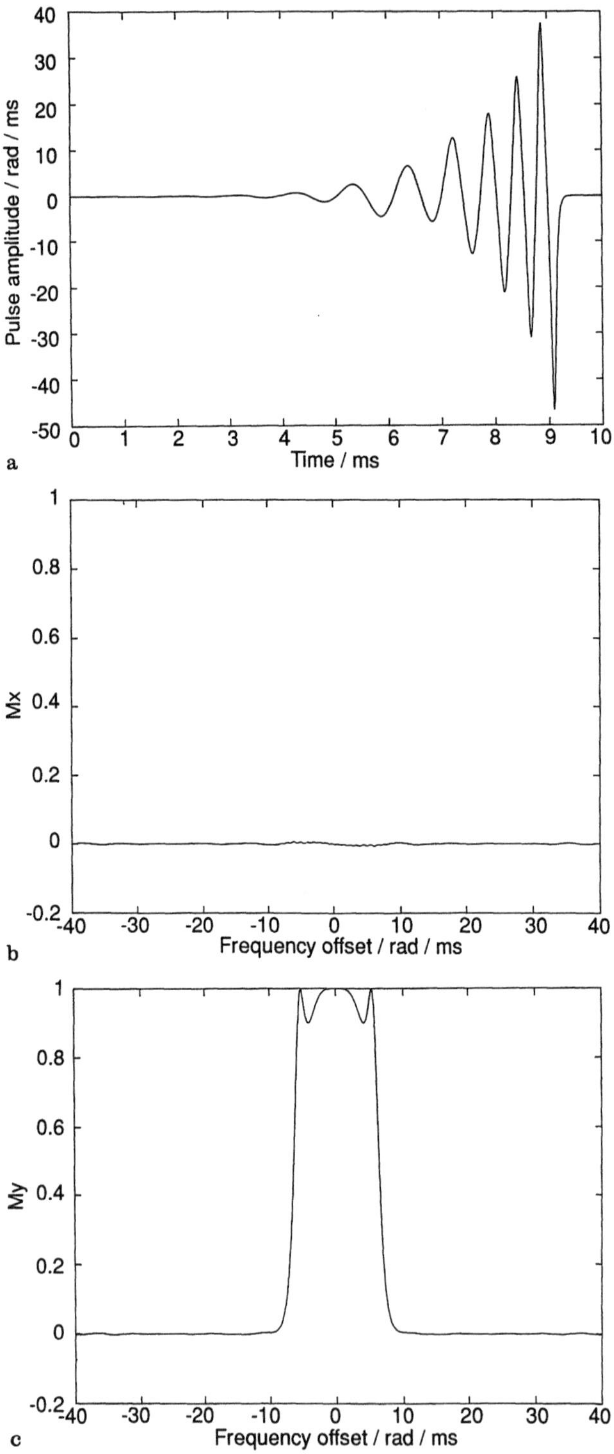

Fig. 4. *(Continued)*

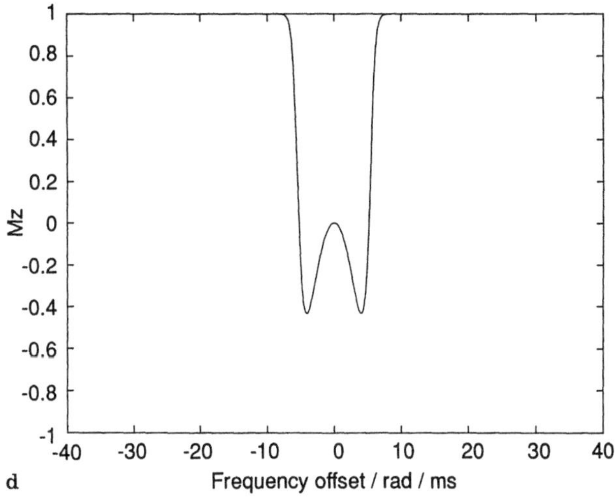

d

Fig. 4a–d. The real phase-compensated RF pulse which is the analytical solution of the Bloch equations for a top hat magnetization profile approximated by a 20th order rational function **a**. Magnetization profiles M_x(**b**), M_y(**c**) and M_z(**d**) corresponding to this pulse. (From [18])

especially because it is effective in reaching the 'global minimum'. (Complicated systems may have many local minima in which optimization schemes based on the monotonic minimization of a single error functional are prone to entrapment.) Simulated annealing was first used to optimize the layout of integrated circuits, but derives its name by analogy with the process in which a substance finds its global energy minimum when cooled. This requires the cooling process to be sufficiently slow: for example, if molten silicon dioxide is cooled slowly, it will form a crystalline structure, whereas, if it is cooled too rapidly, it will form glass—a less ordered state of higher energy.

The optimization procedure involves the definition of a cost function which is a measure of how close the response is to the desired solution. This cost is the quantity which is minimised and is analogous to the energy of the system. The optimization proceeds by varying a degree of freedom by a small, random amount and recalculating the cost. If this is reduced, the change is accepted. If it has increased, the probability of acceptance is determined by the ratio of the increase in cost to the current temperature. (Acceptance of a small detrimental change to the cost function is analogous to the thermal energy which permits a structure to 'jump out' of a local minimum.) This is carried out for each degree of freedom in turn and the process is repeated a number (typically 20) of times at a given 'temperature'. The temperature is then reduced so that the magnitude of detrimental changes to the cost function is reduced (the permissible 'thermal' fluctuation is less).

The initial temperature, time spent at each temperture and number of temperature steps are not strictly defined. Usually, the initial temperature is chosen so that 50% of the changes to the degrees of freedom are accepted at the start. In a typical optimization, each of the degrees of freedom might be adjusted 20 times at each temperature, and the temperature reduced through 40 steps in a linear decreasing fashion. Although conceptually simple and straightforward to implement, simulated annealing can be very costly in computation time, taking several hours or days to run even on a mainframe computer. This procedure is illustrated in Fig. 5 for the case of a phase-compensated $\pi/2$ pulse. Three full simulated annealing cycles were applied. Although not constrained to do so, the cost can be seen to decrease monotonically through the first two cycles of optimization. (The small positive fluctuations are masked by passing through the degrees of freedom 20 times at each temperature.) However, a small increase in the cost function can be discerned near the start of the third cycle. If the degrees of freedom correspond to the

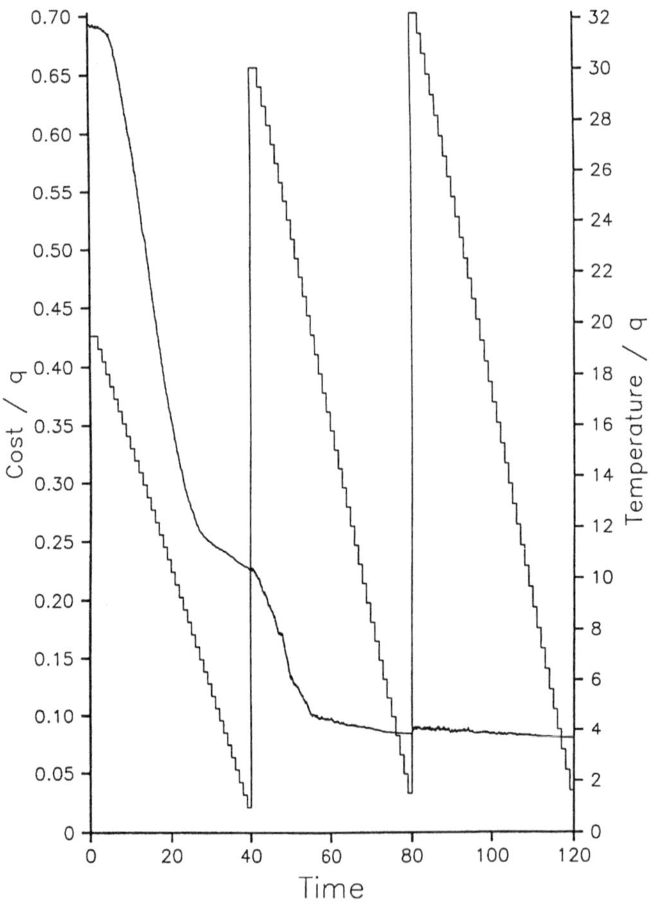

Fig. 5. The cost function and temperature during three cycles of optimization by simulated annealing of a prefocused $\pi/2$ selective pulse. (From [12])

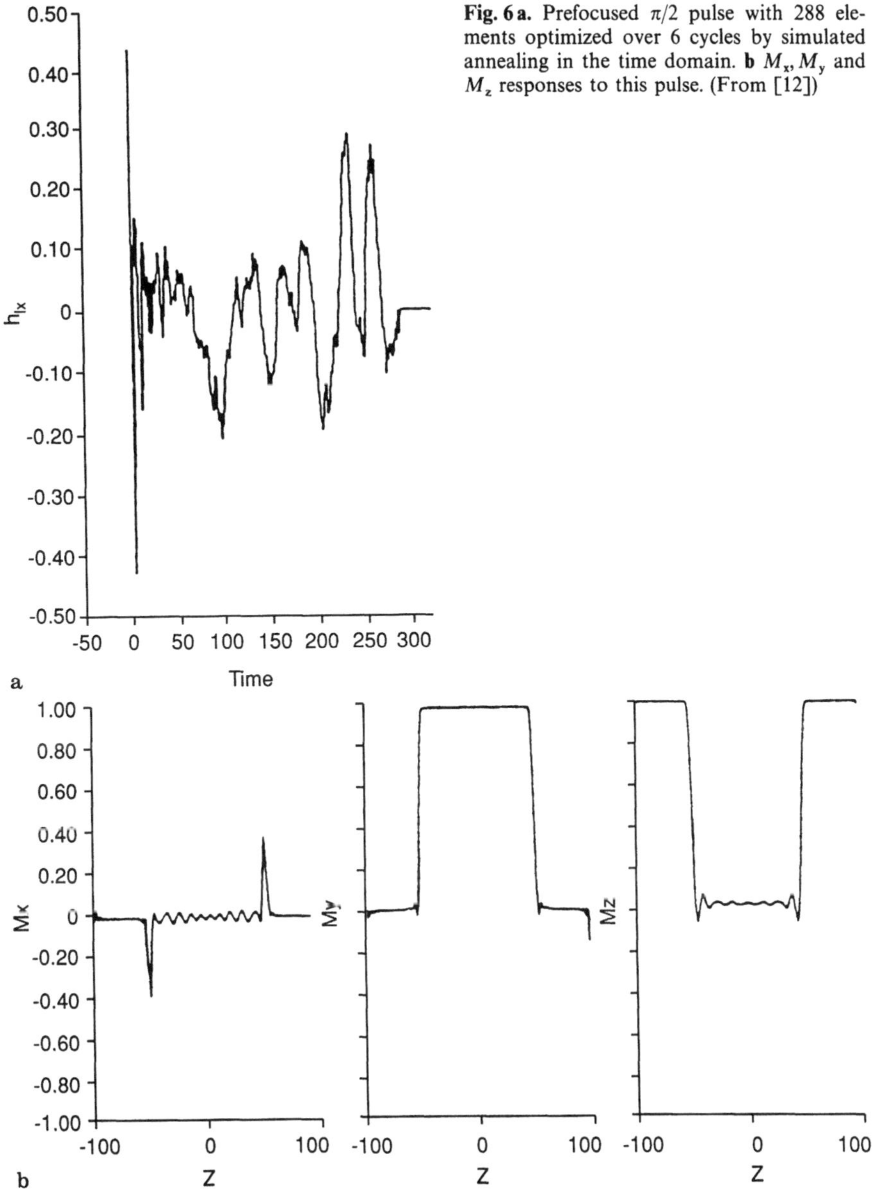

Fig. 6a. Prefocused $\pi/2$ pulse with 288 elements optimized over 6 cycles by simulated annealing in the time domain. **b** M_x, M_y and M_z responses to this pulse. (From [12])

amplitudes of the RF pulse envelope (i.e. the optimization is in the time domain), the optimized pulse generally has a good theoretical response but tends to oscillate wildly at its origin as shown in Fig. 6 [22]. This tendency can be circumvented by optimizing the pulse in the Fourier domain, taking as degrees of freedom, a finite series of Fourier coefficients [23]. A typical result is shown in Fig. 7—a much smoother pulse. (The rather poorer spatial response is

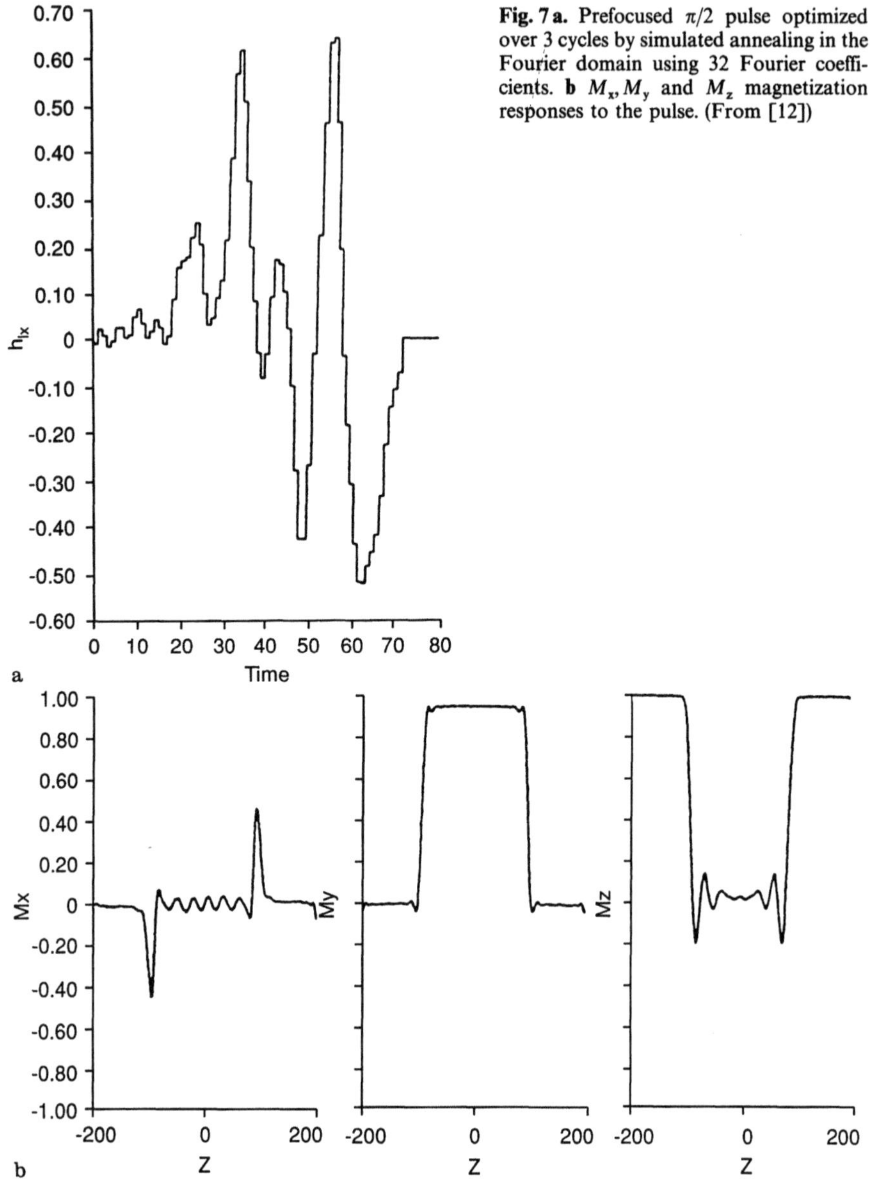

Fig. 7 a. Prefocused $\pi/2$ pulse optimized over 3 cycles by simulated annealing in the Fourier domain using 32 Fourier coefficients. **b** M_x, M_y and M_z magnetization responses to the pulse. (From [12])

probably largely a consequence of the very small number (32) of degrees of freedom used.)

Early attempts to generate two-dimensional selective pulses, based on fixed amplitude rotating gradients (ρ pulses), were fundamentally flawed due to their failure to cover, even approximately, the appropriate region of k-space [11]. The result was the rather poor selectivity and ringing in the two-dimensional selectivity profile of the so-called sombrero, egg carton and stalagmite pulses [24, 25]. Although virtually unusable as designed, successful attempts were made to improve their performance by optimizing both the RF and the gradient waveforms using simulated annealing [26, 27]. As in the case of the one-dimensional phase-compensated pulses, direct application in the time domain led to spikes on the RF amplitude, and worse—on the gradient amplitudes as well. Again the solution was to take as degrees of freedom the amplitudes of the coefficients in a Fourier series expansion of the RF and gradient waveforms. This has the additional merit of automatically limiting the maximum rate of change of gradient (slew rate). The effect on the spatial response of the sombrero pulse (a square π inverting pulse, applied in the presence of a gradient of fixed amplitude rotating through one cycle) was to improve the response from that shown in Fig. 8(a) to that in (b). The RF and gradient waveforms corresponding to the optimized version are illustrated in (c).

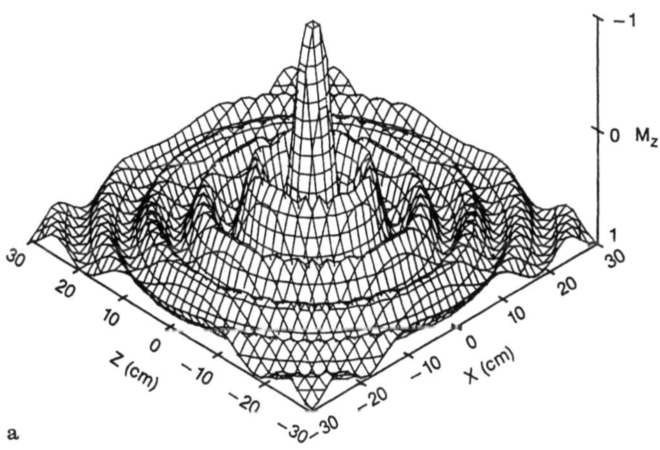

a

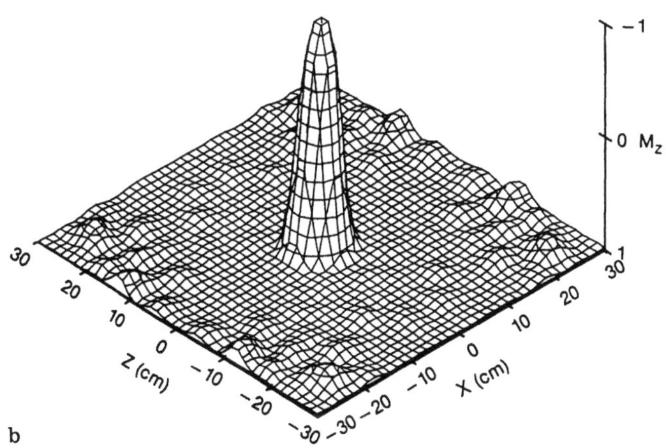

b

Fig. 8. (*Continued*)

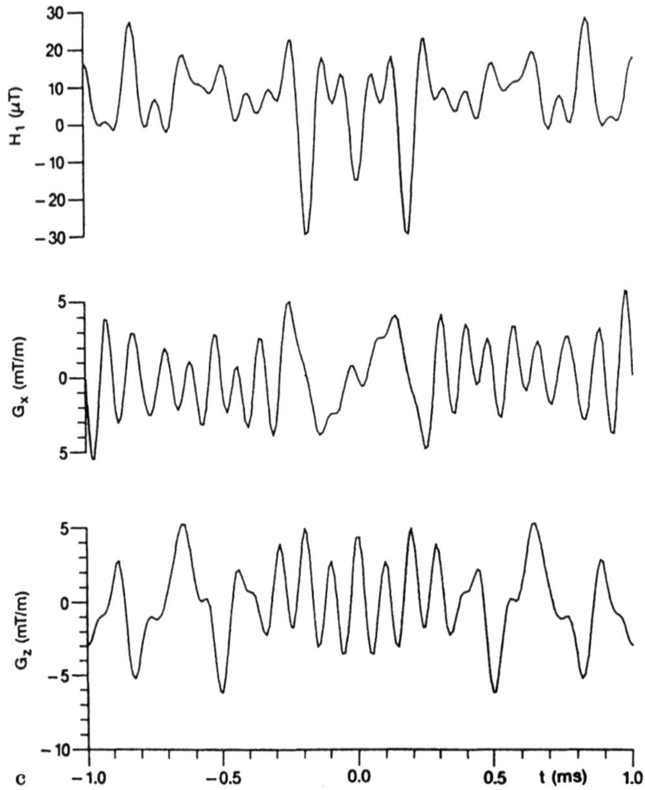

Fig. 8 a. Computed spatial distribution of the z-magnetization, normalized to the equilibrium value in the xz-plane following the application of a 2 ms square π inverting pulse in the presence of a single cycle sine wave x-gradient and cosine z-gradient. **b** Computed spatial distribution of the z-magnetization following application of the optimized two-dimensional ρ pulse shown in **c**. (From [26])

Because of the 'random' nature of the simulated annealing approach, the same optimization, run a second time, is not guaranteed to yield the same result—a feature noted by Hardy et al. [26] and ascribed to nearly degenerate ground state configurations in a 'frustrated' system (one subject to conflicting constraints).

3.2 SPINCALC

SPINCALC is a numerical optimization scheme introduced by Ngo and Morris [28–30]. It uses a switched rotating frame of reference to linearize the Bloch equations, making their inversion possible. This is carried out simultaneously for a number of isochromats, subject to linearity and continuity constraints. As a result, convergence is typically two orders of magnitude faster than for optimal control methods.

In the absence of relaxation, the Bloch equation describing the magnetization $\mathbf{M}(t, q)$ of a single isochromat, labelled q, can be written as:

$$\frac{d}{dt}\mathbf{M}(t,q) = \omega(t,q) \times \mathbf{M}(t,q)$$

where the instantaneous angular velocity $\omega(t, q)$ is given by $-\gamma\mathbf{B}(t, q)$. This can be rewritten in matrix form:

$$\mathbf{M}(t+\delta t, q) = \mathbf{R}(\omega(t,q)\,\delta t)\mathbf{M}(t,q)$$

which can be 'integrated' to give:

$$\mathbf{M}(T,q) = \prod_{0}^{T} \mathrm{R}(\omega(t,q)\,\delta t)\mathbf{M}(0,q) = \mathrm{R}_{\mathrm{tot}}(q)\,\mathbf{M}(0,q)$$

where $R_{\mathrm{tot}}(q) = R_{\mathrm{n}}(q) \ldots R_2(q)\,R_1(q)$, corresponding to the n intervals into which the pulse is divided. Note that, in the real situation, this is not an approximation, since the pulse envelopes are generated digitally via digital-to-analogue converters and hence have exactly this stepwise form. The object is to find, for each q, a new rotation matrix which more closely approximates the desired response

$$R_{\mathrm{tot,new}}(q) = \delta R_{\mathrm{n}}(q) R_{\mathrm{n}}(q) \ldots \delta R_1(q) R_1(q).$$

At this point, the transformation to the switched frame of reference is introduced:

$$\delta R'_{\mathrm{m}}(q) = U_{\mathrm{m}}(q)\delta R_{\mathrm{m}}(q)\,U_{\mathrm{m}}(q)^{-1}$$

$$\text{where } U_{\mathrm{m}}(q) = R_{\mathrm{n}}(q)R_{\mathrm{n-1}}(q)\ldots R_{\mathrm{m+1}}(q) \text{ for } m = n-1, n-2 \ldots 0$$
$$= 1 \qquad\qquad m = n.$$

It can then be shown that:

$$R_{\mathrm{tot,new}}(q) = \delta R_{\mathrm{tot}}(q)\,R_{\mathrm{tot}}(q)$$

where

$$\delta R_{\mathrm{tot}}(q) = \delta R'_{\mathrm{n}}(q) \ldots \delta R'_1(q).$$

If sufficiently small, the corrections are linear, and the product of the small matrices can be replaced with a sum of small vectors:

$$\delta\mathbf{R}_{\mathrm{tot}}(q) = \sum_{\mathrm{m}} \delta\mathbf{R}'_{\mathrm{m}}(q).$$

Only certain degrees of freedom are permitted by the hardware configuration of the imaging device. For a single-phase RF system, the amplitude of each individual pulse element can be varied whereas for a quadrature excitation system both amplitude and phase are available. The $\delta\mathbf{R}'_{\mathrm{m}}(q)$ are then determined for each degree of freedom and a simple least squares fitting procedure is used to assemble the desired response from the responses available. Note that during a particular iteration, the desired response will not normally be the final one since it is important to choose intermediate targets consistent with the linearity requirements (small rotations). Equally it is important to enforce continuity in the rotations applied to adjacent isochromats in order to avoid phase discontinuities. Thus several iterations are normally required to achieve the final response. At each stage, the exact response of the intermediate pulse is determined

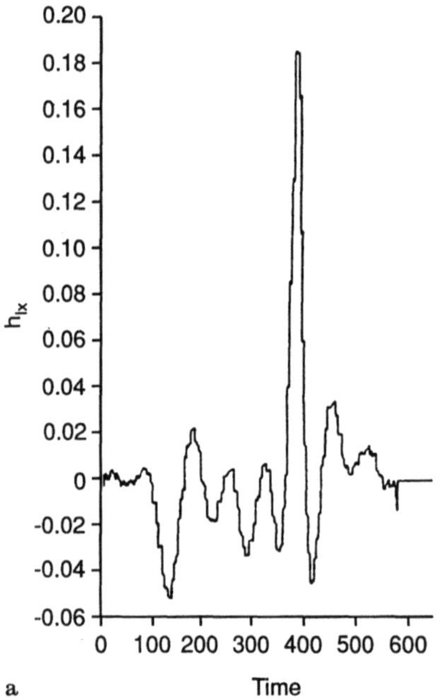

a

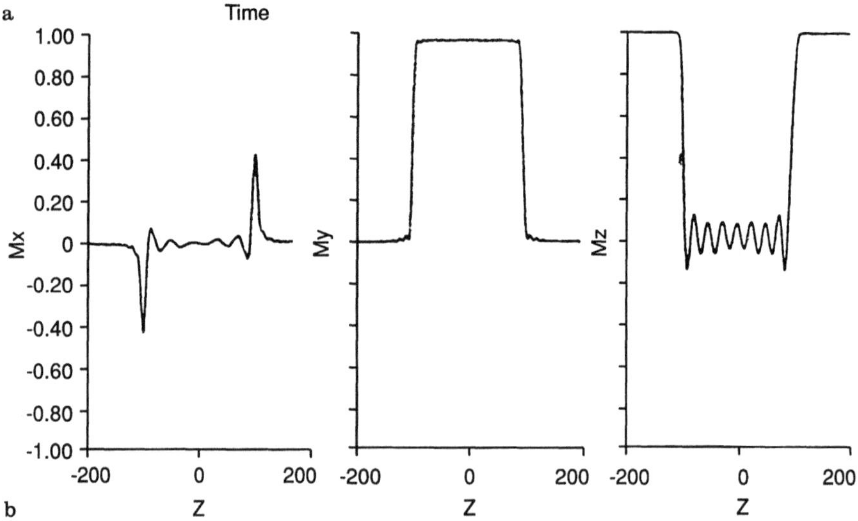

b

Fig. 9 a. Prefocused $\pi/2$ pulse with 576 elements, optimized using SPINCALC. **b** M_x, M_y and M_z responses. (From [31]) from the Bloch equations and is compared with the desired response. There is therefore no question of accumulating error. It is possible to apply further constraints during the least squares fitting. For example, the power levels can be kept below a given maximum in order to limit RF tissue heating effects which are of great concern in high field clinical imaging systems.

The SPINCALC optimization procedure has been used to generate a number of phase-compensated pulses. An example of such a pulse is shown in Fig. 9. This 576-element real phase-compensated $\pi/2$ pulse prepares the excited spins with a small linear phase shift in the negative sense. After a short time, shown to the right of the pulse envelope, during which no RF is applied, the spins will rephase in the existing field gradient. This small deliberate negative phase dispersion permits the gradient to be ramped down to zero as the spins rephase, leaving them ready for the next stage of the MRI/S procedure. The initial starting approximation for the pulse was a selective $\pi/2$-selective π pair, resembling a shortened spin–echo sequence. The final pulse shape retains much of this character. It is also possible to generate such pulses ab initio [12]. That the resulting pulse shape turns out to be different from that shown in Fig. 9, indicates that the method has not found a global minimum and suggests that there is a large family of pulses which are inherently phase-compensated.

4 k-Space Analysis

4.1 Theory

The great success of k-space analysis in producing a unified description of the many different imaging schemes [10], and in suggesting improved variants of them, led Pauly and colleagues to examine its possible application to selective pulse design [11]. In order to cast the Bloch equations in the requisite form, it is necessary to decouple the longitudinal magnetization, and assume that M_z is well approximated by the equilibrium magnetization M_0. This corresponds to the low tip angle approximation and might be thought an unpromising start for any selective pulse design method. However, the results achieved belie any negative preconceptions.

The solution to the Bloch equations for non-interacting spins, with the magnetization initially in the state $(0, 0, M_0)$ and under the above approximation is [11]:

$$M_{xy}(\mathbf{r}) = i\gamma M_0 \int_0^T B_1(t) e^{-i\gamma \mathbf{r} \cdot \int_t^T \mathbf{G}(s)\, ds}\, dt.$$

If a spatial frequency variable is defined as:

$$\mathbf{k}(t) = -\gamma \int_t^T \mathbf{G}(s)\, ds$$

then, with some rearrangement, the expression for the transverse magnetization, M_{xy}, can be rewritten as:

$$M_{xy} = i\gamma M_0 \int_{\mathbf{K}} p(\mathbf{k}) e^{i\mathbf{r}\cdot\mathbf{k}} \, d\mathbf{k},$$

indicating that the resulting transverse magnetization is just the Fourier transform of the weighted k-space trajectory. The weighting factor, $p(\mathbf{k})$, can be factored into two terms, $W(\mathbf{k}) = \dfrac{B_1(t)}{|\gamma \mathbf{G}(t)|}$ and $S(\mathbf{k})$, corresponding to a spatial weighting function and a sampling function, determining the extent and density of the k-space representation. Thus:

$$M_{xy}(\mathbf{r}) = i\gamma M_0 \int_{\mathbf{k}} W(\mathbf{k}) S(\mathbf{k}) e^{i\mathbf{r}\cdot\mathbf{k}} \, d\mathbf{k}.$$

If the coverage of k-space is uniform and complete, the sampling function can be ignored and the magnetization profile and spatial weighting function become simply a Fourier transform pair.

4.2 Phase-Compensated Pulses

Conventional slice-selective $\pi/2$ pulses, such as a truncated sinc pulse, require a negative gradient lobe to refocus the linear phase shift introduced during the course of the RF pulse (see Fig. 3). In the k-space interpretation, (one-dimensional) k-space is traversed linearly and the origin is reached when the remaining gradient integrates to zero. This occurs halfway through the (symmetric) RF pulse and again following the negative gradient lobe. At the latter point the RF weighting is centered in k-space and symmetric about the origin. The slice profile, which is the Fourier transform of this weighting, is in phase. This interpretation immediately suggests other means of achieving phase-compensation in one or multiple dimensions: what is required for automatic phase-compensation, is that the RF and gradient waveforms terminate at the middle of a symmetric weighting of k-space [11].

4.3 Two-dimensional Pulses

The k-space analysis also indicates the reason for the poor results obtained for two-dimensional pulses based on rotating gradients of constant amplitude. What is required is uniform coverage of the two-dimensional k-space. If the pulse is to be inherently phase-compensated then the traversal of k-space must also terminate at the origin. The trajectory which most readily meets these requirements is a (decreasing) spiral, an example of which is shown in Fig. 10

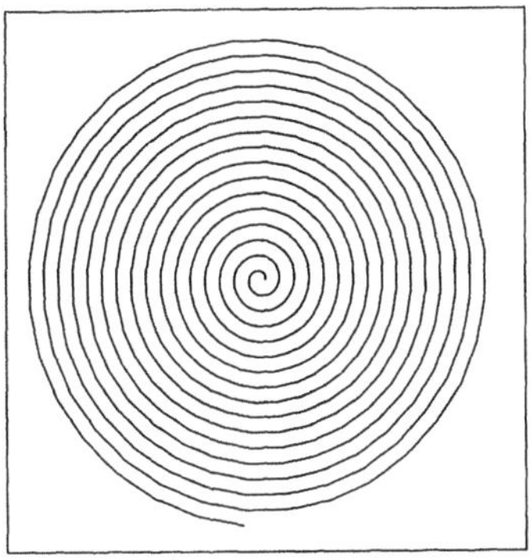

Fig. 10. A 16 turn k-space spiral with 512 time intervals. (From [12])

and which can be described mathematically by:

$$k_x(t) = A(1 - t/T)\cos 2\pi nt/T$$

$$k_y(t) = A(1 - t/T)\sin 2\pi nt/T,$$

where A is a constant and the spiral makes a total of n turns through two-dimensional k-space in a time T. Several choices present themselves for the RF envelope. Pauly et al. [11] illustrated the application of their new theoretical treatment by generating a circularly symmetric Gaussian excitation. In this case $W(\mathbf{k})$ is also a circularly symmetric Gaussian function and the required B_1 waveform is simply the product of $W(\mathbf{k})$ and the gradient waveform (see definition of **k** in Sect. 4.1).

Hardy and Cline [32] have used k-space theory to develop a variety of two-dimensional pulses which give radially symmetric excitation. Examples are shown in Fig. 11. Design principles (based on a Green's function method) were also discussed by these authors for two-dimensional pulses of square and arbitrary cross-sections.

Although such pulses represent a great improvement over previous two-dimensional pulses, they still display artefacts due to approximations inherent in the low tip angle analysis and also due to the non-uniform coverage of k-space. Hardy et al. [33] have demonstrated that the latter effect can be allowed for in a straightforward extension of k-space theory. The former effects are more problematic. At the present state of the art they are probably best 'cleaned up' by numerical optimization methods. An example is shown in Fig. 12 of the use of the SPINCALC algorithm to optimize a radially symmetric two-dimensional π pulse.

The practical application of two-dimensional pulses on standard whole-body imaging systems is restricted by constraints on gradient amplitudes and (especially) slew rates. These problems can be alleviated to some degree by using optimization methods to limit these parameters [34]. Two-dimensional pulses can be further shortened by splitting the coverage of k-space over many (typically 16) such pulses. In one such implementation the normal spiral trajectory

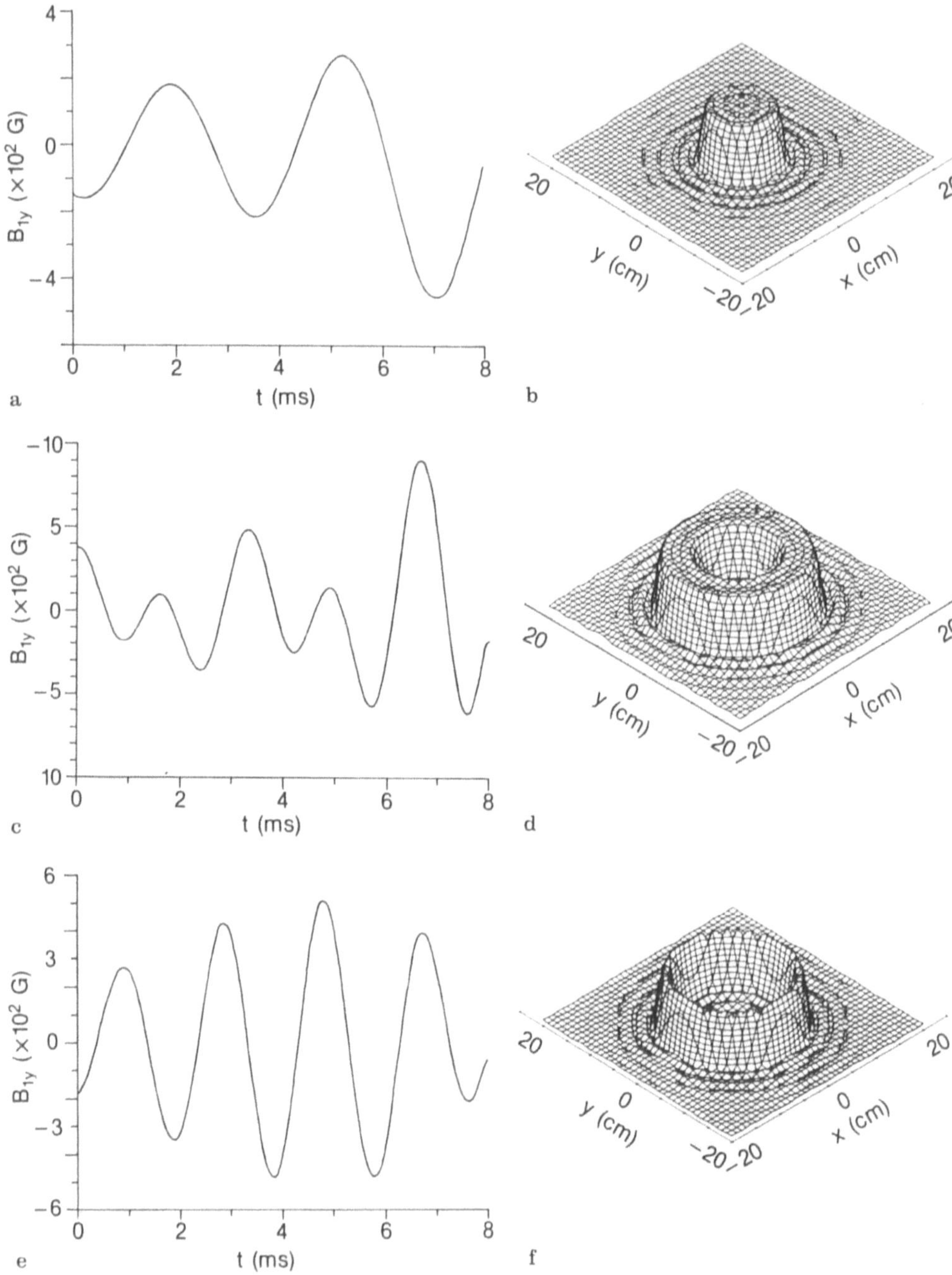

Fig. 11. Three radially symmetric magnetization profiles with their respective RF excitation waveform envelopes displayed on the left. The gradient waveforms correspond to a 16-turn spiral trajectory through k-space. (From [32])

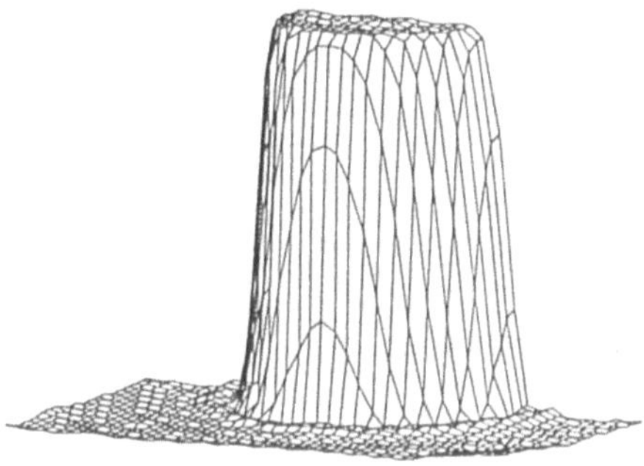

Fig. 12. The M_z response (shown inverted) to a two-dimensional π pulse optimized using SPINCALC. (From [6])

is replaced by spiral arms resembling a pinwheel. Although overcoming the slew rate constraint, susceptibility to motional artefact is reintroduced. Nevertheless, such pulses have been used to record accurately localized ^{31}P spectra of high quality across the wall of the human heart [35].

5 Conclusions

Selective pulses are central to MRI and MRS, yet commercial systems still rely on first generation designs based on an intuitive Fourier approach. Although these 'do a job' they are far from ideal and can be greatly improved upon. The recent progress in designs of phase-compensated pulses should benefit applications in systems with short T_2s. They will also allow selective pulses to be more widely applied in high resolution applications.

The recent developments in k-space theory have greatly clarified design principles and have spurred a renewal of interest in multidimensional pulses. Effective two-dimensional pulses are already available, though their practical application is limited by their requirement for gradient amplfiers with high slew rates. This limitation applies even more so to three-dimensional pulses which are the ultimate goal of the in vivo spectroscopist. Nevertheless, it is clear that acceptable compromises will be possible.

6 References

1. Garroway AN, Grannell PK, Mansfield P (1974) J Phys C 7: L457
2. Mansfield P, Morris PG (1982) NMR imaging in biomedicine. Academic, New York.
3. Morris PG (1986) NMR imaging in medicine and biology. Oxford University Press, Oxford
4. Aue WP (1986) Rev Magn Reson Med 1: 21
5. Morris PG (1988) Ann Rep NMR Spectrosc 20: 1

6. Morris PG, McIntyre DJO, Rourke DE, Ngo JT (1989) Magn Reson Med Biol 11: 167
7. Locher PR (1980) Phil Trans R Soc Lond B 289: 537
8. Conolly S, Macovski A (1985) Proc 4th Meeting Soc Magn Reson Med 958
9. Lent AH, Kritzer MR (1985) Proc 4th Meeting Soc Magn Reson Med 1015
10. Ljunggren S (1983) J Magn Reson 54: 338
11. Pauly J, Nishimura D, Macovski A (1989) J Magn Reson 81: 43
12. Morris PG, McIntyre DJO, Rourke DE, Ngo JT (1989) NMR in Biomed 2: 257
13. Mansfield P (1977) J Phys C 10: L55
14. Coxon R, Ordidge RJ, Mansfield P (1987) Proc 6th Meeting Soc Magn Reson Med 226
15. Ordidge RJ, Connelly A, Lohman JAB (1986) J Magn Reson 66: 283
16. Baum J, Tycko R, Pines (1983) J Chem Phys 79: 4643
17. Silver MS, Joseph RI, Chen C-N, Sank VJ, Hoult DI (1984) Nature 310: 681
18. Rourke DE, Morris PG (1991) Proc 10th Meeting Soc Magn Reson Med
19. Zakharov VE, Shabat AB (1972) Soviet Phys JETP 34: 62
20. Kirkpatrick S, Gelatt Jr, CD, Vecchi MP (1983) Science 220: 671
21. Metropolis N, Rosenbluth AW, Rosenbluth MN, Teller AH, Teller E (1953) J Chem Phys 21: 1087
22. Rourke DE, McIntyre DJO, Morris PG (1989) Proc 8th Meeting Soc Magn Reson Med 863
23. Morris PG, McIntyre DJO, Rourke DE, Ngo JT (1989) NMR in Biomed 5/6: 257
24. Bottomley PA, Hardy CJ (1987) J Magn Reson 74: 550
25. Bottomley PA, Hardy CJ (1987) J Appl Phys 62: 4284
26. Hardy CJ, Bottomley PA, O'Donnell M, Roemer P (1988) J Magn Reson 77: 233
27. Hardy CJ, Bottomley PA, Roemer P (1988) J Appl Phys 63: 4741
28. Ngo JT, Morris PG (1986) Biochem Soc Trans 14: 1271
29. Ngo JT, Morris PG (1987) Magn Reson Med 5: 217
30. Ngo JT, Morris PG (1986) TAMU Newsletter 337: 38
31. Morris PG, Rourke DE, Ngo JT, McIntyre DJO (1988) Proc 7th Meeting Soc Magn Reson Med 656
32. Hardy CJ, Cline HE (1989) J Magn Reson 82: 647
33. Hardy CJ, Cline HE, Bottomley PA (1990) J Magn Reson 87: 639
34. Hardy CJ, Cline HE (1989) J Appl Phys 66: 1513
35. Hardy CJ, Bottomley PA (1991) Magn Reson Med 17: 315

Rf Pulses for Multiple Frequency Excitation: Theory and Application

S. Müller

MR Center of the University, Klingelbergstr. 50, CH-4031 Basel, Schweiz

Table of Contents

Frequency-selective radiofrequency pulses are of high importance in biological and clinical applications of Magnetic Resonance since they provide the basis for localized imaging and spectroscopy. In this contribution an extension of frequency-selective irradiation is described: by simple technical modifications standard selective radiofrequency pulses can be converted into pulses which contain multiple frequencies. These pulses select multiple regions for simultaneous multiregional investigations. They avoid many basic limitations of standard imaging and spectroscopy techniques: e.g. the number of available slices for multislice imaging is not restricted by the repetition time TR of the imaging sequence. Another application is cardiac imaging in which multiple cardiac cross sections can be measured simultaneously at the same cardiac phase. Many other imaging and spectroscopy experiments obtain more flexibility and a higher efficiency in terms of signal to noise ratio per measurement time.

Multifrequency-selective radiofrequency pulses are described and different spatial encode/decode schemes are discussed. Using the Hadamard encoding scheme, imaging and spectroscopy applications are presented which demonstrate the potential of multifrequency radiofrequency excitation.

NMR Basic Principles and Progress, Vol. 26

1 Introduction

One of the most important techniques in biological and medical Magnetic Resonance (MR) is the spatially selective radiofrequency (rf) irradiation. With this tool the MR signals originating from a large heterogeneous sample like the human body can be focussed to a small area of interest to perform refined chemical and anatomical investigations. Nowadays the technique has a broad range of applications: it defines an imaging-slice in two-dimensional (2D) MR imaging, it selects volumes for 3D imaging, it predefines regions for spectroscopic imaging, it saturates areas to avoid flow and motion artefacts and it prepares volumes for localized spectroscopy.

In principle, the spatial focussing within the sample can be performed in several ways. The best established techniques for MR systems with B_0 gradient coils are based on frequency selective rf pulses in the presence of B_0 field gradients. This technique spatially limits the resonance condition of the MR experiment to a slice within the sample. The slice is oriented perpendicular to the magnetic field gradient and under appropriate conditions the whole MR experiment takes place only in this predefined slice. The slice selection process can be repeated with a different gradient orientation for a second slice selection. In this case the combination of both experiments leads to the selection of a rod (the intersecting volume of two nonparallel slices). Finally, a parallelepiped (e.g. a cube) may be selected with a third gradient direction and selective irradiation.

In this article we will describe experimental conditions under which these volume selection experiments can act on multiple locations simultaneously. With this trick several volumes are irradiated at the same time and the subsequent experiments are performed simultaneously on all the selected volumes.

Advantages of this technique are:

- —Multislice imaging can be performed even with very short repetition times
- —Cardiac images of multiple slices can be measured simultaneously at the same cardiac phase
- —Multiple well-separated volumes may be prepared for high-resolution 3D imaging or spectroscopic imaging
- —Localized spectroscopy can be performed on multiple volumes for spectral quantification

Different approaches for multivolume experiments will be reviewed. To introduce the concepts the frequency-selective rf pulses which provide the basis for single volume preparation will be discussed first. Then the step to multi-frequency-selective rf pulses will be made and computer simulations will demonstrate the response of the spin system in this case. In the next section, we will discuss different encode/decode schemes for the separation of the MR information of the multiple volumes. Some experimental results obtained with multilocal preparation will demonstrate the feasibility and the power of the technique. Finally advantages and drawbacks of multiregional examinations will be discussed in general.

2 Theoretical Aspects

2.1 Single- and Multifrequency-Selective Rf Pulses

Let us consider an extended sample containing H_2O in a homogeneous external magnetic field B_0. The hydrogen protons of this sample give rise to a splitting of the energy levels for nuclear spins parallel and antiparallel to the applied magnetic field. The frequency difference ω_L of the splitting is:

$$\omega_L = \gamma B_0 \tag{1}$$

where γ is the gyromagnetic ratio, ω_L is called Larmor frequency.

An externally applied rf field of frequency ω will induce transitions between the energy levels as soon as $\omega = \omega_L$. This is called the resonance condition.

Consider an rf irradiation of the sample with a fixed frequency ω_a. As long as $\omega_a \neq \omega_L$ no transitions will occur in the sample. However, if a magnetic field gradient in z direction:

$$\partial B_0/\partial z = G_z \tag{2}$$

is introduced, then the magnetic field will become spatially dependent:

$$B(z) = B_0 + G_z \cdot z \tag{3}$$

at the same time the Larmor frequency ω_L in the sample will become a linear function of z:

$$\omega_L(z) = \gamma B(z) = \gamma(B_0 + G_z \cdot z). \tag{4}$$

If somewhere in the sample $\omega_L(z)$ is equal to the frequency ω_a of the external rf field, resonance transitions will occur at this position. From

$$\omega_a = \omega_L(z) \tag{5}$$

follows

$$z = z_a = (\omega_a/\gamma - B_0)1/G_z \tag{6}$$

i.e. the resonance condition is only fulfilled for $z = z_a$, which is in a plane perpendicular to the gradient direction. This is illustrated in Fig. 1a.

In a similar way it can be verified that if the rf irradiation is the sum of two single frequency rf waves of frequency ω_a and ω_b, then the resonance condition will be fulfilled simultaneously at two positions (Fig. 1b):

$$z_a = (\omega_a/\gamma - B_0)1/G_z \tag{7}$$

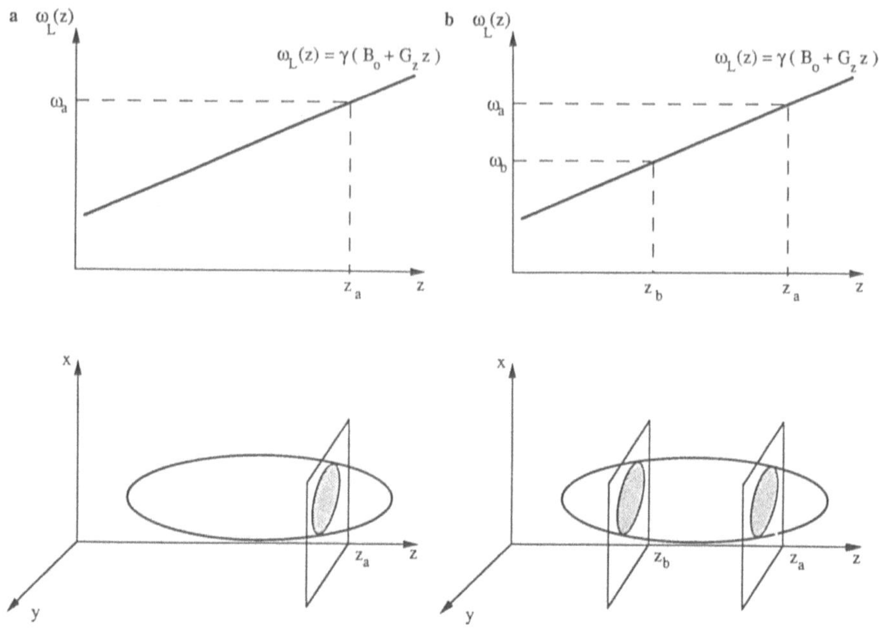

Fig. 1 a. *Top*: Larmor frequency ω_L plotted as a function of position along the field gradient direction z. For an rf wave of frequency ω_a resonance transitions between the Zeeman levels occur at position $z = z_a$. This position corresponds to a plane perpendicular to the direction of the gradient in the sample (*bottom*). **b** Double frequency irradiation using an rf wave with frequencies ω_a and ω_b. Transitions occur simultaneously at positions z_a and z_b. This corresponds to two parallel planes in the sample

and

$$z_b = (\omega_b/\gamma - B_0)1/G_z. \tag{8}$$

Rf waves which contain only a single frequency (e.g. ω_a) are infinitely long. As soon as the rf irradiation is reduced in duration, it will automatically contain more frequency components. They can be determined by Fourier transformation of the rf pulse. Nowadays pulse MR has become the method of choice in most MR experiments, and the durations of the rf pulses are usually in the range of μs or ms. By using the Fourier Transform (FT) it can be shown that an rf pulse of duration τ_p with a carrier frequency ω_a has a frequency distribution centered at ω_a with a spread of about $2\pi/\tau_p$.

Therefore, if such an rf pulse is applied to the sample described above, the resonance condition will lead to transitions in a slice centered at $z = z_a$ with thickness Δz_a where:

$$\Delta z_a \approx 2\pi/\gamma G_z \tau_p. \tag{9}$$

The same happens for an rf pulse containing frequencies ω_a and ω_b. Such a pulse can be generated by adding together the time signals of two rf pulses

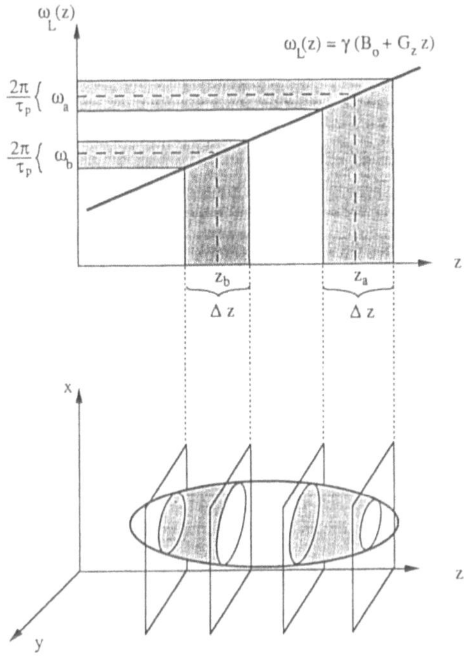

Fig. 2. Double-slice selection using multi-frequency-selective rf pulses. The finite length of the rf pulse causes a frequency distribution around the carrier frequencies. This is transformed via the B_0 gradient into the spatial irradiation of two slices centered at $z = z_a$ and $z = z_b$ with thicknesses Δz

with carrier frequencies ω_a and ω_b. Due to the linearity of the Fourier Transform this will lead to a frequency distribution, whose shape is duplicated at the center frequencies ω_a and ω_b. Therefore the resonance condition is fulfilled at two slice positions z_a and z_b with the same thickness Δz (Fig. 2).

Basically any frequency distribution $I(\omega)$ can be Fourier transformed for the generation of a time domain rf pulse. Via the gradient of the magnetic field this frequency distribution is then transformed into a spatial irradiation distribution in the sample. $I(\omega)$ may include e.g. a comb of frequencies generating transitions in multiple parallel slices where each slice may be individually designed in position and shape.

Note that until now only the resonance condition was discussed which causes transitions between the Zeeman energy levels. So far no attention was paid on the spin response which generates the MR signal following these transitions. Unfortunately, the spin systems are strongly nonlinear with regard to their response to perturbations. Therefore the linear arguments mentioned so far may fail for more quantitative discussions.

A simple example illustrates the problem: we regard the effect of a short rectangular rf pulse of duration τ_p and amplitude vector B_1. Suppose B_1 is perpendicular to B_0. If the carrier frequency of the rf pulse is identical to the Larmor frequency ω_L of the spin system, then the Bloch equations tell that the rf pulse causes a nutation of the spins around an axis parallel to the direction

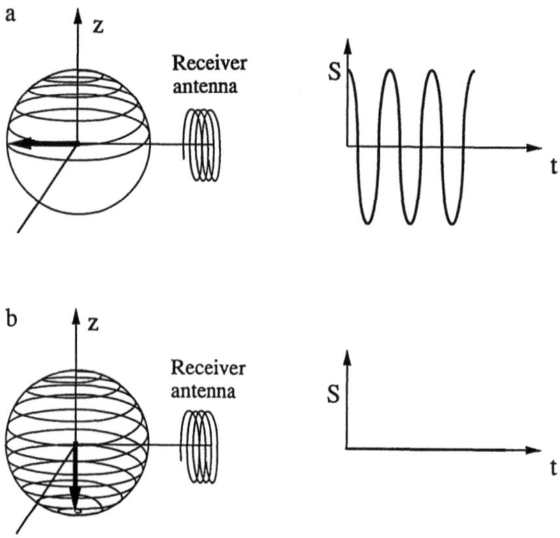

Fig. 3 a, b. Nonlinearity of the MR response: comparison of the trajectories of the magnetization vector during a 90° pulse **a**) and a 180° pulse **b**. After the 90° pulse the magnetization processes in the transverse plane and an intense MR signal S is measured. After the 180° pulse the magnetization is aligned along-z and does not precess in the transverse plane any more. No signal is picked up with the receiver coil

of B_1. The nutation frequency $\omega_1$is:

$$\omega_1 = \gamma B_1. \tag{10}$$

This nutation is superimposed to a precession of the spins around the main magnetic field direction with the (much higher) frequency ω_L. Therefore the dip of the spin vector performs a spiral pathway over the surface of a sphere during the rf pulse. At the end of the pulse the nutation stops and only the precession around B_0 persists. This precession corresponds to a time dependent magnetic field whose magnetization vector rotates in a plane perpendicular to B_0 with frequency ω_L. It can be measured with an rf receiver antenna and is equivalent to the MR signal which is acquired in a simple pulse MR experiment (Fig. 3).

It is evident, that if the nutation angle is 90° at the end of the rf pulse, i.e. either τ_p or B_1 are adjusted such, that:

$$\omega_1 \tau_p = \gamma B_1 \tau_p = \pi/2 \tag{11}$$

then the spins are in the transverse plane after the pulse and a high oscillatory signal S is sensed with the receiver pickup coil (Fig. 3a).

On the other hand after a nutation of 180°, i.e. for τ_p or B_1 are adjusted such that:

$$\gamma B_1 \tau_p = \pi \tag{12}$$

the spins are antiparallel to B_0 after the pulse and no precession is measured (Fig. 3b).

This example has shown that doubling the B_1 field strength (the input) does not double the MR signal. On the contrary, twice the rf input voltage as in the case of the 180° pulse leads to zero MR signal, a strongly nonlinear effect. Thus we have seen that adding together two rf pulses of the *same* carrier frequency does not result in the sum of the corresponding single frequency MR responses. In this article multifrequency selective rf pulses are described which are linear combinations of frequency selective rf pulses with *different* carrier frequencies. A basic requirement for these techniques to work is that the MR signal after such a multifrequency irradiation is the same as the linear combination of the corresponding single frequency MR signals. Thus the question arises, under which conditions we may add together rf pulses of different carrier frequencies and rely on the required linearity arguments for the spin response.

Let us first discuss the simple case in which two rf pulses of different carrier frequencies ω_a and ω_b are added together.

If we call the rf pulse with carrier frequency ω_a:$B_a(t)$ and the second pulse $B_b(t)$ (carrier frequency ω_b), then each pulse has a frequency distribution $I_a(\omega)$ and $I_b(\omega)$, and due to the linearity of the FT it is clear the sum $B_{a+b}(t)$ of both rf pulses:

$$B_{a+b}(t) := B_a(t) + B_b(t) \tag{13}$$

has a frequency distribution $I_{a+b}(\omega)$ corresponding to the sum of the individual distributions:

$$I_{a+b}(\omega) := I_a(\omega) + I_b(\omega). \tag{14}$$

One might argue, that the spin response S_{a+b} of this pulse will be the sum of the responses S_a and S_b as soon as the relevant parts of $I_a(\omega)$ and $I_b(\omega)$ do not overlap. Simulations of the spin behaviour during an rf pulse show, however, that especially during the first part of the rf pulse spins with very different Larmor frequencies are influenced as well and only towards the end of the rf pulse the frequency selectivity corresponding to the FT of the rf pulse is obtained. This astonishing fact can be intuitively understood even with spectral arguments, keeping in mind that for the spins the rf pulse has an unknown future. As long as the pulse is not finished it might also stop at any time which would result in a broad frequency distribution of the rf pulse and a corresponding broadband spin irradiation. Therefore this FT argument based on the nonoverlapping frequency distributions of the rf pulses might be insufficient, since it does not fully treat the effects during the rf pulse.

Another way of treating such a double frequency irradiation is given with the Bloch equations. In this picture the rf pulse $B_b(t)$ with a carrier frequency ω_b creates at the frequency ω_a a nutation around an effective axis ω_{eff}:

If $\omega_b = \omega_a$, the on resonance condition as described earlier holds: $\omega_{eff}(t) = \gamma B_b(t)$, while for strongly separated frequencies ω_{eff} is almost aligned along B_0 and only minor transverse magnetization is generated by $B_b(t)$ at

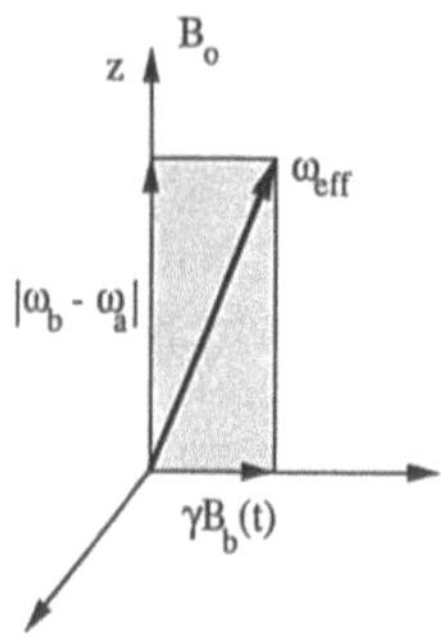

Fig. 4. Off-resonance effect using the Block equations. An rf pulse of frequency ω_b introduces at the frequency ω_a a nutation around the axis ω_{eff}. For strongly separated frequencies ω_b and ω_a, ω_{eff} is almost aligned along B_0 and only minor transverse magnetization is generated by the ω_b rf frequency

frequency ω_a. The higher the frequency difference $|\omega_b - \omega_a|$ the less signal is generated by the off-resonance rf pulse, a similar result as obtained before.

This analysis is quite convenient for rectangular rf pulses where $B_1(t)$ is constant in time. In the situation of specifically designed rf pulses, $B_1(t)$ may be more complex. In addition the carrier frequency may change during the rf pulse in which case ω_{eff} follows a complicated pathway.

In general therefore simulations, which calculate the spin behaviour using the Bloch equations for each point in time and space provide more insight into the spin response.

2.2 Simulations

Let us first compare a single frequency selective excitation with a simultaneous double frequency excitation.

In Fig. 5a the time evolution of the transverse magnetization during a frequency-selective rf pulse is shown. For that purpose a one-dimensional phantom containing spins aligned along the magnetic field gradient direction was prepared for the simulation. During a sinc-shaped rf pulse the Bloch equations were solved for each spin at each point in time. The resulting profiles of transverse magnetization are plotted as a function of time yielding the surface plot in Fig. 5a.

Note that the evolution of the selective slice excitation is only complete at the end of the rf pulse, while at the beginning of the time evolution a broader range of spins is excited than the FT of the rf pulse would suggest. This effect was already mentioned in Sect. 2.1.

A corresponding two slice excitation is shown in Fig. 5b where the rf pulse was replaced by the sum of two sinc rf pulses centered at two well-separated carrier frequencies. All other parameters were kept the same as before. Note the evolution of two slices with a similar shape as in the previous example.

The number of slices can be increased to any feasible number, an example of an 8-slice excitation is shown in Fig. 6.

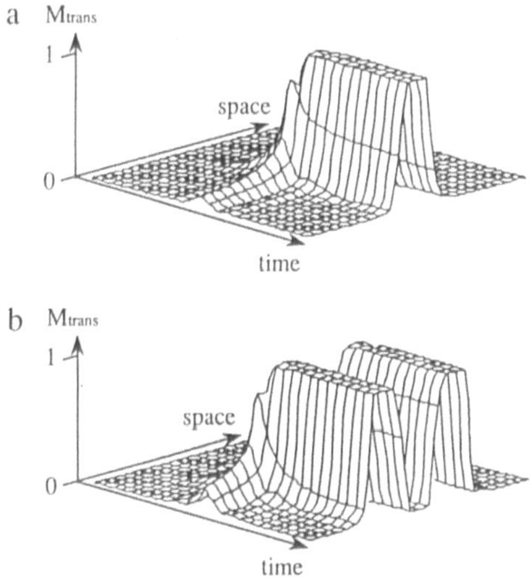

Fig. 5 a, b. Comparison of a single-slice and a double-slice sinc excitation using numerical simulations of the Bloch equations. For the *top figure* a since excitation was applied in a slice selection gradient. The spins were mathematically arranged in a one-dimensional slab phantom (256 discrete spins). For the *bottom figure* a double-slice selection using a multifrequency sinc rf pulse was applied otherwise using the same parameters. During the rf pulse (duration 5.12 ms, 256 sampling points) the development of the spatial slice profiles can be observed. For details of the simulations see Ref. [2]

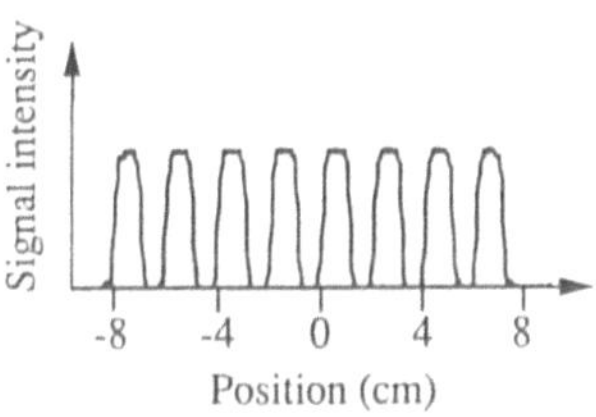

Fig. 6. Simultaneous excitation of 8 slices. Using the same simulation parameters as in Fig. 5 an 8-slice multifrequency rf pulse was applied to the phantom and the resulting signal intensity plotted as a function of position

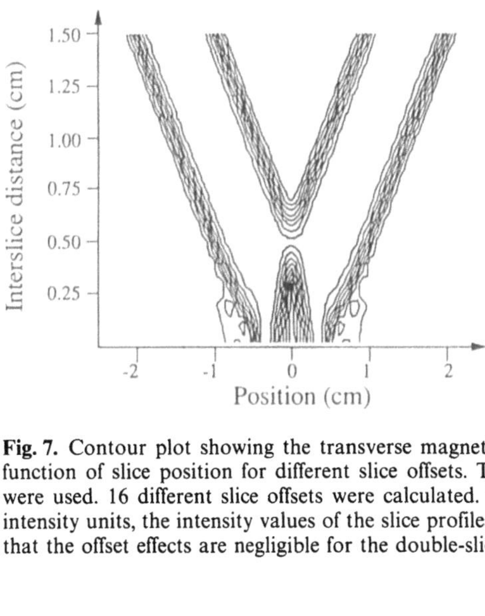

Fig. 7. Contour plot showing the transverse magnetization after a two slice sinc excitation as a function of slice position for different slice offsets. The same simulation parameters as in Fig. 5 were used. 16 different slice offsets were calculated. The contour lines are equally spaced by 50 intensity units, the intensity values of the slice profiles range from 0 to 430 units. The figure shows that the offset effects are negligible for the double-slice sinc excitation until the slices merge

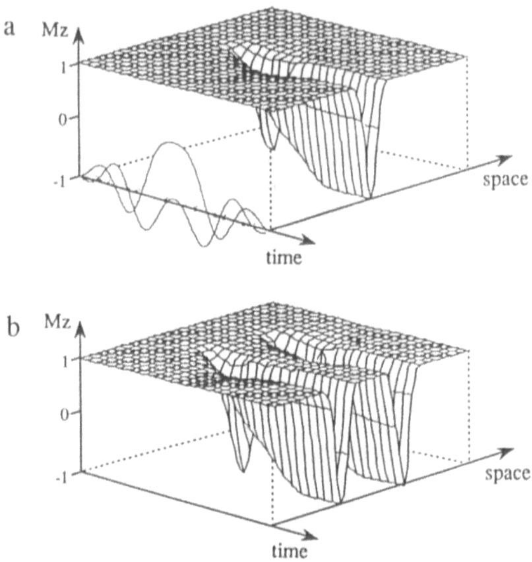

Fig. 8 a, b. Comparison of a single-slice and a double-slice hypersecant inversion using numerical simulations of the Bloch equations. The evolution of the longitudinal magnetization is shown as a function of time and space. For the *top figure* a hypersecant inversion was applied to a one-dimensional slab phantom mathematically constructed from 256 discrete spins. A slice selection gradient was applied to obtain a spatial profile of the inversion. For the *bottom figure* a double-slice selection using a multifrequency hypersecant rf pulse was applied otherwise using the same parameters as in the *top figure*. During the rf pulse (duration 5.12 ms, 256 sampling points, real and imaginary part are indicated on the *top figure*) the development of the spatial slice profiles can be observed. For details of the simulations see Ref. [2]

The off-resonance behaviour for double-slice sinc excitation is demonstrated in Fig. 7. Note that even for adjacent slices a precise multislice selection is possible.

An interesting question is also, whether frequency-modulated adiabatic rf pulses could be used for multifrequency irradiation.

Under certain constraints this is indeed the case and an example of a two slice adiabatic inversion pulse of the hypersecant type [1] is shown in Fig. 8 together with the corresponding single slice evolution.

It can be shown that some of the features of the single-slice adiabatic inversion pulse are lost in such a multislice experiment, e.g., the B_1 insensitivity of the double-frequency inversion is not as pronounced as in the single-frequency case. The limits of applicability of the multifrequency ideas have been discussed by Hafner [2] and Goelman [3] for a number of rf pulse shapes.

2.3 *Encoding/Decoding of Spatial Information*

In the previous section it was demonstrated that multifrequency rf pulses usually generate MR signals which are a linear combination of the corresponding

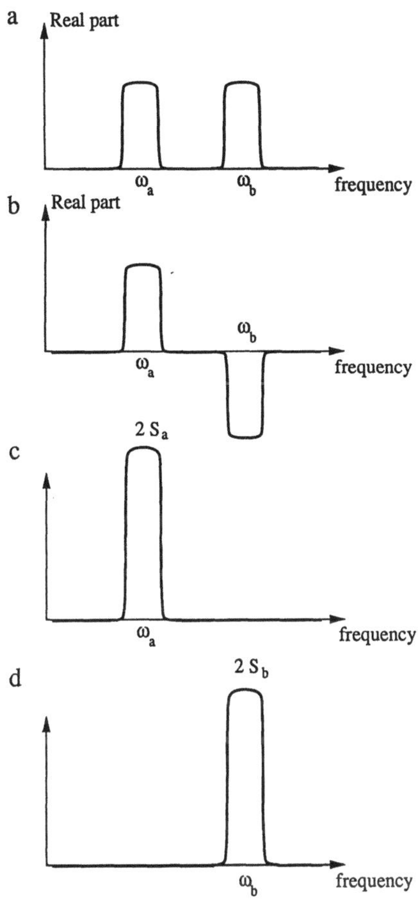

Fig. 9 a–d. Example of a simple encode-decode procedure for two slices at frequencies ω_a and ω_b. **a)** Real part of the spectrum of a (+ +) excitation pulse. This pulse generates a similar excitation profile and a similar shape of the transverse magnetization in the sample. **b)** Real part of the (+ −) excitation. **c)** After summing the results of excitation **a)** and **b)** the signals from frequency ω_b cancel and the signal at ω_a is obtained twice. **d)** The decoding of the signals from the slice at ω_b is obtained by subtracting the results from **a)** and **b)** yielding twice S_b

single-frequency results. However, after the multifrequency irradiation the MR signals of *each* spatial component are of interest and have to be sorted out from the composed multifrequency MR response. Therefore, somehow positional information has to be introduced into the MR response such that the signals from each spatial component can be identified. An elegant and sensitive way to do that is to label the phase of the MR signal with spatial information. This can be done *during* the rf pulse and leads to multilocal techniques with optimum efficiency.

A simple example will illustrate this idea: suppose we want to measure two slices *a* and *b* in an object simultaneously and then decompose the signals into the two original signals S_a and S_b again. We apply the following procedure: in a first experiment a double-slice excitation pulse is created whose Fourier transform is shown in the Fig. 9a.

Such a pulse corresponds to the sum of two identical sinc pulses with carrier frequencies ω_a and ω_b. We may call this pulse a (+ +) excitation. The spin

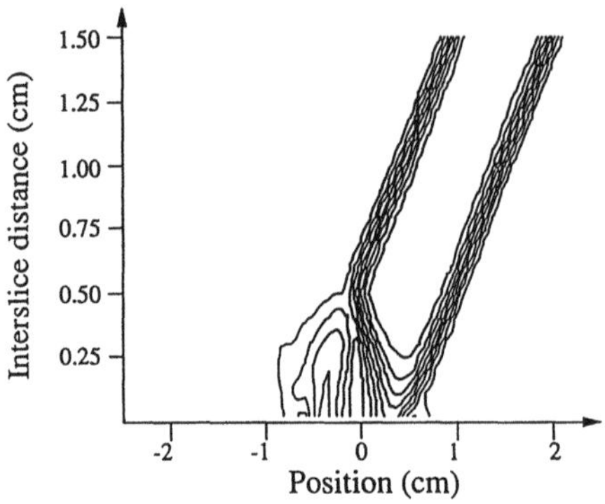

Fig. 10. Contour plot showing the transverse magnetization after sinc excitation as a function of slice position for different slice offsets. The same simulation parameters as in Fig. 7 were used, however the encode/decode procedure of fig. 9 was applied to the simulated MR responses in order to obtain the signal of one single slice alone. The figure shows that the slice offset has a negligible effect on the quality of the slice selection even after decoding

response will have a similar frequency distribution for the transverse magnetization.

In a second experiment the excitation pulse is constructed such that the pulse with carrier frequency ω_b is inverted, (a (+ −) excitation). This pulse has a frequency distribution as shown in Fig. 9b. Since MR is a phase-sensitive technique, the nutation of the spins at ω_b will be backwards instead of forward and the phase of the MR signal will be inverted.

In the first experiment the signal S_1 therefore is:

$$S_1 = S_a + S_b \tag{15}$$

and in the second experiment:

$$S_2 = S_a - S_b. \tag{16}$$

To obtain the signals from slice a and b alone, the signals S_1 and S_2 have to be added and subtracted:

$$S_1 + S_2 = 2S_a, \tag{17}$$

twice the signal at frequency ω_a alone, and

$$S_1 - S_2 = 2S_b, \tag{18}$$

twice the signal at frequency b (Figs. 9c and 9d).

In Fig. 10 a simulation of this procedure is shown using the results of Fig. 7. It can be noted that the decoding of the information from slice b is independent of the slice-to-slice distance.

This was a simple example of an encode–decode process which allows to separate the combined MR signals after the multivolume excitation: using two experiments the signals from slice a and slice b were obtained each with a sensitivity corresponding to two acquisitions. If slice a and b would have been measured one after the other, the sensitivity after both experiments would only correspond to one excitation for each slice. This illustrates a basic advantage of the multifrequency experiment.

Note that the above ideas can be generalized for the excitation of two rods or cubes and to a higher number of locations. Note further that in order to use the phase information for decoding, the MR experiments which follow the multilocal irradiation have to preserve the phase information such that the imposed preparation of the spins is transferred into the observation period. Most experiments fulfill this condition.

In the following we will discuss different encode–decode procedures. For simplification we will regard the example of four simultaneously measured slices and discuss different possibilities for this situation. Let $S_a(t), S_b(t), S_c(t)$, and $S_d(t)$ be the MR signals arising from slices a, b, c, and d after frequency-selective rf pulses $B_a(t), B_b(t)$, $B_c(t)$, and $B_d(t)$. $B_a(t)$ is for example an rf pulse with carrier frequency ω_a.

A general simultaneous multifrequency excitation for phase encoding may have the following form:

$$B_1(t) = e_{1a}B_a(t) + e_{1b}B_b(t) + e_{1c}B_c(t) + e_{1d}B_d(t) \tag{19}$$

with phase factors $e_{1a}, \dots e_{1d}$ being complex numbers of absolute value $\leqq 1$.

With the arguments mentioned earlier the MR signal $S_1(t)$ after such a multifrequency excitation can therefore be written:

$$S_1(t) = e_{1a}S_a(t) + e_{1b}S_b(t) + e_{1c}S_c(t) + e_{1d}S_d(t). \tag{20}$$

It is evident that to sufficiently encode four positions, at least four experiments are necessary with different linear combinations of the rf (remember e.g. the example of the two-slice experiment where at least two experiments were required to separate the signals from each slice).

Thus the result of a complete encoding may be written as:

$$\begin{aligned}
S_1(t) &= e_{1a}S_a(t) + e_{1b}S_b(t) + e_{1c}S_c(t) + e_{1d}S_d(t)\\
S_2(t) &= e_{2a}S_a(t) + e_{2b}S_b(t) + e_{2c}S_c(t) + e_{2d}S_d(t)\\
S_3(t) &= e_{3a}S_a(t) + e_{3b}S_b(t) + e_{3c}S_c(t) + e_{3d}S_d(t)\\
S_4(t) &= e_{4a}S_a(t) + e_{4b}S_b(t) + e_{4c}S_c(t) + e_{4d}S_d(t).
\end{aligned} \tag{21}$$

where $S_i(t)\,(i = 1, \dots, 4)$ was generated by the rf pulse:

$$B_i(t) = e_{ia}B_a(t) + e_{ib}B_b(t) + e_{ic}B_c(t) + e_{id}B_d(t). \tag{22}$$

In vector notation the above results can be written as:

$$\begin{pmatrix} S_1(t)\\ S_2(t)\\ S_3(t)\\ S_4(t) \end{pmatrix} = \underbrace{\begin{pmatrix} e_{1a} & e_{1b} & e_{1c} & e_{1d}\\ e_{2a} & e_{2b} & e_{2c} & e_{2d}\\ e_{3a} & e_{3b} & e_{3c} & e_{3d}\\ e_{4a} & e_{4b} & e_{4c} & e_{4d} \end{pmatrix}}_{\mathbf{E}_4} \cdot \begin{pmatrix} S_a(t)\\ S_b(t)\\ S_c(t)\\ S_d(t) \end{pmatrix} \tag{23}$$

The encoding matrix $\mathbf{E}_4$ consists of complex elements with absolute value $\leqq 1$. If $\mathbf{E}_4$ can be inverted such that $\mathbf{E}_4^{-1}$ exists with:

$$\mathbf{E}_4^{-1} \cdot \mathbf{E}_4 = \mathbf{1} \text{ (the unity matrix)}, \tag{24}$$

then the above equations can be solved:

$$\mathbf{E}_4^{-1} \cdot \begin{pmatrix} S_1(t) \\ S_2(t) \\ S_3(t) \\ S_4(t) \end{pmatrix} = \mathbf{E}_4^{-1} \cdot \mathbf{E}_4 \cdot \begin{pmatrix} S_a(t) \\ S_b(t) \\ S_c(t) \\ S_d(t) \end{pmatrix} = \begin{pmatrix} S_a(t) \\ S_b(t) \\ S_c(t) \\ S_d(t) \end{pmatrix}. \tag{25}$$

In other words, the MR signals $S_1(t), \ldots, S_4(t)$ can be used for the calculation of the signals from each slice: $S_a(t), \ldots, S_d(t)$.

$\mathbf{E}_4^{-1}$ represents the decoding matrix. In an optimum encoding technique in terms of SNR/measurement time, the encoding matrix is proportional to a unitary matrix [44] which implies that:

$$\mathbf{E}_4^{-1} \propto \mathbf{E}_4^{*\mathrm{T}}, \tag{26}$$

where $\mathbf{E}_4^{*\mathrm{T}}$ is the hermitian conjugate of $\mathbf{E}_4$. (If $\mathbf{E}_4$ contains only real elements, $\mathbf{E}_4^{*\mathrm{T}}$ is the same as $\mathbf{E}_4^{\mathrm{T}}$, the transpose of $\mathbf{E}_4$).

Therefore:

$$\mathbf{E}_4^{-1} \propto \begin{pmatrix} e_{1a}^* & e_{2a}^* & e_{3a}^* & e_{4a}^* \\ e_{1b}^* & e_{2b}^* & e_{3b}^* & e_{4b}^* \\ e_{1c}^* & e_{2c}^* & e_{3c}^* & e_{4c}^* \\ e_{1d}^* & e_{2d}^* & e_{3d}^* & e_{4d}^* \end{pmatrix} \tag{27}$$

(*denotes the complex conjugate). Such a matrix pair leads to the well known $\sqrt{n}$ SNR enhancement for n times signal averaging.

There are numerous unitary matrices which can be used for optimum encoding. Three examples will be given below:

a) Fourier

$$\mathbf{F}_4 = \begin{pmatrix} 1 & 1 & 1 & 1 \\ 1 & i & -1 & -i \\ 1 & -1 & 1 & -1 \\ 1 & -i & -1 & i \end{pmatrix} \tag{28}$$

i is $\sqrt{(-1)}$ and corresponds to a 90° phase shift of the rf signal.

b) Hadamard

$$\mathbf{H}_4 = \begin{pmatrix} 1 & 1 & 1 & 1 \\ 1 & -1 & 1 & -1 \\ 1 & 1 & -1 & -1 \\ 1 & -1 & -1 & 1 \end{pmatrix} \tag{29}$$

or

$$\mathbf{H}_4^\$ = \begin{pmatrix} -1 & 1 & 1 & 1 \\ 1 & -1 & 1 & 1 \\ 1 & 1 & -1 & 1 \\ 1 & 1 & 1 & -1 \end{pmatrix} \tag{30}$$

c) An example of an encode/decode technique with suboptimum efficiency is the sine scheme

$$\mathbf{S}_4 = \begin{pmatrix} \sin(\pi/5) & \sin(2\pi/5) & \sin(2\pi/5) & \sin(\pi/5) \\ \sin(2\pi/5) & \sin(\pi/5) & -\sin(\pi/5) & -\sin(2\pi/5) \\ \sin(2\pi/5) & -\sin(\pi/5) & -\sin(\pi/5) & \sin(2\pi/5) \\ \sin(\pi/5) & -\sin(2\pi/5) & -\sin(2\pi/5) & -\sin(\pi/5) \end{pmatrix} \tag{31}$$

here the SNR gain is 1.58 which is less than the optimum $\sqrt{4}$.

Note that the Fourier encoding can in principle also be performed by B_0 gradients which cause a similar (but continuous) phase evolution. This type of encoding is realized, e.g., in conventional 2D FT imaging with the phase–encode gradient.

2.4 *The Hadamard Scheme*

In the following we will concentrate on the Hadamard encoding technique since it is experimentally simple to implement (only 1, −1 coefficients) and it yields optimum efficiency. In the Appendix some properties of the Hadamard matrices of an arbitrary order are listed. Here we will concentrate on the simultaneous excitation of *four* slices with the Hadamard encode/decode scheme.

Several equivalent Hadamard matrices exist. They can be transformed into each other by multiplication of a column or row by −1, by permutation of rows or by permutation of columns. Therefore, we first have to decide which special realization of $\mathbf{H}_4$ should be used. Several arguments favor the matrix $\mathbf{H}_4^\$$.

First: Remember that the multifrequency rf pulse is synthesized with a linear combination of single frequency rf pulses, whose coefficients are the elements of $\mathbf{H}_4$ in each row. Therefore the peak amplitude of the combined rf pulse $B_1(t)$ may be identical to the sum of the 4 peak amplitudes of each component, if the *normalized* $\mathbf{H}_4$ is used. The rf pulses $B_2(t)$, $B_3(t)$, and $B_4(t)$ for the generation of $S_2(t)$, $S_3(t)$, and $S_4(t)$ are much lower in amplitude (c.f. Fig. 11a). On the transmitter side this requires a high linearity of the rf amplifier in order to transmit the rf pulses with the correct amplitudes. In addition, the $B_1(t)$ pulse may exceed the voltage limits of the transmitter path. (Note that the *power* of the rf pulses is identical for each version of the Hadamard matrix, therefore the power deposition in a biological sample cannot be optimized by selecting appropriate Hadamard matrix variants).

Second: A similar problem with the normalized $\mathbf{H}_4$ matrices exists in the observation period: Since $S_1(t)$ is the sum of the signals from each slice:

$S_a(t) + S_b(t) + S_c(t) + S_d(t)$, the receiver sensitivity has to be adapted to the sum of the signals from each slice. To preserve linearity this receiver setting is maintained for $S_2(t), S_3(t)$, and $S_4(t)$ which may have lower signal amplitudes. In this way again the dynamic range of the receiving path may not be optimally used.

Most of these problems can be overcome by using $\mathbf{H}_4^\$$ which generate on the transmitter and on the receiver side almost equal signal amplitudes for all excitations (Fig. 11b).

The rf pulses required for complete $\mathbf{H}_4^\$$ encoding are:

$$\begin{aligned} B_1(t) &= -B_a(t) + B_b(t) + B_c(t) + B_d(t) \\ B_2(t) &= B_a(t) - B_b(t) + B_c(t) + B_d(t) \\ B_3(t) &= B_a(t) + B_b(t) - B_c(t) + B_d(t) \\ B_4(t) &= B_a(t) + B_b(t) + B_c(t) - B_d(t). \end{aligned} \tag{32}$$

Figure 11 illustrates, that encoding with $\mathbf{H}_4^\$$ yields similar peak amplitudes for all four rf pulses $B_1(t), \ldots, B_4(t)$, while in the normalized Hadamard scheme considerable variations in amplitudes occur.

These rf pulses $B_i(t)$ generate MR signals $S_i(t)$ which are linear combinations of the signals of the excited slices:

$$\begin{aligned} S_1(t) &= -S_a(t) + S_b(t) + S_c(t) + S_d(t) \\ S_2(t) &= S_a(t) - S_b(t) + S_c(t) + S_d(t) \\ S_3(t) &= S_a(t) + S_b(t) - S_c(t) + S_d(t) \\ S_4(t) &= S_a(t) + S_b(t) + S_c(t) - S_d(t). \end{aligned} \tag{33}$$

The combined signals $S_i(t)$ can now be decoded using the inverse of $\mathbf{H}_4^\$$. Since $\mathbf{H}_4^\$$ is a symmetric matrix, encode $\mathbf{H}_4^\$$ and decode $(\mathbf{H}_4^\$)^{-1} = (\mathbf{H}_4^\$)^{\mathrm{T}}$ matrices are identical in this special case and decoding follows exactly the same scheme:

$$\begin{aligned} 4S_a(t) &= -S_1(t) + S_2(t) + S_3(t) + S_4(t) \\ 4S_b(t) &= S_1(t) - S_2(t) + S_3(t) + S_4(t) \\ 4S_c(t) &= S_1(t) + S_2(t) - S_3(t) + S_4(t) \\ 4S_d(t) &= S_1(t) + S_2(t) + S_3(t) - S_4(t) \end{aligned} \tag{34}$$

which can easily be verified by substitution of the $S_i(t)$ as given above.

Therefore, in four experiments four signals $S_1(t), \ldots, S_4(t)$ were acquired which could be decomposed in signals $4S_a(t), \ldots, 4S_d(t)$. Since the noise during four acquisitions adds up as $\sqrt{4}$, the well-known $\sqrt{4}$ SNR improvement resulted for each of the four slices.

3 Applications

In the following three examples are given which demonstrate the feasibility and potential of simultaneous multilocal investigations.

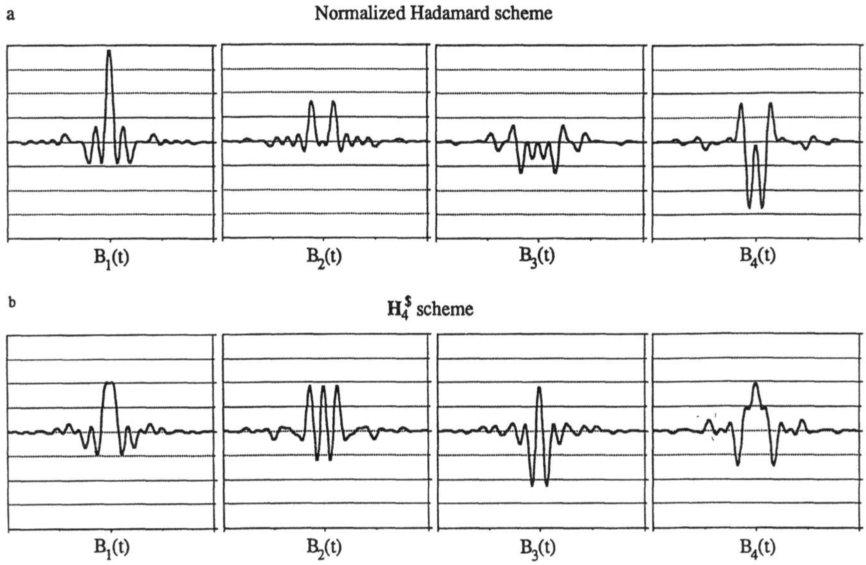

Fig. 11 a, b. Comparison of the rf pulses for two different Hadamard encoding schemes. **a**) Multifrequency-selective rf pulses with the normalized Hadamard matrix. **b**) Multifrequency encoding using the $H_4^{\$}$ scheme. It can be noted, that the peak voltage of the rf pulses in **a** is higher as in **b** and that the scheme b) should be used with preference

3.1 Cardiac Imaging

In MR imaging the simultaneity of the multislice excitation is of advantage if a physiological parameter triggers the image acquisition. An example is ECG gated spin echo imaging of the heart: conventional imaging schemes provide an inefficient mapping of the two-dimensional slice position-cardiac phase space as shown in Fig. 12. Therefore anatomic information from several slice positions at well-defined cardiac phases is not available and has to be gathered using inefficient cycling of the whole scheme.

SIMUSIM (SImultaneous MUltiSlice IMaging) represents a more useful imaging scheme as shown in Fig. 13 for the case if the cardiac anatomy during end-diastole and systole is of interest (Fig. 13):

Cardiac crosssections obtained using SIMUSIM in 24 min are shown in Fig. 14 [5].

3.2 Multislice Imaging with Short Repetition Time

In conventional time interleaved multislice imaging the number of slice position which can be integrated in the same measurement cycle, is limited by the repetition time TR of the imaging sequence. An extreme case is Snapshot FLASH

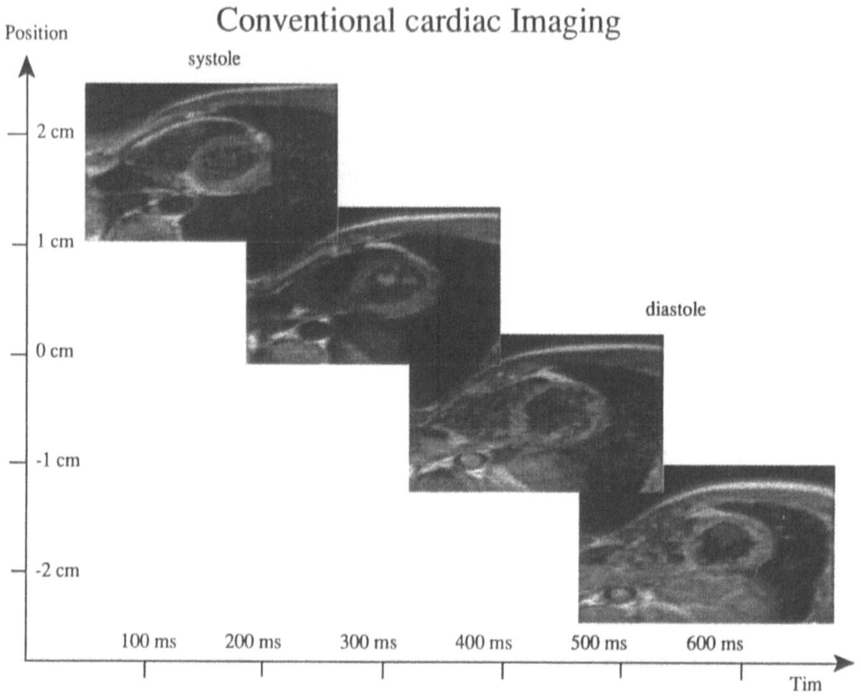

Fig. 12. Schematic figure demonstrating the slice position/cardiac phase dependence in conventional Spin Echo multislice imaging. It is not possible e.g., to measure four cardiac cross sections at early systole within a single step. The ony solution to measure different slice positions at a predefined cardiac phase is to cycle the slice time order in subsequent experiments. This however results in a sensitivity loss for each image, since the number of acquisitions contributing to the SNR of each image is lower than the total number of acquisitions

where TR minimized to the lowest technically possible limit. In this way subsecond imaging times result, however only one single slice can be measured per experiment.

The slice selective rf pulses of Snapshot FLASH can easily be replaced by SIMUFREX (Simultaneous Multifrequency) pulses together with the previously described Hadamard scheme. Such SIMUSIM images from the head of a volunteer are shown in Fig. 15 [6].

3.3 Multilocal Spectroscopy

In in vivo spectroscopy a volume selection step usually precedes the acquisition of the spectroscopic information. With SIMUVOSP (SImultaneous MUlti-VOlume SPectroscopy) [7] the volume selection part can be extended to the simultaneous measurement of several volumes. In Fig. 16 the ^1H spectrum from a brain tumor is measured simultaneously with a spectrum from the healthy

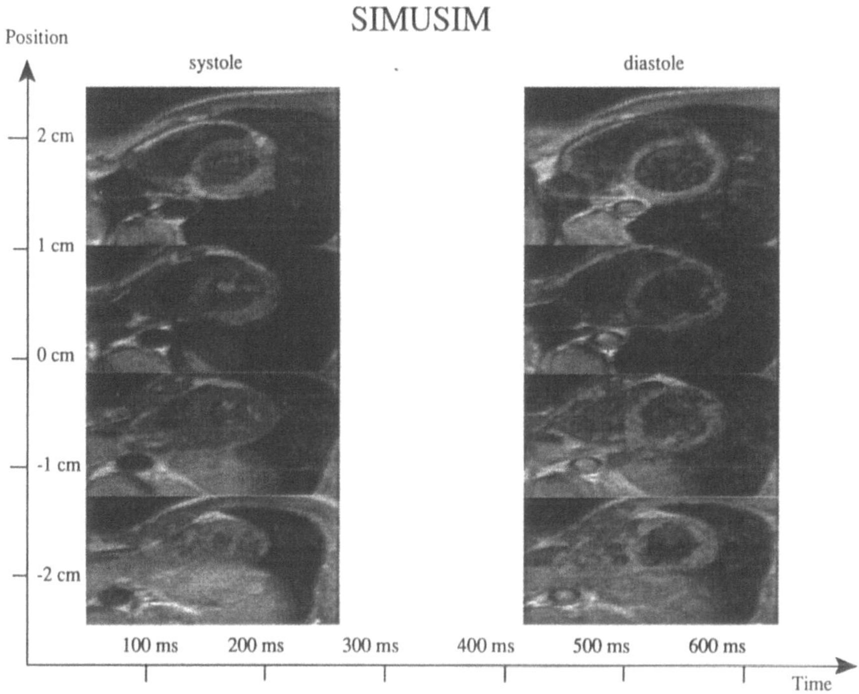

Fig. 13. Slice position/cardiac phase dependence for SIMUSIM cardiac imaging. Note that the simultaneity of the excitation of each slice results in higher efficiency in terms of SNR per measurement time compared to the conventional scheme in Fig. 12

contralateral side. In this way reference data are available for spectral quantification [8].

4 Discussion

Simultaneous multislice excitation was first proposed by Maudsley [9] for multiple line-scanning spin density imaging. Maudsley proposed the $\mathbf{H}_4^{\$}$ Hadamard matrix for encoding and decoding. In Ref. [10] the same technique was reviewed and compared with a number of different FT MRI methods. Later multislice regional imaging and several encode/decode matrices were discussed with respect to efficiency and SNR gain [4]. In the same paper, first multislice images were shown obtained with Fourier- and Hadamard rf encoding and reconstruction from projections. It was also mentioned that the Fourier encoding can be performed either with rf pulses or with an evolution period in a B_0 gradient. These two versions of encoding were discussed later by Maudsley [11] with respect to sensitivity in Fourier imaging.

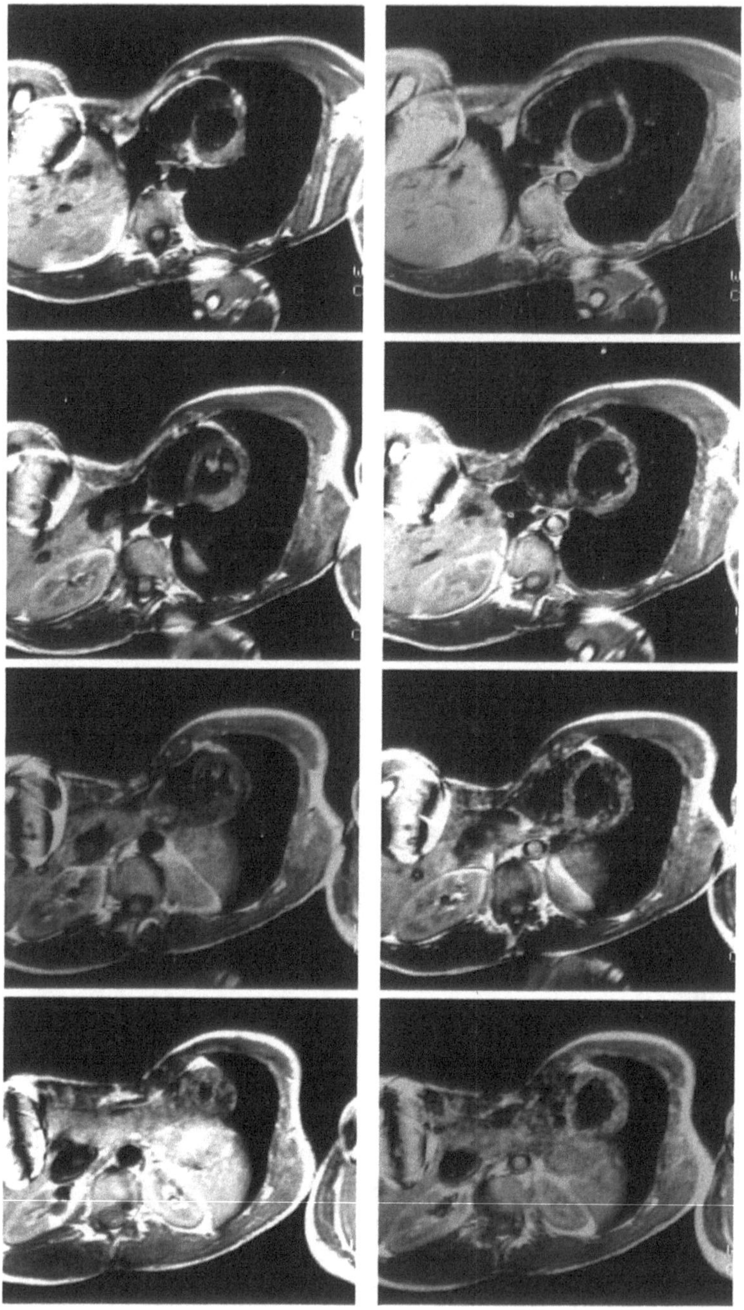

Fig. 14. Cardiac short axis cross sections using the SIMUSIM scheme. The four images on the left side were simultaneously acquired during early systole, the images on the right at end diastole. The whole set of Spin Echo images were obtained in 24 min image parameters: slice thickness: 8 mm, slice gap: 12 mm, TR ≈ 1 s (ECG gated), TE: 22 ms

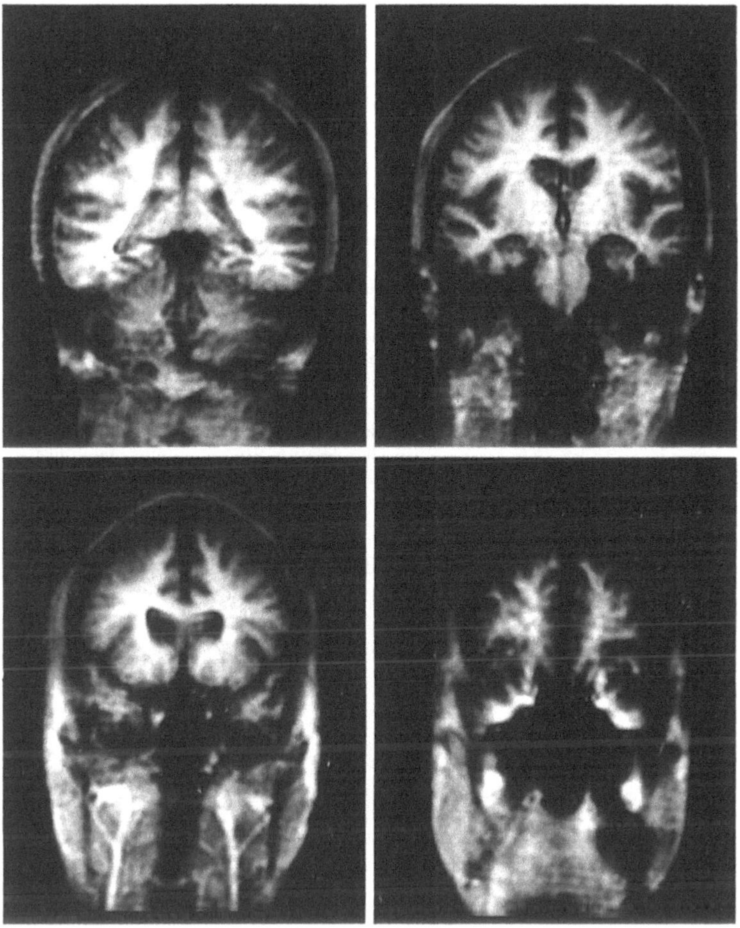

Fig. 15. Coronal snapshot FLASH images simultaneously acquired from four slice positions using SIMUSIM. The image contrast was induced by a 180° pulse 300 ms prior to the image acquisition. Image parameters: TR: 10 ms, TE: 4.9 s, acquisition matrix: 128 × 256, 4 × 1 acquisitions, slice thickness: 10 mm, slice gap: 10 mm, total measurement time: 6 s

First images obtained with standard 2D FT imaging and a commercial clinical whole-body system were described [12]. The specific advantages of the simultaneity of the multislice excitation for cardiac imaging were theoretically and experimentally demonstrated by Müller [5].

A similar multislice imaging technique termed SIMA was implemented in a whole body system and compared with 3D FT imaging [13]. It was found that especially the interslice interferences of 3D FT imaging are greatly alleviated by Hadamard multislice encoding.

A different encode/decode idea was proposed recently and was dubbed POMP (Phase Offset Multi Planar) [14]. POMP is also based on simultaneous

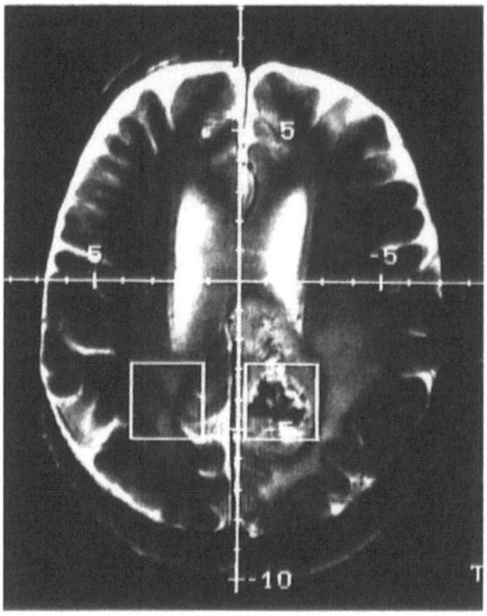

Fig. 16. Axial ^{1}H scout image of a patient suffering from a glioplastoma. Two regions of interest were chosen for simultaneous localized ^{1}H spectroscopy. The resulting spectra are indicated in the bottom, they were simultaneously acquired during 6.5 min. A modified STEAM SIMUVOSP sequence was used for this purpose. Volume sizes: $(2.5\,\text{cm})^3$, TR = 1.5 s, TE = 50 ms, 256 acquisitions. The reference spectrum on the left has a normal appearance, revealing resonances from *N*-acetyl aspartate (NAA), phosphocreatine and creatine (PCr/Cr), choline (Cho), and inositole (Ins.) The spectrum from the tumor region has an inverted choline to PCr/Cr ratio and only marginal NAA signals

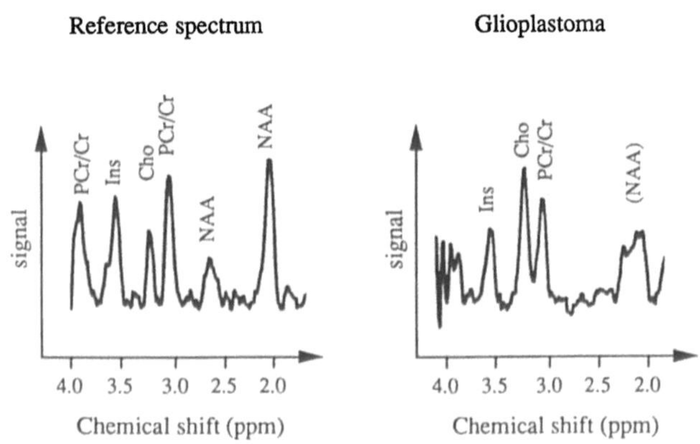

multislice excitation but can be used if a phase-encoding dimension is used after the volume selection, as for example in 2D FT imaging. If $B_a(t)$ is the rf modulation envelope of slice *a* and $B_b(t)$ the envelope for slice *b*, then the POMP rf pulse for both slices: $B_{a+b}(t)$ becomes:

$$B_{a+b}(t) = B_a(t) + B_b(t)e^{ik_y y_0} \tag{35}$$

where k_y is the view-dependent phase-encoding value in the imaging process and y_0 a choosable parameter which defines the image displacement of slice *b*. Since the imaging phase-encoding $e^{ik_y y}$ is then superimposed to the rf phase encoding of slice *b*, a complete phase factor $e^{ik_y(y+y_0)}$ results in the MR signal which after conventional two-dimensional Fourier transform results in an offset of slice *b* by amount y_0 in the phase-encoding direction.

Applications of simultaneous multivolume selection for localized spectroscopy were first proposed in Refs. [15, 16]. Experimental results on phantoms and in vivo showed the feasibility of the technique and demonstrated the increase in efficiency over single volume techniques. Later Bolinger and Leigh [17] described extensions of these multivolume ideas called Hadamard Spectroscopic Imaging (HSI). In that report the question was addressed how conventional spectroscopic imaging using B_0 gradient phase encoding compares with rf phase encoding of comparable spatial resolution. They found that using specially designed rf pulses with very sharp transition zones, HSI compares favorably with respect to volume definition and flexibility. The approach to multi-cube selection using ISIS was discussed in detail [18] leading to similar ideas as in HSI and SIMUVOSP. The SIMUVOSP results were reviewed and applications for ^{1}H, ^{31}P, and ^{13}C spectroscopy were shown [7]. Recently Hafner [8] showed spectra simultaneously acquired from healthy and diseased tissue in the brain of a tumor patient using a STEAM SIMUVOSP technique.

Multifrequency-selective excitation and inversion has been carefully studied using simulations of the Bloch equations [2]. It was found that sinc excitation of up to eight slices yields spin responses which can be completely decoded and have optimum sensitivity. This was verified for slice gaps between zero and twice the slice thickness. In the same study two-slice hypersecant inversion pulses were investigated as they can be, e.g., used in ISIS multivolume experiments. Again slice profiles comparable in quality and selectivity were obtained as in single-slice experiments. However the rf power insensitivity of the double-slice hypersecant pulse is less pronounced than in the single-slice case. For example, for a slice gap equal to the slice thickness the rf field intensity can only be increased to about 2.5 times the threshold value for inversion, for higher intensities the slice profiles deteriorate.

A different approach for the analysis of the multifrequency excitation was undertaken by Goelman et al. [3]. Using average Hamiltonian theory conditions were described under which the spin response can be described as the Fourier transform of the rf excitation. These criteria were used to construct rf pulses for transverse HSI.

Standard multislice and multivolume experiments on commercial MR systems are usually performed in a time-sequential way, i.e., the time gaps between the acquisition of each volume are filled with the measurement of neighboring volumes [19]. Even in localized spectroscopy this is possible under certain conditions (SLIT DRESS [20], STABLE [21]). In general, however, the repetition time TR determines an upper limit for the number of slices or volumes to be measured in this time-multiplexed way. If more volumes are needed, either only partial interleaving is possible, negating the efficiency advantage, or lengthening TR which again is either inefficient in terms of signal gain per measurement time, or deterimentally alters the contrast in the images.

Especially with the advent of gradient echo imaging, where the TRs may be chosen as short as technically possible, standard multislice imaging is reduced to the acquisition of one or two slices only. Also in T_2 weighted or inversion recovery imaging, the long echo and inversion times strongly limit the number of slices available for conventional multislice imaging. In addition, in all experiments where the selectivity of the slice selection is not preserved in the course of the experiment (ISIS [22], FROGS [23], VSE [24]) this time-interleaved method fails.

Thus in a number of clinical situations conventional multislice imaging requires compromise of either throughput (efficiency), tissue discrimination or spatial coverage. In these cases the *simultaneous* excitation and acquisition of multiple volumes improves the flexibility of the experiments, since this limitation in number of available volumes is absent. Snapshot FLASH images combined with simultaneous multislice excitation illustrate these possibilities (Fig. 15). Another advantage of simultaneous multislice imaging was illustrated with ECG gated cardiac imaging (Fig. 14). There conventional multislice imaging is inefficient if the anatomy of the heart has to be displayed in multiple sections at one or several cardiac phases.

An alternative method that avoids the short TR problem and offers significant advantages in ease of implementation, stability, and quality of the results is gradient phase encoding. As in any volume excitation method, acquisition of signal from a large fraction of the noise generating volume yields improved SNR, provided that the receiver has sufficient dynamic range. Contiguous thin slices can be sampled and the data postprocessing is straightforward even for large numbers of slices. These properties may make 3D FT imaging or spectroscopic imaging the method of choice for many situations. However, the limitations of gradient phase encoding are also obvious:

1. Slice profiles obtained with gradient phase encoding have a sinc-shaped sensitivity distribution. This is a fundamental artefact produced by the limited number of samples in Fourier-based gradient phase encoding. It manifests itself by leakage of signals between adjacent volumes which may be very pronounced in areas of large volume-to-volume signal variations. The effect may be somewhat alleviated by low pass filtering although at some cost in spatial resolution.

2. Since the phase-encode gradient and hence the phase of the spins are a continuous function of space, the encoding results in a phase dispersion of the signals within each resolved voxel. This results in a reduced observable magnetization and a signal loss of 13% from the theoretical maximum [11].

3. A more practical limitation of gradient phase encoding is the fact that contiguous slices of equal thickness are not always desirable in clinical applications. For instance, in the studies of symmetrical structures (temporomandibular joints, knee, breast, shoulder, hip..) unequal interslice gaps can be very useful. Similarly, high-resolution spectroscopic imaging of a volume in a diseased region in the brain could be compared with the same volume at the contralateral side if unequal intervolume distances could be produced with gradient phase encoding.

4. The phase encoding with B_0 gradients requires a time interval in the pulse sequence, during which the transverse magnetization evolves in the B_0

gradient. In 2D FT MR imaging this time interval can usually be implemented in the sequence without increasing the time between excitation and acquisition. In spectroscopic imaging, however, especially of ^{31}P [25] and ^{13}C [26] the time delay caused by phase evolution in the gradient produces loss of signal of short T_2^* components together with phase and baseline artefacts.

Most of these limitations of gradient phase encoding are not present in rf phase-encoding techniques.

1. Slice profile definition and phase dispersion within each voxel can be freely designed, in principle the rf phase encoding allows a discrete definition of phase and amplitude of each resolved point in the sample. Thus, leakage between slices can be minimized to zero and phase encoding from rf excitation theoretically avoids any phase dispersion within the voxels.

2. Position and shape of each excited volume can be freely designed with rf encoding. Contiguous slices as shown in HSI [17] or well-separated slices with equal of unequal slice parameters may be measured with rf phase encoding.

Since many gradient phase-encoding methods like 3D FTor spectroscopic imaging are preceded by a volume-selective part (to avoid aliasing), rf phase encoding in this case might enlarge the spectrum of applications of the gradient-based techniques: For instance, in a study of symmetrical structures two volumes can be simultaneously prepared using rf encoding and then spatially resolved using gradient phase encoding.

3. It is clear that nowadays the time delays necessary for gradient phase evolution and the eddy current problems have been minimized through recent advantages in technology (e.g., actively shielded gradients, high gradient power..). Nevertheless, rf phase encoding is performed *during* the selective rf pulse and evolution delays are absent. A special feature of this fact is, that very high-order phase encoding can be achieved in this way without any special demands. Thus, POMP [14], which is a combination of rf and gradient phase encoding, simulates very high gradient values in slice-encoding direction via rf phase encoding. In this way overlap free reconstruction of multiple slices displaced in phase-encode direction can be performed.

Aside from these advantages of rf phase-encoding limitations exist which may make gradient phase-encoding techniques the method of choice in some situations.

1. With gradient phase encoding almost any MR experiment in which a transverse evolution period exists, can be converted into a spatially resolved experiment. This is a very convenient way to obtain spatial mapping in one or more dimensions. In rf phase encoding the spatial resolution can usually only be introduced through a volume/slice selection part of the experiment.

2. Slice selection with frequency-selective rf pulses and B_0 gradients are usually subject to chemical shift artefacts. This is the same for selective single- or multivolume irradiation. This artefact can be minimized by using strong B_0 gradients, but in gradient phase encoding it is completely absent.

3. Simultaneous excitation/inversion of multiple slices requires broad-band selective rf pulses. This results in higher rf peak voltage compared to

single-volume techniques and may prohibit certain sequences due to rf power considerations.

5 Conclusions

In this article a technique was reviewed which has the potential of increasing the versatility and efficiency of almost every slice- or volume-selective experiment. The idea is to modify the volume- or slice-selective part in the sequence such that instead of one single volume, several volumes are measured simultaneously. The rf pulse provides a positional encoding via the phase of the MR signal and allows the separation of the signals from each volume.

Several different encode/decode schemes were discussed. Each scheme provides an individual phase alternation for the volumes in successive experiments which is the basis for decoding. The Hadamard scheme which is optimally efficient and easy in implementation was used for some illustrative examples in the article.

The discussion showed that rf phase encoding and gradient phase encoding may be alternative techniques, but may also supplement each other as in multivolume selection preceding spectroscopic imaging, 3D FT imaging or in POMP. In comparison with conventional single slice/volume methods the flexibility, e.g., of SIMUSIM is higher since the number of available slices for multislice imaging is not limited by TR. Thus, short TR gradient echo imaging obtains multislice capability. Cardiac imaging is another field of application for SIMUSIM, since the simultaneity of the multislice imaging provides the basis for efficient and precise volume measurements in the heart.

Rf pulses for multilocal excitation were so far mainly designed using linear arguments. Optimization of pulse shapes with respect to spin response, rf power, pulse duration etc. open the field for new pulse designs, especially if the rf pulses are combined and synchronized with time dependent B_0 gradients. Thus the field of multiple frequency-selective rf pulses is still open for further developments.

6 Appendix

Hadamard matrices **H** are known for orders 1, 2 and nearly any multiple of 4. Equivalent Hadamard matrices can be obtained by multiplication of a column or row by -1, by permutation of rows or by permutation of columns. Thus numerous Hadamard matrices exist for any order. Higher order matrices may be derived from lower orders by means of Kronecker products of two Hadamard matrices: If $\mathbf{H}_n$ and $\mathbf{H}_m$ are Hadamard matrices of order n and m, respectively, then $\mathbf{H}_{nm} = \mathbf{H}_n \otimes \mathbf{H}_m$. Hadamard matrices of order $2n$ are most efficient for encoding/decoding. $\mathbf{H}_{2n}$ can be iteratively generated from $\mathbf{H}_{2n-1}$ by:

$$\mathbf{H}_{2n} = \begin{pmatrix} \mathbf{H}_{2n-1} & \mathbf{H}_{2n-1} \\ \mathbf{H}_{2n-1} & -\mathbf{H}_{2n-1} \end{pmatrix} \tag{36}$$

or with a different approach the element h_{jk} of a Hadamard matrix $\mathbf{H}_m$ may be written as:

$$h_{jk} = (-1) \sum_{i=0}^{n-1} ((j-1)_i(k-1)_i) \tag{37}$$

where $n = \log_2 m$ and $(j)_i$ is the i-th bit in the binary representation of j. Finally, the rows of the Hadamard matrices are the digitized Walsh functions.

Since Hadamard matrices may be transformed into each other, one can construct a normalized Hadamard matrix from each set of equivalent matrices, whose elements are $+1$ in the first row and in the first column. Some normalized Hadamard matrices are given below, they can be conveniently generated with Eq. (36):

$$\mathbf{H}_1 = (1) \tag{38}$$

$$\mathbf{H}_2 = \begin{pmatrix} 1 & 1 \\ 1 & -1 \end{pmatrix} \tag{39}$$

$$\mathbf{H}_4 = \begin{pmatrix} 1 & 1 & 1 & 1 \\ 1 & -1 & 1 & -1 \\ 1 & 1 & -1 & -1 \\ 1 & -1 & -1 & 1 \end{pmatrix} \tag{40}$$

$$\mathbf{H}_8 = \begin{pmatrix} 1 & 1 & 1 & 1 & 1 & 1 & 1 & 1 \\ 1 & -1 & 1 & -1 & 1 & -1 & 1 & -1 \\ 1 & 1 & -1 & -1 & 1 & 1 & -1 & -1 \\ 1 & -1 & -1 & 1 & 1 & -1 & -1 & 1 \\ 1 & 1 & 1 & 1 & -1 & -1 & -1 & -1 \\ 1 & -1 & 1 & -1 & -1 & 1 & -1 & 1 \\ 1 & 1 & -1 & -1 & -1 & -1 & 1 & 1 \\ 1 & -1 & -1 & 1 & -1 & 1 & 1 & -1 \end{pmatrix}. \tag{41}$$

$\mathbf{H}_4$ has an experimentally useful variant $\mathbf{H}_4^{\$}$ which can be written as:

$$\mathbf{H}_4^{\$} = \begin{pmatrix} -1 & 1 & 1 & 1 \\ 1 & -1 & 1 & 1 \\ 1 & 1 & -1 & 1 \\ 1 & 1 & 1 & -1 \end{pmatrix}. \tag{42}$$

Since the elements of $\mathbf{H}_n$ are real, the inverse $\mathbf{H}_{n-1}$ is identical to the transpose $\mathbf{H}_n^T$, therefore decoding is very simple.

7 References

1. Silver MS, Joseph RI, Hoult DI (1984) J Magn Reson 59: 347
2. Hafner HP, Müller S, Seelig J (1990) Magn Reson Med 13: 279
3. Goelman G, Subramanian VH, Leigh JS (1990) J Magn Reson 89: 437
4. Oh CH, Park HW, Cho ZH (1984) IEEE Trans Med Ima Mi-3(4): 170
5. Müller S (1989) Magn Reson Med 10: 145
6. Müller S (1990) Magn Reson Med 15: 497
7. Müller S, Hafner HP, Beckmann N (1989) NMR in Biomed 2(5/6): 209
8. Hafner HP, Radü E, Seelig J (1990) Magn Reson Med 15: 135
9. Maudsley AA (1980) J Magn Reson 41: 112
10. Cho ZH, Kim HS, Song HB, Cumming J (1982) Proc IEEE 70(10): 1152
11. Maudsley AA (1986) J Magn Reson 68: 363
12. Müller S (1988) Magn Reson Med 6: 364
13. Souza SP, Szumowski J, Dumoulin CL, Plewes DP, Glover G (1988) J Comp Ass Tomogr 12(6): 1026
14. Glover GH, Shimakawa A (1988) Proc 7th Ann Meeting Soc Magnetic Resonance in Medicine, 20–26 August 1988. San Francisco, USA, p 241
15. Müller S, Sauter R, Weber H, Seelig J (1987) Proc 6th Ann Meeting Soc Magnetic Resonance in Medicine, 17–21. August, New York City, USA, p 936
16. Müller S, Sauter R, Weber H, Seelig J (1988) J Magn Reson 76: 155
17. Bolinger L, Leigh JS (1988) J Magn Reson 80: 162
18. Ordidge RJ, Bowley RM, McHale G (1988) Magn Reson Med 8: 323
19. Crooks LE, Ortendahl DA, Kaufman L, Hoenninger J, Arakawa M, Watts J, Cannon CR, Brant-Zawadzki M, Davis PL, Margulis AR (1983) Radiology 146: 123
20. Bottomley PA, Smith LS, Leue WM, Charles C (1985) J Magn Reson 64: 327
21. Ernst T, Hennig J (1989) Proc 8th Ann Meeting Soc Magnetic Resonance in Medicine, Amsterdam, 1: 247
22. Ordidge RJ, Connelly A, Lohman JAB (1986) J Magn Reson 66: 283
23. Sauter R, Müller S, Weber H (1987) J Magn Reson 75: 167
24. Aue WP, Müller S, Cross TA, Seelig J (1984) J Magn Reson 56: 350
25. Brown TR, Kincaid BM, Ugurbil K (1982) Proc Natl Acad Sci 79: 3523
26. Müller S, Beckmann N (1989) Magn Reson Med 12: 400

Spectrum Analysis

Conventional data processing in Fourier transform (FT) spectroscopy involves a Fourier transformation of the time domain data after some preprocessing (exponential or Gaussian filtering, baseline correction etc.). The intensities are then determined by integration of the spectrum or by spectral deconvolution. These procedures have been reviewed in the literature [1, 2] and will therefore not be discussed in this volume. Instead, the focus will be on an alternative approach, in which model parameters, such as resonance frequency and intensity, are derived from the original data in the measurement domain.

There are several reasons for data analysis in the time domain. Depending on the quality of the raw data a discrete or fast Fourier transformation (FFT) yields an estimate of the true spectrum. If data sampling is incomplete (e.g. missing initial data points, since data collection was started only after gradient ring-down) or non-uniform, the FFT spectrum may deviate considerably from the 'true' spectrum. Any preprocessing (filtering, removal of broad spectral components) will add to these deviations. In addition, signal-to-noise in FFT may not be optimal. The analysis of NMR data by time domain fitting procedures is discussed by de Beer and van Ormondt.

An important aspect in in vivo MRS is the evaluation of the absolute concentration of metabolites in the undisturbed biological environment. However, many problems arise with such procedures. Aspects such as coil loading, partial saturation, inhomogeneous RF excitation, definition of the volume of interest have to be considered for the data collection step. Deconvolution or time domain fitting procedures have to be applied for accurate determination of intensity factors, which are proportional to the number of resonating nuclei. These and other aspects associated with absolute quantitation of in vivo spectra are discussed by Cady.

1. Lindon JC, Ferridge AG (1980) Progr NMR Spectrosc 14: 27
2. Ernst RR, Bodenhausen G, Wokau A (1987) Principles of Nuclear Magnetic Resonance in One and Two Dimensions (Oxford University Press) pp 91

Analysis of NMR Data Using Time Domain Fitting Procedures

R. de Beer and D. van Ormondt

Applied Physics Laboratory, Delft University of Technology, P.O. BOX 5046, 2600 GA Delft, The Netherlands

Table of Contents

This chapter concerns quantification of data directly in the measurement domain. In the case of in-vivo NMR, this principle amounts to processing in the time domain. Among other things, measurement domain processing obviates correction for artefacts that arise when incomplete or nonuniformly sampled data are subjected to FFT. Two basically different signal-processing approaches are described. 1) Black-box (or noninteractive) methods which require only minimal involvement and knowledge of the user, but cannot exploit prior knowledge. 2) Interactive methods which do have the capability to exploit prior knowledge and by this very nature require a trained user. Applications of both methods to real-world in-vivo signals in the fields of spectroscopy, spectroscopic imaging, and inversion recovery relaxation time measurements are given.

List of Abbreviations

BCF	Baseline Cosine Fitting
CLEAN	Deconvolution method
CR	Cramér–Rao (lower bounds)
ESPRIT	Estimation of Signal Parameters via Rotational Invariance Techniques (related to HSVD)
HSVD	Hankel Singular Value Decomposition (State Space)
LP	Linear Prediction
LPSVD	Linear Prediction and Singular Value Decomposition

LP/VARPRO	Linear-Prediction and VARiable PROjection (Evans-Fischl)
LS	Least Squares
MEM	Maximum Entropy Method
MLE	Maximum Likelihood Estimation
SVD	Singular Value Decomposition
TLS	Total Least Squares
VARPRO	VARiable PROjection

1 Introduction

1.1 Time Domain Versus Frequency Domain

1.1.1 Model Parameter Estimation or Spectrum Estimation

The beginning of this chapter sets out our main objective and how we intend to achieve it. As the title indicates, the content is concerned with parameter estimation in the time domain. Literally this is correct, but the term Time Domain is to be interpreted as *Measurement* Domain. Since in-vivo NMR measurements take place in the time domain, these two terms imply the same thing in practice. However, the intended interpretation should be kept in mind.

Why is it important to process the data in the measurement domain? This has to do with the wish to 'quantify' the parameters of the physical model function that describes the NMR measurement. Given this goal one should try to avoid preprocessing of the data to which the model function is to be fitted. The particular form of preprocessing alluded to is discrete (or fast) Fourier transformation (DFT or FFT) and attendant windowing, the well-known method for estimating the spectrum associated with a measurement. If the data are incomplete, or have been sampled at nonuniformly distributed times, then the FFT estimate of the spectrum deviates from the 'true' spectrum [1–6]. In addition, if the noise in the data is other than Gaussian, then the signal-to-noise ratio (SNR) of the Fourier coefficients as estimated by FFT may not be optimal [7].

The remarks just made are by no means meant to belittle the powerful technique which FFT is. In fact, a spectrum is often an indispensable tool for pre-viewing and interpreting time domain data, and for procuring starting values. Moreover, methods to reduce deviations from the 'true' spectrum have been devised; see e.g. Refs. [3, 8–10], and Sect. 1.1.3. However, in spite of the latter, the highest precision is usually attainable only if the ultimate step of the estimation of parameters is applied to the original data. Note that in statistical treatment one usually wields the word 'estimation' rather than 'quantification'. (We shall use both.)

In summary then, our main objective is to estimate the parameters of the model function that describes the original (i.e. time domain) data, rather than to first estimate a 'true' spectrum and then proceed by quantifying model parameters in the frequency domain. Various alternative approaches to time domain model function fitting will be treated. Also, some recent methods to improve the spectrum will be briefly reviewed.

1.1.2 Time Domain and Frequency Domain Model Functions

In this subsection, some possible forms of the time domain model function and their associated Fourier transforms will be reviewed. For convenience only free induction decays (FIDs) are considered, although analysis of echoes is also feasible. Among other things, we intend to indicate that an NMR time domain model function is often simpler to handle than its Fourier transform.

Note that the two domains have a common serious problem, namely that the available model functions may be inadequate to describe the data (or spectrum). Should this be the case, there seems to be no other option than to restrict oneself to only partial quantification, for instance to estimating only the net amplitude (time domain) or area (frequency domain) of a spectral feature, see Sect. 2.5.2.

1.1.2.1 Exponential Damping (Lorentzian Lineshape)

The FID model function most often used is a sum of exponentially damped complex-valued sinusoids, sampled at uniformly distributed times t_n, $n = 0, 1, \ldots, N-1$, i.e.:

$$\hat{x}_n = \sum_{k=1}^{K} c_k \exp[i(\varphi_0 + \varphi_k)] \exp[(\alpha_k + i\omega_k)t_n], \tag{1}$$

where $c_k, \alpha_k, \omega_k, \varphi_k$ are respectively the real-valued amplitude, damping factor, angular frequency ($\omega = 2\pi\nu$), and phase of sinusoid number k. Furthermore, $t_n = t_0 + n\Delta t$, in which t_0 is the time lag between the effective time origin and the first data point to be included in the analysis; φ_0 is the zero order phase and $i = \sqrt{(-1)}$. Note that t_0 is equivalent to the first order phase, whereas the φ_k represents extra degrees of freedom that may be required under certain experimental conditions (usually all φ_k are zero). Note also that $\alpha_k \leqq 0$. The caret on x denotes that this quantity represents the model function rather than the actual measurement.

Equation (1) is to be fitted to the data using one of the available approaches (see below). We emphasize that φ_0 and t_0 can easily be included in the fit procedure. Thus, absence of initial data points, for some reason or another, poses no problems. In addition, the mathematical properties of the numbers $z_k \overset{\text{def}}{=} \exp[(\alpha_k + i\omega_k)\Delta t]$ enable one to reduce the computational load considerably [11]. For example, it is easy to derive closed-form expressions for the sums $\Sigma_n n^l (z_k z_{k'}^*)^n$, including the special case $(z_k z_{k'}^*) \to 1$, where $*$ denotes complex conjugation.

In the frequency domain one could use the *continuous* Fourier transforms of the exponentially damped ($\alpha_k \leqq 0$) sinusoids in Eq. (1), i.e.:

$$\hat{S}(\omega) = \sum_{k=1}^{K} c_k \frac{\exp[i(\varphi_0 + \varphi_k)]}{\alpha_k + i(\omega_k - \omega)} \{\exp[(\alpha_k + i(\omega_k - \omega))t_{N-1}] - \exp[(\alpha_k + i(\omega_k - \omega))t_0]\}, \tag{2}$$

and fit these to the DFT of the data. If $t_0 = 0$ and the terms $\exp(\alpha_k t_{N-1})$, $k = 1, \ldots, K$, are negligible, then Eq. (2) is simply a sum of Lorentzian lines.

However, complications can be seen to arise when the latter conditions are not satisfied. In this context it should be pointed out that Abildgaard et al. were able to derive a closed-form expression for the DFT of the (exponentially damped) model function of Eq. (1) [12] including the effects of truncation. Thus, an exact frequency domain counterpart of Eq. (1) is available which in turn implies that fitting Lorentzians in the frequency domain without having to resort to approximations is in fact possible. See Ref. [13] for reduction of the computational burden attendant to fitting Lorentzian functions.

Matters become more complicated in 2D experiments; see e.g. Ref. [14], page 324 and references therein. There the model function comprises products of sinusoids, defined in an unprimed and a primed dimension, i.e.:

$$\hat{x}_{nn'} = \sum_{k=1}^{K} \sum_{k'=1}^{K'} c_{kk'} \exp[i(\varphi_0 + \varphi_{kk'})] \exp[(\alpha_k + i\omega_k)t_n)] \cdot \exp[(\alpha'_{k'} + i\omega'_{k'})t_{n'}]. \tag{3}$$

Fitting Eq. (3) to 2D data poses no problem [15], apart from the increased computational burden. However, in the frequency domain one is confronted with a mixture of absorption and dispersion. Various ways of dealing with this aspect have been devised [14].

1.1.2.2 Other Forms of Damping

In various experiments the damping function may be more complicated than $\exp(\alpha t)$. In the time domain it is simple to extend the damping with a *Gaussian* contribution, i.e. $\exp(\alpha t) \rightarrow \exp(\alpha t + \beta t^2)$. The latter function can easily be incorporated in Eq. (1), and β_k can then be included in the fit in natural fashion. Note however, that the option to use a fast algorithm which exploits the favourable properties of exponential damping, is then lost. In the frequency domain, the function $\exp(\alpha t + \beta t^2 + i\omega_0 t)$ leads to a *Voigt* lineshape which is described by a complicated formula, the real part of which is proportional to:

$$\hat{S}(\omega) \propto \int_{-\infty}^{\infty} \frac{\exp(-\omega'^2/4\beta)}{(\omega - \omega' - \omega_0)^2 + \alpha^2} d\omega'. \tag{4}$$

The fitting of such a function to spectral lines is treated in e.g. Refs. [16, 17].

Finally, we point out that it is often relatively easy to include experimental complications into the measurement domain model function. A very recent example of this is Ref. [18] which treats the magnetization-induced change of the net static magnetic field B_0 as a result of an exciting pulse. It is shown in Ref. [18] that this particular effect can be accounted for by simply multiplying each damped sinusoid by a factor $\exp[-i\Delta\omega\psi(t)]$, in which $\Delta\omega$ is the initial resonance shift due to the magnetization change of B_0 and $\psi(t)$ governs the return of the resonance shift to zero. In principle, one can include this term in a time domain fit without much effort. However, the corresponding model function in the frequency domain may be complicated and therefore difficult to implement in a fit procedure.

1.1.3 *Deconvolution of Spectral Artefacts Due to Missing Data Points and/or Nonuniform Sampling*

Most often the spectrum attendant to a set of time domain samples is obtained by applying fast Fourier transformation (FFT). In order to yield the 'true' spectrum, FFT requires that no samples be missing at either end of the measurement and also that the time lags between successive samples be constant (uniform sampling). In actual practice, however, these conditions can not always be met. A typical case of missing initial data points occurs with ^{31}P spectroscopic imaging, where sampling can begin only after the field gradient is over; see e.g. Ref. [19]. Another example is that initial data points are sometimes deliberately truncated prior to FFT with the aim to remove broad spectral features; see e.g. Refs. [1, 10, 20, 21]. Finally, it may be desirable to increase the time lags between successive samples exponentially in an attempt to enhance the spectral resolution (nonuniform sampling) [3, 4, 22].

The present subsection indicates briefly some methods devised to deconvolve artefacts pertaining to FFT of data belonging to one of the categories just mentioned. Two main approaches are distinguished: a) a time domain method, namely Linear Prediction, and b) frequency domain methods such as maximum entropy (version 2), CLEAN, and baseline cosine fitting (BCF).

Before proceeding we stress that the problem at hand is in principle non-existent when analysing in the measurement domain. This is because the model function pertaining to this domain simply accepts any value of the running variable. Moreover, note that even the Nyquist sampling rule need not be observed, provided the pertinent prior knowledge be incorporated in the starting values of the model parameters.

a) Linear Prediction [23–25, 2, 5]

One approach to deconvolution is to reconstruct missing data points using Linear Prediction (LP), and then append these to the available data points [19, 26–29]. LP is typically a time domain method and therefore it is justified to devote some space to deconvolution based on it. Essentially, LP amounts to estimating a data point from a linear combination of past or future ones. For socalled *back*ward prediction one writes:

$$\hat{x}_n = \sum_{m=1}^{M} a_m x_{n+m}. \tag{5}$$

The entities $a_m, m = 1, \ldots, M$ are called prediction coefficients. Note that they are independent of n. When noise is absent and the model function of Eq. (1) is correct, then Eq. (5) is exact and $M = K$ prediction coefficients are needed. When noise is present, and/or the damping is non-exponential, then Eq. (5) is an approximation and it is advisable to make M much larger than K.

The prediction coefficients can be estimated from the available data with the aid of a linear least squares (LS) fit procedure [23] or by applying the Burg algorithm [30, 31]. The Burg algorithm is faster than LS [32a], but it should be kept in mind that strictly speaking it is intended for stationary signals (see e.g. Refs. [30–33] and references therein). Once the prediction coefficients have been determined, missing initial data points can be reconstructed by repeated use of Eq. (5). Interestingly, the actual model parameters entering in Eq. (1) are not required in this process. The 'true' spectrum is finally obtained by applying FFT to the data appended with the reconstructed part. An example of this LP-based deconvolution is shown in Fig. 1c. In the context of this paragraph, it should be mentioned that the linear least squares procedure for estimating the prediction coefficients has been statistically refined, resulting in so-called Total Least Squares (TLS) [34–36]. Very recently, this was successfully applied to NMR for the first time [37].

b) Other methods

Various other methods have been devised to construct the 'true' spectrum from incomplete or otherwise non-ideal data. Since these methods are outside the scope of this chapter we confine ourselves to a brief characterization of three approaches.

The best-known approach is the Maximum Entropy Method (MEM2) [38–41]. The extension '2' is intended to distinguish this MEM from that of Burg, see e.g. Refs. [26, 33] and point a). The Burg algorithm is indicated by MEM1, and is actually a form of linear prediction. MEM2 is an iterative double optimization method, i.e., simultaneously it maximizes the 'entropy of the spectrum' and minimizes the sum of squared differences between the inverse FT of the spectrum and the available time domain data. MEM2 is not only able to correct for missing data points, but can handle non-uniform time lags between successive samples as well [3]. An example of MEM2 deconvolution is depicted in Fig. 1d.

Next, we mention so-called CLEAN, which originated in the field of astronomy [42]. This method was first applied to solid-state NMR [43] where quantitative analysis required adequate correction for missing of initial data

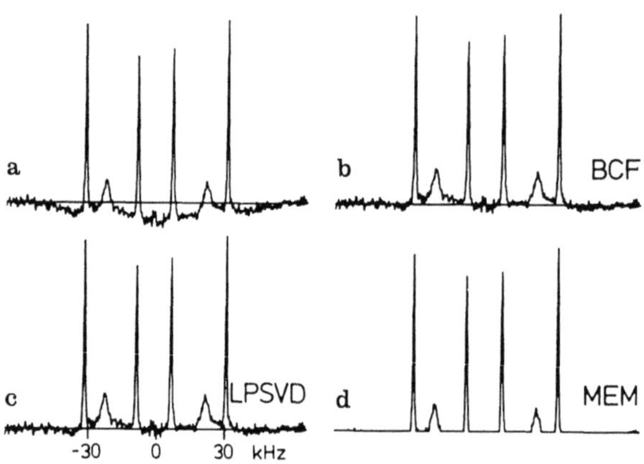

Fig. 1 a–d. Comparison of various methods to correct baseline distortion of an NMR spectrum caused by truncation of initial data points prior to FFT [9]. **a**) No correction. **b**) Correction by baseline cosine fitting [9]. **c**) Correction by linear prediction. **d**) Correction by MEM2. When quantifying in the measurement domain such corrections are not required points due to spectrometer dead time. CLEAN can deconvolve any window function imposed on the time domain data. In addition, like MEM2, it can correct for nonuniformity of the sampling times; this, plus a comparison of CLEAN with MEM2 can be found in Refs. [44, 45]. An application of CLEAN to in vivo NMR was very recently given in Ref. [46].

Finally, the most recently devised approach is so-called Baseline Cosine Fitting (BCF) [9], originating from solid-state NMR. An example of BCF deconvolution is depicted in Fig. 1b. See also Ref. [8].

1.1.4 The Influence of Noise on Spectrum Estimation

The previous section was devoted to spectral artefacts caused by missing of data points or by nonuniformity of the distribution of the sampling times. These artefacts can be considered as systematic errors in the spectrum. In this section we focus on the statistical errors in the spectrum as obtained by FFT [7].

FFT of a series of data amounts to representing these data by a sum of undamped sinusoids with fixed 'equidistant' frequencies ν_n, i.e. $\Sigma_n c_n \exp(i2\pi\nu_n t)$. The coefficients c_n of the expansion, the Fourier coefficients, follow immediately from applying FFT to the data. A plot of the Fourier coefficients yields the spectrum. The important point here is that FFT can be shown to be equivalent to least squares fitting of the sum of sinusoids (with the predetermined values ν_n) to the data. A consequence of this fact for the statistical error in the FFT estimate of the spectrum is indicated in the following paragraph.

It has been proved that when the noise distribution function is Gaussian, the statistical errors pertaining to least squares fitting, approach the fundamental lower bounds [7]. However, should the noise distribution be other than Gaussian, then this property need not be true. For example, if the distribution is Laplacian, the standard deviation of the Fourier coefficients (spectrum) obtained by least squares fitting (FFT) is at least twice as large as the theoretical lower bound [7] (see also Sect. 1.2). Consequently, the statistical errors in parameters estimated from the FFT spectrum are in turn larger than necessary. In the latter case, an alternative estimator (e.g. Maximum Likelihood, see Sect. 1.2.1) should be used to arrive at the spectrum.

In conclusion, we point out that if the SNR is low, it may be worthwhile to ascertain that the noise distribution at hand warrants the use of FFT.

1.2 The Fundamental Limit of the Statistical Error [47–50]

1.2.1 The Likelihood Function

It is desirable to have a standard of precision to which the result of a quantification can be compared. Such a standard is available, in the form of the so-called Cramér–Rao lower bounds (CR bounds). In order to treat this one must first introduce the so-called likelihood function.

Suppose the model function $\hat{x}_n, n = 0, \ldots, N-1$, of Eq. (1) can exactly represent a noiseless measurement $x_n, n = 0, \ldots, N-1$. Adding noise to the data x_n, one can then write $x_n - \hat{x}_n = \varepsilon_n$, where ε_n equals the (complex-valued) noise contribution to data point n. Next, suppose that the distribution function, $p_n(\varepsilon_n)$ of the noise of each data point is given. Strictly speaking, $p_n(\varepsilon_n)$ is the product of two distribution functions, namely one for the real part of the noise and one for the imaginary part (supposedly uncorrelated). Thus, assuming Gaussian noise as an example one may write

$$p_n(\varepsilon_n) = \frac{1}{\sqrt{(2\pi\sigma_r^2)}} \exp\left[-\left(\frac{\varepsilon_{nr}}{2\sigma_r}\right)^2\right] \times \frac{1}{\sqrt{(2\pi\sigma_i^2)}} \exp\left[-\left(\frac{\varepsilon_{ni}}{2\sigma_i}\right)^2\right] \tag{6}$$

where the subscripts i and r denote real and imaginary respectively, and σ is the standard deviation. Since the model function was assumed exact, one may now substitute $\mathrm{Re}(x_n - \hat{x}_n)$ for ε_{nr}, and $\mathrm{Im}(x_n - \hat{x}_n)$ for ε_{ni}.

The joint probability distribution function P of the measurement x as a whole, equals the product of the probability functions of the individual samples, according to:

$$P(\bar{\varepsilon}) = P(\bar{x} - \hat{\bar{x}}) = \prod_{n=0}^{N-1} p_n(x_n - \hat{x}_n), \tag{7}$$

with $\bar{x} = (x_0 x_1 \cdots x_{N-1})^\sim$, $\hat{\bar{x}} = (\hat{x}_0 \hat{x}_1 \cdots \hat{x}_{N-1})^\sim$, $\bar{\varepsilon} = (\varepsilon_0 \varepsilon_1 \cdots \varepsilon_{N-1})^\sim$, and $\sim$ denotes transposition.

From the standpoint of probability theory, Eq. (7) yields the probability of measuring the data $\bar{x}$, given the values of the parameters of the associated model function $\hat{\bar{x}}$. In parameter estimation, however, the situation is reversed. One is given a data set $\bar{x}$ and the task is to find the values of the model parameters comprised in $\hat{\bar{x}}$ that yield the highest probability. In the latter interpretation $P(\bar{x} - \hat{\bar{x}})$ is called the likelihood function, and the values of the model parameters thus found are called Maximum Likelihood estimates.

Often it is convenient to use the log-likelihood function, defined by $L(\bar{x} - \hat{\bar{x}}) \stackrel{\text{def}}{=} \ln(P(\bar{x} - \hat{\bar{x}}))$. Note that the parameter values that maximize $P(\bar{x} - \hat{\bar{x}})$ also maximize $L(\bar{x} - \hat{\bar{x}})$. The advantage of using the log-likelihood function becomes immediately clear when returning to the case of Gaussian noise. For, assuming $\sigma_r = \sigma_i = \sigma$ one finds:

$$L(\bar{x} - \hat{\bar{x}}) = -N \ln(2\pi\sigma^2) - \frac{1}{2\sigma^2} \sum_{n=0}^{N-1} |x_n - \hat{x}_n|^2 \tag{8}$$

Equation (8) implies that, in the case of Gaussian noise, maximizing the log-likelihood function amounts to minimizing the sum of squared residues. This result should comfort those confronted with the maximum likelihood approach for the first time.

Should the noise distribution be other than Gaussian, then maximizing the (log-) likelihood function leads to a different result. For instance, referring to Sect. 1.1.4, a Laplacian noise distribution can be shown [7] to yield optimum parameter estimates by minimizing $\Sigma_n |x_n - \hat{x}_n|$, rather than $\Sigma_n |x_n - \hat{x}_n|^2$!

At this point we are ready to turn to the standard of precision mentioned at the beginning of this section. However, before we proceed note that the standard deviations of maximum likelihood estimates usually approach this standard [47–50]. This constitutes an important reason for choosing a maximum likelihood estimator.

1.2.2 The Cramér–Rao Lower Bounds

It can be shown that the theory of maximum likelihood estimation leads to lower bounds on the statistical errors in the parameter estimates. The existence of such bounds implies that irrespective of the method used to quantify (estimate) the values of model parameters from a data set, there is a lower bound on the precision that can not be superseded. For reasons of space we can only present here the recipe for evaluating these so-called Cramér–Rao lower bounds. In this process we shall indicate the set of J model parameters to be estimated (e.g. amplitudes, frequencies, etc.) by the symbols $\vartheta_1, \vartheta_2, \ldots, \vartheta_J$. Furthermore, the log-likelihood function will be indicated in short-hand notation simply by L.

First one sets up the Fisher Information Matrix, I, whose $J \times J$ elements are given by [49, page 117].

$$(I)_{jl} \stackrel{\text{def}}{=} \mathrm{E}\left[\frac{\partial L}{\partial \vartheta_j} \times \frac{\partial L}{\partial \vartheta_l}\right] = -\,\mathrm{E}\left[\frac{\partial^2 L}{\partial \vartheta_j \partial \vartheta_l}\right], \quad j, l = 1, \ldots, J, \tag{9}$$

where E stands for Expectation Value, which amounts to averaging over all possible noise realisations, while keeping the model parameters fixed at their 'true' values (i.e. pertaining to the noiseless data). In the case of Gaussian noise it is not difficult to work out expressions for the elements of I [47–50, 23, 24b]. Thus, substituting Eq. (8) and assuming uncorrelated noise, Eq. (9) can be rewritten in the simple form (* denotes complex conjugation)

$$(I)_{jl,\,\text{Gauss}} = \frac{1}{\sigma^2} \sum_{n=0}^{N-1} \mathrm{Re}\left[\frac{\partial \hat{x}_n^*}{\partial \vartheta_j} \frac{\partial \hat{x}_n}{\partial \vartheta_l}\right]. \tag{10}$$

It can be shown that subject to some not particularly restrictive conditions [49, page 114], the CR lower bound on the standard deviations of the parameter estimates are given by:

$$\text{St. Dev. } (\vartheta_j) \geqq \sqrt{(I^{-1})_{jj}}, \quad j = 1, \ldots, J. \tag{11}$$

Equations (10) and (11) enable one to evaluate the fundamental lower bounds on the precision of the parameter estimates. For certain simple cases one can derive a closed-form formula that provides useful insight [51]. One result, agreeing with common sense, is that for uncorrelated Gaussian noise the CR lower bounds can be seen to be proportional to σ. As a rule, Eqs. (10) and (11) are worked out numerically by computer.

Note that in an actual experiment (in contradistinction to a simulation) the CR lower bounds can of course not be calculated *exactly* because the estimates

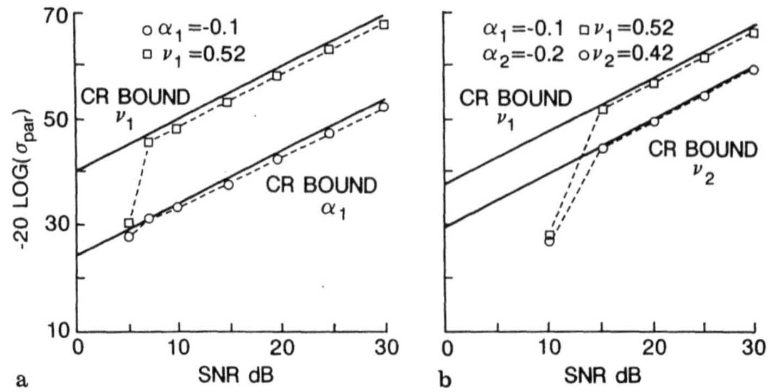

Fig. 2 a, b. Example of breakdown of an estimation method below a certain SNR [21], demonstrated by plotting the standard deviation σ_{par} of estimated parameters as a function of the SNR. 25 data points were used, corrupted with 500 different noise realizations (Monte Carlo experiment). **a**) $\sigma_{\alpha 1}$ and $\sigma_{\nu 1}$ for a single sinusoid; $\alpha_1 = -0.1$, $\nu_1 = 0.52$, $\Delta t = 1$. **b**) and $\sigma_{\nu 2}$ for two sinusoids with $\alpha_1 = -0.1, \nu_1 = 0.52, \alpha_2 = -0.2, \nu_2 = 0.42, \Delta t = 1$. At high SNR the standard deviation of the Monte Carlo experiment (*dashed line*) is near the Cramér–Rao lower bound (*drawn line*). As the SNR is lowered a deviation occurs. This is called breakdown. The breakdown SNR increases as the frequencies of two sinusoids approach each other

of the model parameters are affected by the measurement noise. Nevertheless, the CR lower bounds are very useful for, among other things, predicting the attainable precision of a quantification experiment, using chosen values of the attendant model parameters and experimental variables. Should the predicted precision be insufficient, one can redesign the experiment by computer, i.e. one can vary the noise level (related to the measurement time), number of sample times, and distribution of sample times in such a manner that the precision becomes acceptable [4, 22]. Further improvement can be achieved by redistributing the number of scans averaged at each sample time [22]. Such approaches, called experimental design, may yield considerable saving of expensive spectrometer time. See also Ref. [52].

Finally, we point out that any parameter estimation procedure eventually breaks down as the SNR is decreased. The SNR at which this takes place is called breakdown SNR. See Fig. 2 for an example.

1.3 The Importance of Prior Knowledge

In the previous section we have seen that there is a fundamental lower bound on the precision of a quantitative analysis. The only way to get around this lower bound is to invoke prior knowledge about the model parameters. As an example, it is reasonable to assume that the signal of a proton coupled to a rotating —CH_3 group consists of four distinct sinusoids with equal phases and amplitude ratio $c_1{:}c_2{:}c_3{:}c_4 = 1{:}3{:}3{:}1$. In addition, one may assume $\nu_1 - \nu_2 = \nu_2 - \nu_3 = \nu_3 - \nu_4 = \Delta\nu$ and possibly $\alpha_1 = \alpha_2 = \alpha_3 = \alpha_4$ might apply as

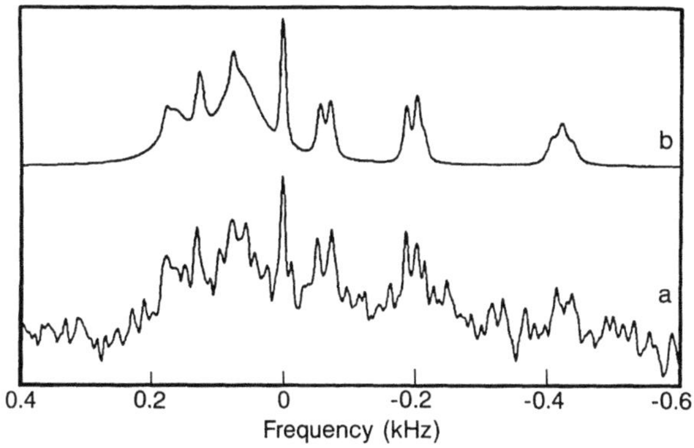

Fig. 3 a, b. Example of model function fitting at low SNR with the aid of prior knowledge. **a)** FFT of an in-vivo ^{31}P-NMR spectrum of a human brain. **b)** FFT of a time domain model function fitted with all relative frequencies fixed, most relative damping factors fixed, and all relative amplitudes of ATP fixed. Taken from lecture notes by R. de Beer; measurement by Philips Medical Systems

well. Imposing such relations on the model function usually yields rather beneficial effects on both the precision and the breakdown SNR. An excellent means to quantitatively investigate the increase of the precision is of course provided by the Cramér–Rao theory treated above.

An extreme case of prior knowledge is when all frequencies and damping factors are known, as shown in Ref. [51]. Figure 3 shows an example where all relative frequencies and most relative damping factors were known. Other examples of the usefulness of prior knowledge are given in e.g. Refs. [6, 53–57].

This section is ended with the remark that the most basic form of prior knowledge is the model function itself. Should a model function be lacking then much of the above is not applicable and one is forced to resort to special methods. One such method, pertaining to the frequency domain, is quantification of a peak area by integration. A possible method for the time domain is treated in Sect. 2.5.2.

1.4 Noninteractive (Black-Box) Versus Interactive Parameter Estimation

In this section a distinction is made between estimation approaches on the basis of the amount of user involvement. As pointed out in Sect. 1.3, prior knowledge, including that of the model function, may be lacking. Also, an expert capable of judging and implementing prior knowledge may not be available. In the latter events one seeks a procedure that requires little or no involvement of the user. Such procedures exist and we indicate them by the adjectives Noninteractive or Black-Box. Examples in the time domain are Singular Value Decomposition (SVD)-based methods, and Evans–Fischl-based methods.

Both black-box methods mentioned are restricted to the model function of Eq. (1). Should this model be inadequate, the quantification is mathematical rather than physical (see e.g. Sect. 2.5.2), but the fit can most often be made quite good. A price to pay for the relative ease of operation is that the breakdown SNR is higher than that of the more elaborate interactive methods.

Implementation of prior knowledge may require extensive involvement of the person running the attendant computer program. Therefore such an approach is indicated by the adjective Interactive. Tasks to be performed may include choosing the proper model function, providing good starting values, gathering prior knowledge, and judging the convergence to the proper (global) minimum. Clearly, this may be rather expensive in terms of human effort. Note however that once prior knowledge has been made available, its implementation is quite simple; see e.g. the penultimate paragraph of Sect. 3.3. In addition, imposition of prior knowledge usually has a beneficial effect on the stability of the convergence of an iterative process. Examples of interactive parameter estimation, applicable in both the time and the frequency domain, are Maximum Likelihood Estimation (defined in Sect. 1.2.1) and Bayesian Estimation. The former will be treated in Sect. 3, the latter will only be very briefly described in Sect. 3.1.2.

2 Black-Box Methods

2.1 General Aspects

In Sect. 1.4 of this chapter two approaches to quantification were introduced, namely a) noninteractive (black-box) processing which requires little user involvement but lacks the possibility to invoke prior knowledge, and b) interactive processing which may demand heavy user involvement but provides the opportunity to specify the model function and invoke prior knowledge. This Sect. 2 is devoted to various alternatives within the black-box category.

Within the black-box category we may distinguish between *non*iterative methods based on Singular Value Decomposition (SVD) and iterative methods based on the Evans–Fischl algorithm [58–62] which combines Linear Prediction (LP) and VARiable PROjection [63] (VARPRO) and will be indicated by LP/VARPRO. Note that LP/VARPRO, although iterative, asks no starting values from the user. As for precision, we point out that LP/VARPRO has Maximum Likelihood properties (see Sect. 1.2), whereas SVD-based methods do not.

Distinguishing further, SVD-based methods can be subdivided into, among others, Linear Prediction [23–25] (LPSVD) and Vandermonde Matrix Decomposition [64, 65] (HSVD or State Space). LP/VARPRO can be subdivided into variants whose computational burden is proportional to either the square [59–61] or the first power [62] of the number of data points.

Of the mentioned black-box variants we shall treat SVD-based methods in some detail. Section 2.2 briefly reviews various facts about SVD of a matrix. In addition, the notion of the rank of a matrix is introduced and the capability of SVD to determine the rank is pointed out. Section 2.3 concerns Vandermonde decomposition of data matrices pertaining to 1D and 2D measurements. Like SVD, Vandermonde decomposition can also reveal the rank. In addition, it immediately yields the wanted signal poles. Note that a method capable of *direct* Vandermonde decomposition of a data matrix has yet to be invented. Therefore a (fast) method to transform an SVD into a Vandermonde decomposition is derived in Sect. 2.4 (HSVD or State Space [64, 65]). Section 2.6 briefly reviews LPSVD and compares this technique with HSVD. Finally, applications to various NMR signals are added in Sects. 2.5 and 2.6. Treatment of LP/VARPRO is beyond the scope of this chapter; for an application of LP/VARPRO to 2D NMR the reader is referred to Ref. [15].

2.2 *Singular Value Decomposition* [34, 66–68]

2.2.1 *General Aspects*

In signal processing one often forms a matrix from the data and then subjects this matrix to singular value decomposition (SVD). SVD can be looked upon as a method to reduce a general, rectangular matrix to diagonal form. Symbolically this process can be indicated as:

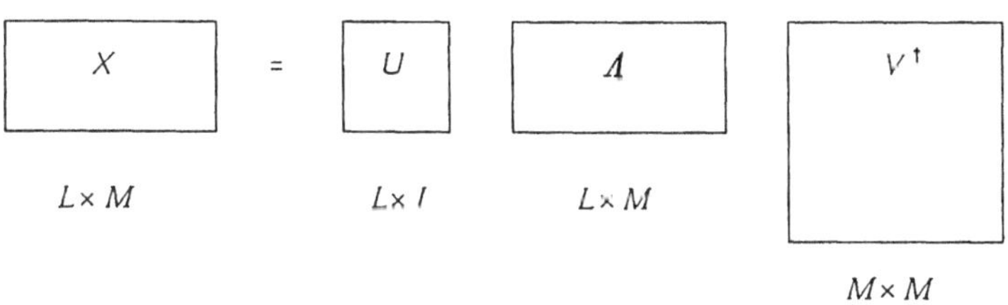

(12)

where X is a general (possibly complex-valued) $L \times M$ matrix, U is an $L \times L$ unitary matrix, Λ is a real-valued $L \times M$ diagonal matrix whose diagonal elements are non-negative, and V is an $M \times M$ unitary matrix . † denotes Hermitian conjugation. In the chosen example L is smaller than M. The diagonal elements of Λ are called singular values, indicated by λ_l, $l = 1, \ldots, L$, and the columns of U and V are called left and right singular vectors, respectively. As a rule, the singular values are arranged in descending order.

The following might be helpful for the insight of spectroscopists: When X is Hermitian (hence square), one may compare SVD to the familiar procedure of diagonalization of the Hamiltonian matrix (zero trace). Conspicuously, SVD yields only nonnegative singular values, whereas diagonalization yields positive *and* negative eigenvalues (zero trace). This seeming disagreement can be understood when realizing that U and V in Eq. (12) need not be equal even when X

is chosen square. The latter property provides the opportunity to adjust the signs of diagonal elements in Λ by reversing the signs of the associated right or left singular vectors.

The decomposition of Eq. (12) is *always possible.* One way to achieve it, called the normal equations approach, is to form the $L \times L$ matrix $XX^{\dagger}$ and the $M \times M$ matrix $X^{\dagger}X$. If $L \leqq M$, then the singular values are equal to the square roots of the eigenvalues of $XX^{\dagger}$, and the left singular vectors are equal to the eigenvectors of $XX^{\dagger}$. The right singular vectors are equal to the eigenvectors of $X^{\dagger}X$. The normal equations approach is computationally efficient for SVD of matrices that are devoid of any structure, see e.g. Ref. [24c]. See also Ref. [69] for further aspects of the normal equations approach.

2.2.2 Exploitation of Special Matrix Structure

In Sect. 2.3 we shall encounter matrices that possess Hankel structure, i.e., all elements on an antidiagonal are equal. It will be clear that SVD of such a highly symmetric matrix by means of a method intended for general matrices, such as the normal equations approach and others, entails a waste of computer time. (Recall that the computational burden of diagonalizing a general matrix is proportional to the cube of its size.) Therefore, mathematicians are searching for ways to exploit the Hankel symmetry. A final solution has not yet been found, but improvement can be achieved by combining the following measures: a) Use of the so-called Lanczos algorithm to perform the SVD [70]. This yields the capability to compute only those singular values and vectors that contain information about the signal [71] (see also Sect. 2.4). b) Exploitation of the fact that a Hankel matrix can be expressed as a submatrix of a circulant matrix. (In a circulant matrix all rows are derived from the first one by applying successive cyclic permutations of the elements.) This provides the capability to execute matrix-vector products attendant to the Lanczos algorithm efficiently by means of FFT [72]. c) Use the van Kats/van der Vorst method [73] for proper orthogonalization of the singular vectors.

The usefulness of 'Hankel/Lanczos' SVD is illustrated by the fact that it is capable of reducing the computational cost attendant to the example to be given in Sect. 2.5.1, by a factor 3 to 5, depending on the number of spectral components to be quantified. Another useful application is the parametrization of the water contribution to an in vivo ^{1}H-NMR spectrum (Sect. 2.5.2). In addition to reduced computation time, Hankel/Lanczos SVD usually requires less computer memory than general SVD.

2.2.3 The Rank of a Matrix

If only K singular values are non-zero, then the remaining $L-K$ singular values and their associated singular vectors do not contribute to the matrix product in Eq. (12). Consequently, U, Λ, and V may be stripped of the corresponding rows and columns without affecting the outcome of the product. After this stripping (truncation) only the shaded regions shown below in Eq. (13) remain.

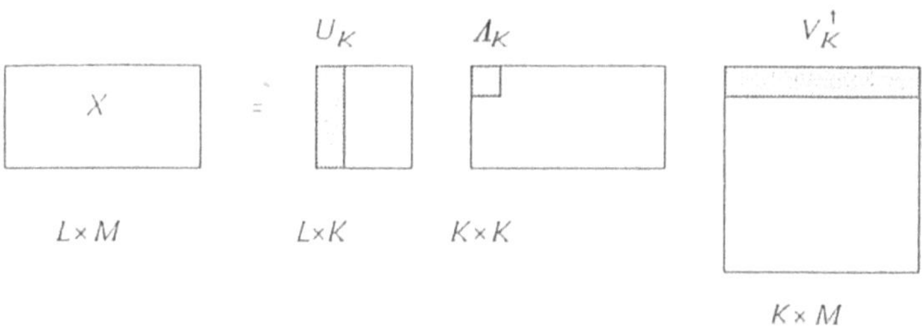

(13)

In Eq. (13) Λ_K is a $K \times K$ diagonal matrix with positive diagonal elements, whereas U_K and V_K are both matrices with K orthonormal columns. Now we recall the following matrix properties:

a) The number of linearly independent rows and linearly independent columns of a matrix are equal to each other. This number is called the rank of the matrix, indicated by the symbol R. If the rank is maximal, then it is said to be full. For example the rank of X is full if $R(X) = L(L < M)$.

b) The rank of the product of a $J \times K$ matrix A and a $K \times M$ matrix B is bounded by the relation:

$$R(A) + R(B) - K \leqq R(AB) \leqq \min([R(A), R(B)], \tag{14}$$

known as Sylvester's inequality [74].

Applying then Sylvester's inequality to the successive products in Eq. (13), one can infer another property·

c) The rank of a matrix equals the number of nonzero singular values.

In the next section the rank of matrices made up from NMR signals comprising exponentially damped sinusoids is considered. We shall see that the rank of such matrices plays an important role in signal processing.

2.3 The Data Matrix and Its Vandermonde Decomposition

2.3.1 1D NMR

In 1D NMR one measures a series of complex-valued samples $x_n, n = 0, \ldots, N-1$. In order to be able to process noninteractively (black-box), the sample times t_n should be chosen equidistant. However, there are no restrictions with respect to t_0 and t_{N-1}, i.e., t_0 need not coincide with the effective time origin of the experiment, and the signal need not have decayed into the noise at t_{N-1}.

The crux of the method lies in the formation of an $L \times M$ data matrix X from 1D data $x_0, x_1, \ldots, x_{N-1}$, according to:

$$X \stackrel{\text{def}}{=} \begin{pmatrix} x_0 & x_1 & x_2 & \cdot & \cdot & \cdot & x_{M-1} \\ x_1 & x_2 & \cdot & \cdot & & & \cdot \\ x_2 & \cdot & \cdot & & & & \cdot \\ \cdot & \cdot & & & & & \cdot \\ \cdot & \cdot & & & & & \\ \cdot & \cdot & & & & & \\ x_{L-1} & \cdot & \cdot & & & \cdot & x_{N-1} \end{pmatrix}, \tag{15}$$

Note that X has the Hankel structure alluded to in Sect. 2.2. Next, we assume temporarily that a) noise is absent and b) the exponentially damped model function of Eq. (1) describes the data exactly. It can easily be checked by writing out that under these assumptions X can be decomposed as follows

$$X = \begin{pmatrix} 1 & \cdots & 1 \\ z_1^1 & \cdots & z_K^{-1} \\ \vdots & \vdots & \vdots \\ z_1^{L-1} & \cdots & z_K^{L-1} \end{pmatrix} \begin{pmatrix} c_1'' & & & \\ & \cdot & & 0 \\ & & \cdot & \\ 0 & & \cdot & \\ & & & c_K'' \end{pmatrix} \begin{pmatrix} 1 & z_1^1 & \cdots & z_1^{M-1} \\ \vdots & \vdots & \cdots & \vdots \\ 1 & z_K^1 & \cdots & z_K^{M-1} \end{pmatrix}, \tag{16}$$

$$= \qquad \zeta_{LK} \qquad\qquad C \qquad\qquad \tilde{\zeta}_{MK}\,, \tag{16'}$$

in which ζ_{LK} and ζ_{MK} are respectively $L \times K$ and $M \times K$ (both full rank) so-called *Vandermonde* [75] matrices, and C is a (full rank) diagonal amplitude matrix whose elements are given by $(C)_{kk} = c_k'' = c_k \exp[i(\varphi_0 + \varphi_k)]$. (A double prime is used rather than a single one, the latter being reserved to denote a second dimension in Sect. 2.3.2.) The symbol ~ denotes transposition. Both L and M should be chosen greater than K in order that the rank of the Vandermonde matrices be equal to K. Given this condition, application of Sylvester's inequality to the successive products in Eq. (16) shows that the rank of a data matrix built from a noiseless 1D signal consisting of a sum of K exponentially damped sinusoids *is exactly equal to* K. Note that a (Hankel) matrix comprising noise usually has full rank ($=\min(L, M)$). However, so long as the SNR of a noise-corrupted signal is above a certain threshold the rank property derived above is a good approximation.

We call Eq. (16) the *Vandermonde decomposition* of the data matrix. Comparing Eq. (16) with Eq. (13), the reader may surmise a close relation between the two decompositions. Such a relation is dearly wanted because direct Vandermonde decomposition of a data matrix has not been achieved so far. The relation does indeed exist but its treatment is deferred until Sect. 2.4. First we briefly consider the case of 2D NMR.

2.3.2 2D NMR

In contrast to 1D NMR data, 2D NMR data $x_{nn'}$, $n = 0, \ldots, N-1$, $n' = 0, \ldots, N'-1$, are arranged in the form an $N \times N'$ matrix X from the start, i.e. $(X)_{nn'} = x_{nn'}$. Quantities in the second dimension are given a prime. Invoking the 2D exponentially damped model function of Eq. (3), we can immediately perform a Vandermonde decomposition, which now has the form

$$X = \begin{pmatrix} 1 & \cdots & 1 \\ z_1^1 & \cdots & z_K^1 \\ \vdots & \vdots & \vdots \\ z_1^{N-1} & \cdots & z_K^{N-1} \end{pmatrix} \begin{pmatrix} c''_1 & \cdots & c''_{1K'} \\ \vdots & \vdots & \vdots \\ c''_{K1} & \cdots & c''_{KK'} \end{pmatrix} \begin{pmatrix} 1 & z'^1_1 & \cdots & z'^{N'-1}_1 \\ \vdots & \vdots & \vdots & \vdots \\ 1 & z'^1_{K'} & \cdots & z'^{N'-1}_{K'} \end{pmatrix}, \tag{17}$$

$$= \qquad \zeta_{NK} \qquad C \qquad \tilde{\zeta}'_{N'K'}\ , \tag{17'}$$

where ζ_{NK} and $\zeta'_{N'K'}$ have the same meaning as in Eq. (16), except that the former is defined in the unprimed domain and the latter in the primed domain. The amplitude matrix C need no longer be diagonal: nondiagonal elements $(C)_{KK'} = c''_{kk'} = c_{kk'} \exp[i(\varphi_0 + \varphi_{kk'})] \neq 0$ can be associated with those cross-peaks whose 2D widths are equal to the widths of the associated parent peaks. Since $R(X) = R(C)$ (a consequence of Sylvester's inequality) various complications may arise in the 2D case. First, if $K \neq K'$, then the number of nonzero singular values is only $\min(K, K')$ at most. Second, if in addition rows (and columns) of C are linearly dependent, then there are even fewer nonzero singular values. As a result there are fewer signal singular values and vectors than spectral components. This complication constitutes a fundamental difference between 1D and 2D SVD-based signal processing. However, by rearranging the 2D data prior to executing SVD, it is possible to resolve the rank problem completely [15]. The cost of the solution is that SVD has to be performed twice. We emphasize that in both SVD rounds all data are used simultaneously, avoiding separate row and column processing or averaging of rows or columns. An example of a simulated signal that could easily be quantified [15] via SVD is displayed in Fig. 4. Results for one of the two smallest of the 12 components in the signal are included in the caption. Further treatment is beyond the scope of this chapter. See Refs. [15, 24a, 76–83] for other approaches to 2D time domain quantification. 2D frequency domain quantification is treated in Ref. [84].

At this point we have considered the data matrices attendant to 1D and 2D NMR measurements and derived the associated Vandermonde decompositions. To date, a method for direct Vandermonde decomposition of a data matrix has not yet been conceived. In signal processing the way to get around this problem is to first perform an SVD and then transform the latter into a Vandermonde decomposition (State Space approach or HSVD [64, 65]). The next subsection reviews this method.

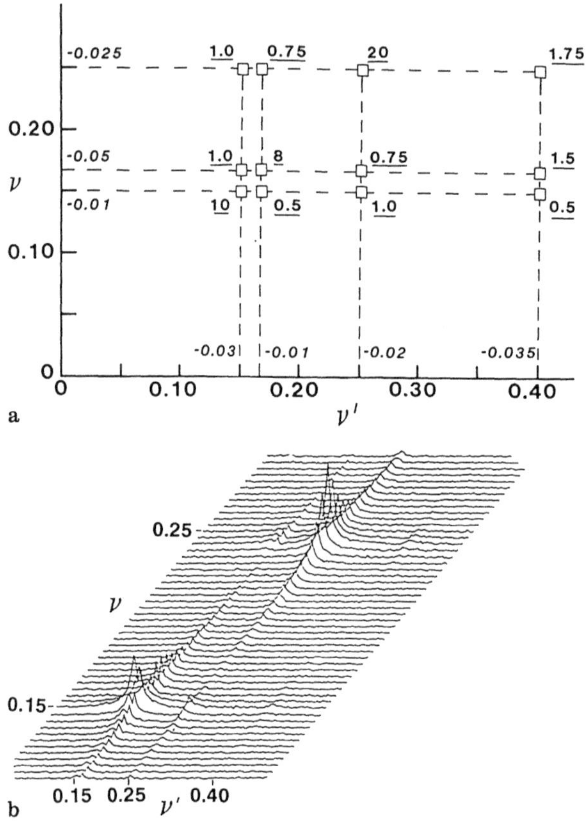

Fig. 4 a, b. 2D HSVD analysis of a simulated signal [15]. **a)** Overview of parameters used to simulate the signal with Eq. (3), $K = 3$, and $K' = 4$, $N = N' = 50$, $\Delta t = 1$. The values of the amplitudes are underlined, and those of the damping factors in italic script. The values of the frequencies can be read from the axes. All phases are equal to zero. **b)** 2D FFT of the simulated signal contaminated with complex-valued Gaussian noise with standard deviation equal to 1 for the real and imaginary part each. Fifty different noise realisations of the same standard deviation were used. 2D HSVD correctly retrieved all 12 components in each trial. For the weak component at $\nu_1 = 0.15$, $\nu'_2 = 0.17$, the estimate of the amplitude, averaged over the fifty trials, was 0.51 with a standard deviation of 0.08, which compares well with respectively the true amplitude value of 0.5 and the Cramér–Rao lower bound which is 0.05

2.4 Theory of HSVD (State Space)

2.4.1 The Principle

Comparing Eqs. (13) and (16) it is immediately clear that the structures of the Singular Value and Vandermonde decompositions are similar: The sizes and ranks of corresponding constituents are seen to match. We concentrate now on finding a relation between U_K and ζ_{LK}. From the equality:

$$U_K \Lambda_K V_K^\dagger = \zeta_{LK} C \tilde{\zeta}_{MK}, \tag{18}$$

it can be inferred that U_K and ζ_{LK} span the same column space. This in turn implies that a full rank $K \times K$ matrix must exist which converts ζ_{LK} into U_K, i.e., $U_K = \zeta_{LK} Q$, and, vice versa $\zeta_{LK} = U_K Q^{-1}$. Now we invoke an important property of Vandermonde matrices, namely that each row is equal to the previous row multiplied from the right by the same diagonal matrix. See e.g. Eq. (16):

$$\mathrm{row}_l(\zeta_{LK}) = (1, 1, \ldots, 1) Z^l, \quad l = 0, \ldots, L-1, \tag{19}$$

where Z is a diagonal matrix, defined by $Z \overset{\text{def}}{=} \mathrm{diag}(z_1, z_2, \ldots, z_K)$. A similar, but slightly more complicated relation can be seen to hold between successive rows of U_K. This follows by writing:

$$\begin{aligned}\mathrm{row}_l(U_K) &= (1, 1, \ldots, 1) Z^l Q = (1, 1, \ldots, 1)(Q Q^{-1}) Z^l Q \\ &= ((1, 1, \ldots, 1) Q)(Q^{-1} Z^l Q) = ((1, 1, \ldots, 1) Q)(Q^{-1} Z Q)^l \\ &= ((1, 1, \ldots, 1) Q) Z'^l.\end{aligned} \tag{20}$$

Apparently, each row of U_K is equal to the previous one multiplied from the right by a nonsingular matrix $K \times K$ matrix Z', defined by $Z' \overset{\text{def}}{=} Q^{-1} Z Q$. In condensed form this can be represented as:

$$(U_K)^{\uparrow} = (U_K)^{\downarrow} Z', \tag{21}$$

in which $(U_K)^{\uparrow}$ and $(U_K)^{\downarrow}$ are $(L-1) \times K$ matrices derived from U_K by respectively deleting the first and last row. If noise is present in the signal, Eq. (21) is of course no longer exact and should be solved in the standard linear least squares sense, according to:

$$Z' = ((U_K)^{\downarrow\dagger}(U_K)^{\downarrow})^{-1}(U_K)^{\downarrow\dagger}(U_K)^{\uparrow}. \tag{22}$$

Using the Sherman–Morrison matrix inversion formula, Eq. (22) can be simplified to:

$$Z' = \left(E + \frac{u u^{\dagger}}{1 - u^{\dagger} u} \right) (U_K)^{\downarrow\dagger} (U_K)^{\uparrow}, \tag{23}$$

where u is the bottom row of U_K, and E the $K \times K$ unit matrix [64, 65]. Subsequently, the quantities $z_k, k = 1, \ldots, K$, which contain the information on the damping factors and frequencies, follow from diagonalizing the complex-valued matrix Z'. Solving Eq. (21) by Total Least Squares (TLS) [34–37] (see Sect. 1.1.3) could yield increased precision. Note, however, that there is much correlation between the errors in $(U_K)^{\downarrow}$ and $(U_K)^{\uparrow}$. This aspect deserves further study.

Finally, the last step of HSVD is to estimate the amplitudes and phases by linear least squares fitting of Eq. (1) to the data, whilst keeping the frequencies and damping factors fixed at the values estimated in the previous step.

Further details of the algorithm have been described in Ref. [64], and will not be treated here for reasons of space. See also the final remark in Sect. 2.4.3.

For a comparison of HSVD with other recently devised techniques such as ESPRIT and Matrix-Pencil, the reader is referred to e.g. Ref. [85].

2.4.2 *Practicalities*

The concept of HSVD described above is very simple. However, some remarks about practical aspects are in order.

a) Often one can discern a significant discontinuity when pursuing a plot of the singular values from the smallest values upwards; see e.g. Fig. 5d. As a rule truncation of singular values and vectors takes place at this discontinuity, or rather, to be on the safe side, at a somewhat higher index [6]. If too many singular values too many are taken into account, this merely results in quantification of some noise features [86].

b) An analysis of the statistical errors in HSVD (and LPSVD) for a single undamped sinusoid has been given in Ref. [87]. It turned out that the *statistical* error in the frequency estimate is minimal (and very nearly reaches the CR lower bound) when $L \approx N/3$ or $2N/3$. However, see Ref. [88] for a treatment of the statistical error of the damping factor. Expressions for the bias (*systematic* error, due to noise) of the HSVD estimates of the damping factor and frequency of a single sinusoid are derived in Ref. [89]. The optimal values of L resulting from the various studies do not coincide. In our opinion, it is recommendable to choose $L \approx N/2$ when a signal is known to comprise many sinusoids.

c) If the signal under investigation does not satisfy the exponentially damped sinusoid model, then HSVD (or LPSVD, for that matter) will in general still accomplish a good fit. However, true to its nature, HSVD achieves this fit by decomposition of each deviating spectral component into a series of exponentially damped sinusoids [90]. By proper combination of the resulting parameters one may then still be able to retrieve physically relevant quantities such as intensities of nondescript spectral features [90]. In addition, the fit may be used for removing strong unwanted background components from a signal. The latter can be a valuable asset when processing spectroscopic imaging signals dominated by a partially suppressed water contribution [91]. See Sect. 2.5.2 and Sect. 2.5.3 for illustrations.

2.4.3 *Summary*

Summarizing, the HSVD procedure for 1D signals comprises the following steps:

a) Form an $L \times M$ data matrix from the N data points. As a rule one chooses $N/3 \leqq L \leqq N/2$. It is advisable to discard corrupted initial data points.

b) Execute SVD of the data matrix.

c) Truncate noise-related singular values and vectors (Eq. (13)). If in doubt where to truncate, retain some extra singular values and vectors.

d) Extract Z' from the truncated left singular vector matrix (Eq. (22)).

e) Diagonalize Z', which yields the frequencies and damping factors.

f) Quantify the amplitudes and phases by fitting Eq. (1) to the data. Truncation of data points is accounted for in the value of t_0. Quantified noise features can often be identified by the following characteristics: small amplitude, a devious phase, and near-zero damping factor.

A FORTRAN 77 (with complex*16 extension) implementation of HSVD on IBM PC formatted floppy disk is obtainable from the authors, on the condition that it be used strictly in a nonindustrial environment.

2.5 *Applications of HSVD to NMR*

2.5.1 ^{13}C *(Enriched) In-Vivo NMR of Rat Liver* [11]

The in-vivo ^{13}C measurement considered here comprises 512 complex-valued data points. Figure 5a shows the region of interest of the spectrum as obtained by FFT of the data. The number of significant spectral components seems to be of the order of twenty, seventeen of which are indicated in Fig. 5c. This number agrees qualitatively with the number that can be inferred from a plot on a logarithmic scale of the singular values ($L = 255$), shown in Fig. 5d. Perusing the plot from the right, the first significant irregularity is indeed discernible at around twenty.

The FFT of the HSVD time domain analysis using $K = 25$ is shown in Fig. 5b. It can be seen that agreement with the original spectrum is satisfactory. Apparently, the heavily overlapping cluster of $k = 1, 2, \ldots, 8$, defined in Fig. 5c, poses no problem to HSVD. The region around $k = 9$ is difficult and would benefit from prior knowledge about relative frequencies and damping factors; however, none was available to the authors. As mentioned before, HSVD, being a black-box method, cannot impose that all phases φ_k be equal. This is evidenced by the detailed results given in Table 1 [11]. The effect of constraining all φ_k to zero, using a fast nonlinear least squares program (see Sect. 3.3), is also given in this Table. In this nonlinear least squares fit the parameters estimated by HSVD were used as starting values. Most errors are reduced by the phase constraint. See also Fig. 5c. Note that at the time of this study, the authors did not yet include φ_0 and t_0 in the fit. Moreover, only the first three data points were discarded.

Another application of HSVD to a real-world signal, namely from ^{1}H in-vitro NMR of blood, is given in Ref. [92].

2.5.2 *Nonexponentially Damped Sinusoids (Simulation)* [90, 93]

In this subsection we address the capability of HSVD to quantify (parametrize) signals that do not satisfy the model function of Eq. (1). By working out an example we show that HSVD achieves a good fit of an arbitrarily damped

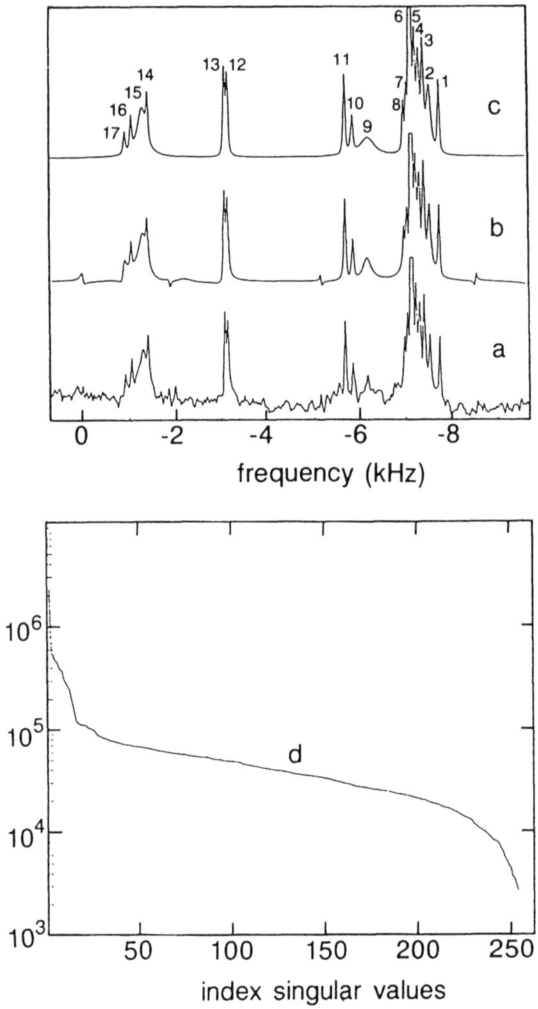

Fig. 5a–d. Quantification of a ^{13}C (enriched) in-vivo NMR signal of the liver of a rat [11]. The number of data points is $N = 512$, and $t = 50\,\mu s$. **a**) FFT of the signal. **b**) FFT of an HSVD fit, using $K = 25$ exponentially damped sinusoids. **c**) FFT of a nonlinear least squares fit with phase constraints, using $K = 17$ exponentially damped sinusoids. The starting values in **c**) were obtained from HSVD. See also Table 1. **d**) Singular values of the data matrix on a logarithmic scale

sinusoid provided a sufficient number of singular vectors is taken into account. Such a fit amounts to decomposition of the deviating function into a limited number of exponentially damped sinusoids. Although the fit thus obtained is mathematical rather than physical, it can still be very useful for at least two reasons, namely a) it yields the total amplitude (intensity) of a spectral component of unknown shape b) it enables one to remove a spectral component

Table 1. Results of quantification of ^{13}C in-vivo (enriched) NMR signal of a rat liver [11]. Upper numbers: HSVD, lower numbers: Nonlinear Least Squares with phase constraint. See Fig. 5c for assignment of k

k	ν_k (kHz)	α_k (kHz)	c_k	φ_k (degr.)
1	−7.7642 ± 0.0038[a]	−0.074 ± 0.024	388 ± 81	−4 ± 12
	−7.7630 ± 0.0023	−0.082 ± 0.023	420 ± 75	0
2	−7.5368 ± 0.0106	−0.193 ± 0.066	650 ± 214	−21 ± 19
	−7.5440 ± 0.0074	−0.335 ± 0.077	1021 ± 178	0
3	−7.4072 ± 0.0076	−0.173 ± 0.048	1181 ± 417	−24 ± 20
	−7.4142 ± 0.0023	−0.115 ± 0.025	678 ± 118	0
4	−7.3320 ± 0.0108	−0.203 ± 0.068	1108 ± 475	35 ± 25
	−7.3158 ± 0.0034	−0.170 ± 0.038	820 ± 143	0
5	−7.2279 ± 0.0041	−0.047 ± 0.026	362 ± 120	−19 ± 19
	−7.2318 ± 0.0018	−0.087 ± 0.020	596 ± 97	0
6	−7.1320 ± 0.0014	−0.112 ± 0.009	2983 ± 254	−14 ± 4
	−7.1365 ± 0.0005	−0.101 ± 0.005	2556 ± 87	0
7	−7.0628 ± 0.0084	−0.173 ± 0.053	909 ± 313	56 ± 20
	−7.0397 ± 0.0032	−0.076 ± 0.032	286 ± 78	0
8	−6.9742 ± 0.0075	−0.024 ± 0.047	124 ± 80	29 ± 36
	−6.9690 ± 0.0031	−0.057 ± 0.032	220 ± 68	0
9	−6.1432 ± 0.0537	−0.651 ± 0.338	666 ± 264	−8 ± 24
	−6.1743 ± 0.0051	−1.159 ± 0.524	945 ± 524	0
10	−5.8401 ± 0.0097	−0.120 ± 0.061	285 ± 112	−5 ± 22
	−5.8417 ± 0.0066	−0.127 ± 0.065	259 ± 100	0
11	−5.6772 ± 0.0039	−0.100 ± 0.025	538 ± 94	4 ± 10
	−5.6762 ± 0.0021	−0.086 ± 0.020	482 ± 75	0
12	−3.1745 ± 0.0047	−0.098 ± 0.030	495 ± 118	−11 ± 13
	−3.1769 ± 0.0028	−0.119 ± 0.027	585 ± 97	0
13	−3.1092 ± 0.0036	−0.083 ± 0.023	535 ± 112	−6 ± 12
	−3.1104 ± 0.0018	−0.064 ± 0.019	430 ± 78	0
14	−1.4206 ± 0.0097	−0.062 ± 0.061	180 ± 129	−13 ± 41
	−1.4221 ± 0.0040	−0.073 ± 0.050	219 ± 112	0
15	−1.3540 ± 0.0365	−0.740 ± 0.230	1613 ± 327	15 ± 12
	−1.3075 ± 0.0202	−0.744 ± 0.177	1588 ± 331	0
16	−1.0660 ± 0.0116	−0.073 ± 0.073	138 ± 93	2 ± 39
	−1.0639 ± 0.0059	−0.091 ± 0.059	190 ± 86	0
17	−0.9049 ± 0.0289	−0.194 ± 0.181	190 ± 140	−56 ± 43
	−0.9256 ± 0.0092	−0.095 ± 0.087	127 ± 78	0

[a] The quoted errors equal 2 × the Cramér–Rao lower bounds pertaining to the estimated parameters

of unknown shape from a signal without affecting the rest. An example of advantage b) will be shown in the next subsection.

The signal to be quantified here is a noiseless simulation, calculated with double precision. This choice is to ensure that the outcome can be compared to 'true' parameters. The chosen signal is [93]:

$$x_n = (1 - i\Delta\omega n - \exp(-i\Delta\omega n))\exp(i\omega n)/(\Delta\omega n)^2, \quad n = 0, 1, \ldots, N-1, \tag{24}$$

which corresponds to a triangular function in the frequency domain, see Fig. 6a. The edge of the triangle is at $\omega/2\pi$, and its width is $\Delta\omega/2\pi$. The first data point equals $x_0 = 0.5$ (= area in frequency domain).

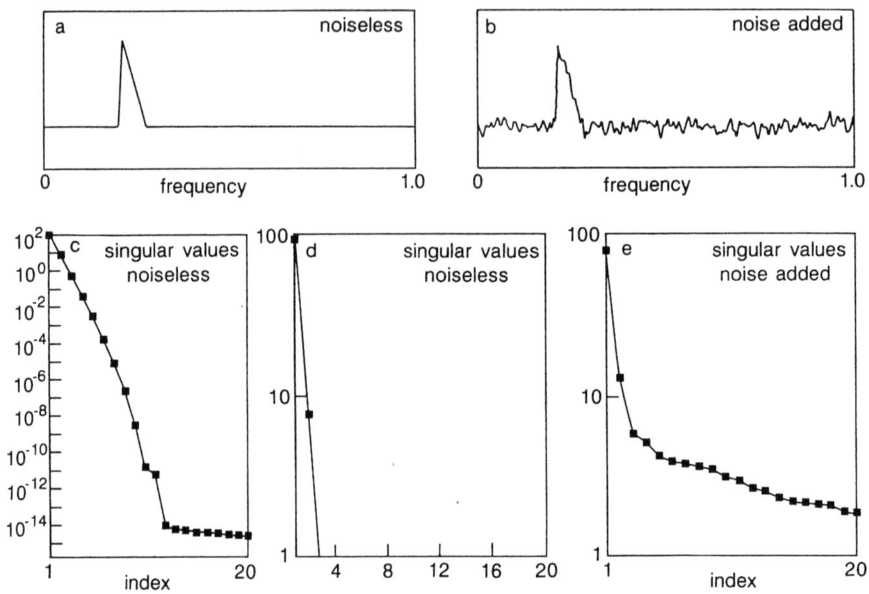

Fig. 6 a–e. SVD of the data matrix of a signal of non-Lorentzian shape, simulated with Eq. (24). **a**) FFT of the noiseless signal. **b**) FFT of the signal with complex-valued noise added. The standard deviation of the noise equals $0.1x_0$. **c**) Singular values of the data matrix of the noiseless signal ($N = 128$). **d**) Same as **c**), but with expanded vertical scale. **e**) Same as **d**), but with noise added to the signal. Note that already the third signal singular value is 'buried' under a noise singular value

First, we consider the singular values, plotted on a logarithmic scale in Fig. 6c and d. Apparently, eleven significant singular values can be distinguished, given the precision of the calculation. Figure 7 shows the precision of the HSVD fit when truncating at respectively $K = 10, 5$, and 2. It follows from these results that quantification (c.q. reconstruction) of nonexponentially damped sinusoids by means of HSVD is feasible. For $K = 2$, the error of the amplitude (peak area) is still less than 5%. The smallness of this error is due to the fact that the magnitude of the singular values decreases quickly as a function of index.

Subsequently, we add (complex-valued) noise with standard deviation equal to 0.05 to the signal, see Fig. 6b. The attendant singular value plot is displayed in Fig. 6e. It is striking that already the third signal singular value of the noiseless signal is dominated by noise singular values. This result indicates that at the given SNR it suffices to truncate at $K = 2$. Quantification of the signal with $K = 2$ yields (for this noise realisation) $x_{0\mathrm{est}} = 0.511$. The deviation from the true value ($x_0 = 0.5$) may be compared with the Cramér–Rao lower bound of the error in the amplitude ($\times 0.5$) of a signal satisfying Eq. (24), which is 0.015, for $N = 128$. Admittedly, in order to be rigorous, this investigation should have been repeated with many different noise realisations (so-called Monte Carlo experiment).

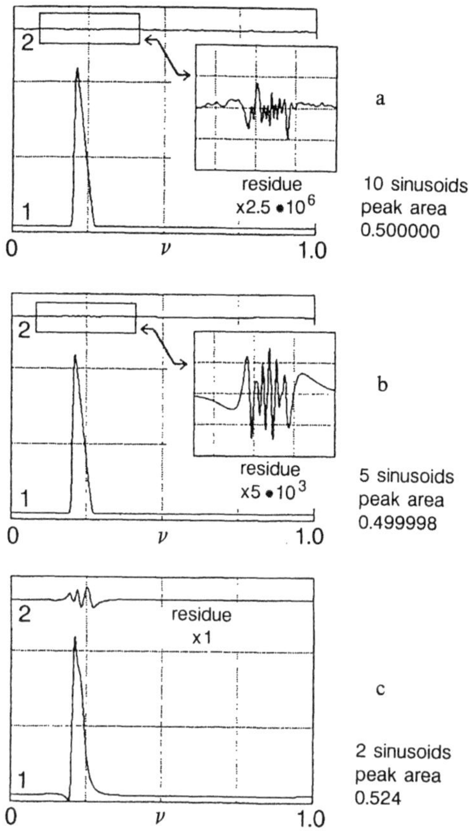

Fig. 7 a–c. HSVD fit of the non-Lorentzian signal of Fig. 6a, simulated with Eq. (24). **a**) FFT of the HSVD fit using $K=10$ exponentially damped sinusoids and FFT of the residue on a greatly expanded scale. The area ($=x_0$) estimated by HSVD is correct to within six decimals. **b**) Same as **a**), but for $K=5$. The area is still correct to within five decimals. **c**) Same as **b**), but with $K=2$. Note that the area is still correct to within 5%

In summary, Sect. 2.5.2 shows that quantification by HSVD of spectral features not satisfying Eq. (1) is feasible, and this without using knowledge about the actual model function of the signal under investigation. It should be realized, however, that when two or more features of unknown shape overlap appreciably (when presented in the frequency domain), then one can in general quantify only their total. Thus, decomposition of overlapping features into the original spectral constituents is not feasible without pertinent knowledge.

2.5.3 Removal of the Water Signal in ^1H In-Vivo Spectroscopic Imaging [91]

In ^1H in-vivo spectroscopic imaging, the contribution of water to the signal constitutes a problem because a) its magnitude is often much larger than that of nearby spectral features of interest and b) its shape is distorted due to suppression techniques. In this subsection we show that quantification by HSVD of a real-world ^1H in-vivo spectroscopic imaging signal works well [91]. After quantification, the contribution of water can be subtracted from the signal.

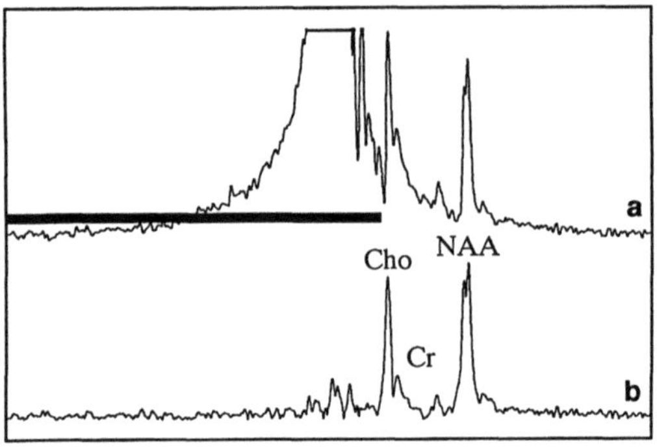

Fig. 8 a, b. Removal of the water contribution to a ^{1}H in-vivo spectroscopic imaging signal of one voxel [91]. **a**) FFT of the signal. The region indicated by the horizontal black bar contains the 'water line'. The lines of interest, indicated by Cho, Cr, and NAA, are disturbed by the 'foothill' of the water line. **b**) FFT of the signal, but now with all components, retrieved and quantified by HSVD in the region of the black bar, subtracted. Components with frequencies outside the region of the black bar are not affected by this procedure. See text for further discussion

Description of the spectroscopic imaging experiment, carried out by Philips Medical Systems, is beyond the scope of this chapter. We mention only that a typical signal spans three dimensions, two pertaining to the field gradient domain (here 16 × 16 points), and the third to the time domain (here 512 points). Thus, the total signal comprises 16 × 16 time domain responses (*echoes*). Prior to the time domain processing considered in this subsection, the signal is transformed from the field gradient domain to the spatial domain by applying 512 2D FFTs.

A typical result of time domain processing in one of the 16 × 16 spatial domain points (voxels) is given in Fig. 8. Figure 8a shows the FFT of the second halve of the echo. The water contribution is seen to dominate the spectral lines of interest. As a result, the baseline of the lines of interest is severely distorted, which hampers further analysis. In our opinion, correction of the baseline by fitting some *polynomial* to the smooth parts of the region of interest and subsequently subtracting this polynomial, is hazardous. This is because criteria for deciding whether or not a polynomial inadvertedly removes parts of the actual lines are difficult to establish. In addition, such polynomial fitting makes no use of the well accessible (high SNR!) information contained in the strong part of the water line. The method applied here brings about a baseline correction by removing a sum of exponentially damped sinusoids whose parameters can be determined with high precision, and whose frequencies are all located in a region where the water line is strong, rather than in the region of interest. In other words, only *tails* of strong Lorentzian lines are subtracted from the region of interest. As a result, the risk of affecting spectral features of interest is reduced.

The result of the subtraction is shown in Fig. 8b. All fitted sinusoids in the region marked by the black bar were removed. It can be seen that some weak features contained in the strong water contribution were overlooked by HSVD, but it is not realistic to suppose that these have a detrimental effect on subsequent analysis of the region of interest. When in doubt, another round of HSVD usually suffices to parametrize the remainder of the water contribution.

Finally, it should be mentioned that subtraction of the water contribution was proposed earlier [54]. However, in that investigation the damping was modelled by a polynomial, rather than by a combination of exponentials. In our opinion, the model function used in this section is a better approximation of physical reality.

2.6 LPSVD

2.6.1 General Aspects and Comparison with HSVD

LPSVD was devised by Kumaresan and Tufts [23] for the purpose of detecting and quantifying damped sinusoids in noise. The technique was first applied to Magnetic Resonance in the field of Electron Spin Echo Envelope Modulation (ESEEM) to reconstruct a substantial initial part of the signal, lost as a result of spectrometer dead time [94], see Fig. 9. The theory of LPSVD has been treated from various angles in Refs. [5, 6, 23, 25, 37, 95]. In view of space limitation a formal treatment is not included in this chapter. Instead we restrict ourselves to recalling the main principles of LPSVD and to pointing at some differences with HSVD. In the next subsection we mention some applications of LPSVD to real-world NMR signals.

The starting point of LPSVD is the linear prediction (LP) equation, Eq. (5), set up for $n = 0, 1, \ldots, N - M - 1$. When the signal is contaminated by noise, the order M of the equation is taken much larger than the number of sinusoids K, as already mentioned in Sect. 1.1.3. Quoting from Ref. [96], the extraneous degrees of freedom thus created 'act as lightning rods' in absorbing noise effects to a substantial degree. The set of LP equations is solved by linear least squares with the aid of SVD, which provides the opportunity to remove a significant noise contribution. This is brought about by putting all noise-related singular

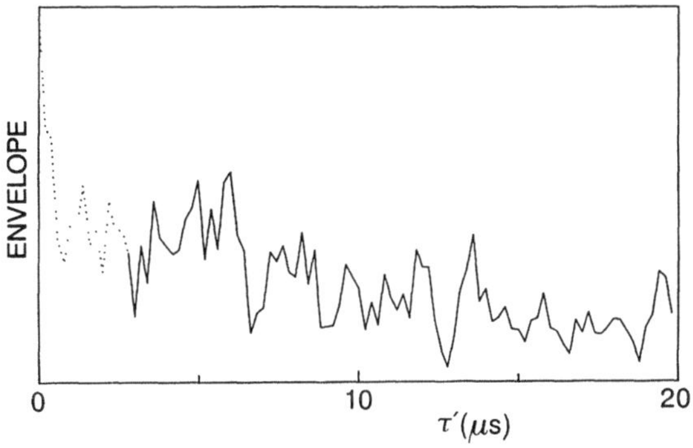

Fig. 9. Reconstruction by means of LPSVD of the missing initial part of a magnetic resonance measurement [94]. The application in question concerned a stimulated electron spin echo envelope modulation measurement. Hence the small time scale

values $\lambda_k, k = K+1, \ldots, M$ to zero, similar to the procedure of HSVD. Note that by doing so, the number of prediction coefficients remains unchanged ($= M \gg K$), but the disturbance of the prediction coefficients by noise is reduced. See Ref. [23] for illustrative examples of this effect.

Once the LP coefficients $a_m, m = 1, \ldots, M$, have been found, the signal poles $z_k = \exp[(\alpha_k + i2\pi\nu_k)\Delta t], k = 1, \ldots, K$, are obtained by rooting the polynomial

$$z^{-M} - a_1 z^{-M+1} - \cdots - a_M z^0 = 0. \tag{25}$$

Equation (25) possesses M roots, and the task is to select from these the proper K signal roots (poles). When the signal is noiseless, this entails no problem since in that case all roots associated with damped sinusoids have absolute value > 1, whereas the absolute values of the remaining, extraneous, roots are < 1. However, when the signal contains noise, the roots (poles) are disturbed. Consequently, as the noise increases, selection of roots (poles) on the basis of magnitude will eventually be jeopardized. By comparing the results of forward and backward prediction this complication can be mitigated however [97]. The underlying idea of the latter approach is that the absolute values of the extraneous roots are insensitive to the direction of the prediction (i.e. backward or forward), whereas the absolute values of the signal roots (poles) are not [23]. An alternative way to alleviate the effect of noise was shown to be the application of a narrow-band prefilter, consecutively tuned to each frequency region where a spectral component is supposedly present ('peak extraction') [98].

Now we turn to similarities and dissimilarities between HSVD and LPSVD [6, 87, 95], see Fig. 10. First, the notion of *extended order*, i.e. $M \gg K$, is present in both. Consequently, the data matrix to be subjected to SVD is the same in the two methods (ignoring for simplicity that the first data point used for the quantification is excluded from the LPSVD data matrix). Problems related to 'intermixing' of signal- and noise-related singular values [86] are therefore identical in both methods. Second, criteria for truncation of singular values are the same in LPSVD and HSVD. After truncation, the methods temporarily take diverging routes. In LPSVD one carries out polynomial rooting and root selection. The former is time-consuming when M is large, whereas the latter may be difficult when the SNR is low, as already mentioned in the preceding paragraph. HSVD circumvents these problems, as discussed in Sect. 2.4, and proceeds in a much simpler manner. Once the signal poles (frequencies and damping factors) have been obtained, LPSVD and HSVD arrive at the amplitudes and damping factors in exactly the same manner.

A detailed comparison of the underlying theories of LPSVD and HSVD is given in Ref. [95]. In addition, the statistical errors in the frequency estimates of LPSVD and HSVD for a single undamped sinusoid in noise are compared in Ref. [87]: It was found that the HSVD estimate is more precise than the LPSVD estimate, but the difference is not dramatic. As for systematic errors, it was proved analytically that the bias of the HSVD damping factor estimate for a single sinusoid in noise is negative (line estimated too wide) [89], whereas the bias of the LPSVD estimate seems to be positive (line estimated too narrow)

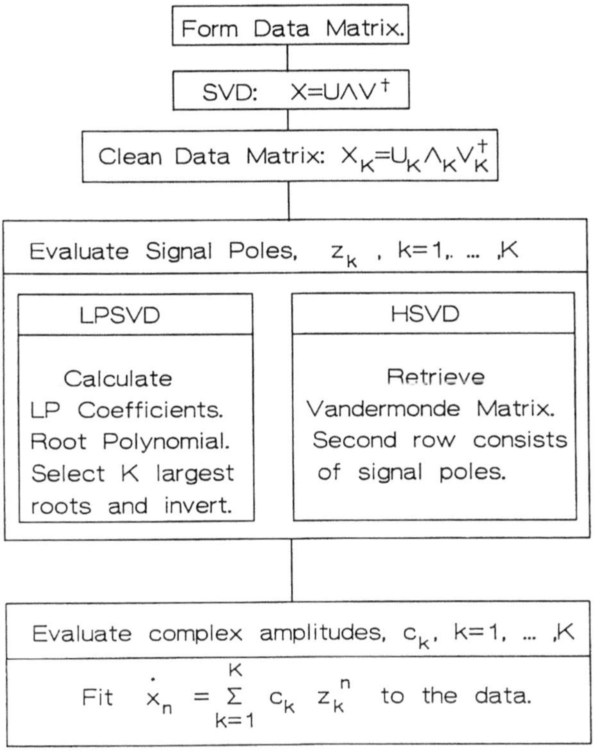

Fig. 10. Principles of LPSVD and HSVD. Both use SVD of the data matrix and then truncate the noise-related part. LPSVD then proceeds by polynomial rooting and root selection, whereas HSVD follows a simpler route. See also Refs. [87, 95]

[23, 24b]. A comparison of LPSVD and HSVD signal detection and quantification applied to in vivo ^{31}P NMR of a tumor in a mouse is presented in Ref. [99].

2.6.2 Applications of LPSVD

In this subsection we mention some applications of LPSVD to real-world NMR signals. Quantification of in vivo ^{31}P NMR on a test animal is treated in Refs. [24a, 99, 100]. A conclusion to be drawn from this work is that truncation of initial data points enables one to alleviate complications caused by the contribution of immobile nuclei. However, at the relatively low SNR at hand, incorporation of prior knowledge would be desirable [6, 53]. Unfortunately, LPSVD is not capable of this feat.

Successful application of LPSVD to in vivo ^{31}P NMR of living cell composition has been reported in Ref. [101]. Here, LPSVD provided intracellular concentration estimates in reduced NMR acquisition time, thus improving time resolution in NMR measurements of intracellular transients.

It has been reported [102] that data sets as large as 8 K originating from 1D ^{1}H and ^{13}C NMR of liquids, such as solutions of small proteins, can be handled by LPSVD. In Ref. [102], the extended order M was made as high as 2400. Naturally, the computing time was accordingly long, namely of the order of several days, but as the authors in question have remarked, this was not prohibitive when compared to a recording time of one day. Note that if only a limited region of a sizeable spectrum is to be quantified, one can resort to a Linear Prediction technique called LP-ZOOM [103].

Quantification of 2D NMR signals of liquids by LPSVD is treated in Refs. [76, 77, 82]. In these studies, phase problems attendant to 2D NMR did not constitute a problem. In addition, truncation of the data sets for reasons of economy could be accommodated without loss of accuracy.

Finally, we mention an application to ^{27}Al zero-field NMR on $GdAl_2$ at 4.2 K [104]. Here, LPSVD was successfully used for processing of ^{27}Al quadrupole echo oscillations.

3 Interactive Quantification Methods

3.1 General Aspects

3.1.1 Advantages of Interactive Methods

Section 2 was concerned with noniterative quantification methods. Such methods are advantageous because they require no starting values of model parameters, but are restricted in the sense that they cannot accommodate prior knowledge. In addition, they are intrinsically based on the exponentially damped sinusoid model function, which on the one hand constitutes a limitation, but on the other hand exempts the user from the obligation to bother about the actual model function.

In this section we consider quantification methods having rather opposite characteristics, such as:

a) A model function which describes the data, including the noise distribution, has to be provided.

b) The procedures are iterative and should converge properly.

c) Starting values of the model parameters have to be provided for the first iteration cycle.

d) Prior knowledge can (and preferably should) be imposed on the model parameters.

Since each of these points may imply significant involvement of the user, we call the methods in question interactive. This user involvement may be viewed as a disadvantage. However, relative to the noninteractive (black-box) methods of Sect. 2 we note the following important advantages.

e) *Improvement of the precision.* The standard deviations of the estimated parameters approach nearer to the (fundamental) Cramér–Rao lower bounds associated with the case at hand (see Sect. 1.2.2). When available prior knowledge is imposed, the Cramér–Rao bounds themselves decrease, which results in further improvement.

f) *Lowering of the threshold SNR.* As a rule, imposition of prior knowledge amounts to a reduction of the number of independent parameters to be estimated. This reduction in turn simplifies the log-likelihood function (see Sect. 1.2.1) and thus favors the proper convergence of the iteration process. Note that the threshold phenomenon is associated here with possible failure of convergence. More fundamental causes of threshold also exist however. With noniterative methods of Sect. 2 a cause of threshold may be 'intermixing' of small signal singular values with noise singular values.

The methods treated in the present section are based on maximizing the log-likelihood function, $L(\bar{x} - \hat{x})$, by varying the independent model parameters comprised in the model function $\hat{x}$ (see Sect. 1.2.1). Recall that this approach is called Maximum Likelihood Estimation (MLE), and that MLE is identical to nonlinear least squares model function fitting when the noise distribution is Gaussian. Moreover, the standard deviations attendant to MLE usually approach very near to the Cramér–Rao lower bounds. In this chapter we adhere to the assumption of Gaussian noise. See e.g. Ref. [7, 50] for examples of the treatment of other noise distributions.

3.1.2 *Alternative Ways to Reach the Global Extremum*

An important aspect of MLE is how to arrive at the global maximum of the log-likelihood function, or, in the case of Gaussian noise, at the global minimum of the sum of squared residues. For a comprehensive treatment of this aspect we refer the reader to Ref. [50]. Here, we restrict ourselves to a mere mention of some alternative approaches.

First, there is of course the brute force approach of simply evaluating $L(\bar{x} - \hat{\bar{x}})$ on a mesh of points in the (multidimensional) model parameter space. Provided the mesh is sufficiently fine, one is bound to find the optimum set of parameter values. It will be clear that the attendant computer time is unacceptably long in most cases.

A much better approach is to resort to so-called Monte Carlo search methods [50] which use a random generator to update the parameter values in consecutive iteration steps. One such approach is 'simulated annealing'. As in any approach each update is judged on its effect on $L(\bar{x} - \hat{\bar{x}})$. However, parameter updates that yield a decrease of $L(\bar{x} - \hat{\bar{x}})$ are not rejected outright, but weighted on the basis of an artificial Boltzmann factor comprising an artificial temperature T. The higher T, the more a decrease of $L(\bar{x} - \hat{\bar{x}})$ is tolerated. Initially, T is high, but as the convergence proceeds, T is lowered according to some protocol and eventually tends to zero. The advantage of simulated annealing is that the parameter space is searched much more efficiently than with the brute force approach, while at the same time the chance of being trapped in some local maximum of $L(\bar{x} - \hat{\bar{x}})$ is small because temporary decreases of this function are allowed. Very recent examples of application of simulated annealing to NMR are, in signal processing Refs. [105, 106], and in RF pulse crafting Ref. [107]. Note that although the convergence is well guided, the threshold phenomenon remains [106]. Another Monte Carlo-based search approach is the genetic algorithm [108]. This has been applied to RF pulse crafting in NMR [109] and to retrieval of sinusoids from noise in simulated signals [108].

A third, and widely applied approach is to derive the iterative parameter updates from the gradients (and second derivatives) of $L(\bar{x} - \hat{\bar{x}})$ with respect to the model parameters. This class of methods can be indicated broadly by the term gradient methods. Usually, gradient methods are in turn much faster than Monte Carlo-based methods. However, the chance of being trapped in a local maximum of $L(\bar{x} - \hat{\bar{x}})$ is of course greater, especially at low SNR. As for the latter aspect, much depends on the quality of the starting values. Many alternative programs are available from various libraries; see e.g. Refs. [50, 110].

In the remainder of this chapter we consider two gradient methods. These are the Variable Projection (VARPRO) method [63, 111] and the Gauss–Newton method [112, 113]. The special feature of VARPRO is that the linear and nonlinear parameters are handled in separate rounds. Favorable consequences of this are that the number of parameters fitted simultaneously is reduced and that starting values of the linear parameters (i.e. amplitudes) need not be provided by the user. The special feature of the Gauss–Newton type method considered below is that use is made of the favorable mathematical properties of the exponentially damped sinusoid model which enable one to analytically work out an important part of the calculations involved. As a result, the remaining computational load can be reduced by a factor of several tens.

Finally, before proceeding, we point at an ascendant, rather different philosophy of parameter estimation called Bayesian Estimation [47, 49, 50, 114–116]. In Bayesian estimation the model parameters are looked upon as stochastic rather than deterministic variables. One consequence of this is that in the initial stage of the estimation one has to provide a *prior probability function* of each model parameter. When the estimation process has terminated, one obtains the *posterior* probability function of each model parameter which contains the wanted information. Another consequence of the Bayesian philosophy is that one need not estimate model parameters that are currently not of interest ('nuisance parameters'). Instead one can 'integrate out' such parameters from the expression for the posterior probabilities; see e.g. Ref. [117]. Like MLE, Bayesian estimation can be used in any domain. Recently, it has been applied to quantification of (in-vivo) NMR data in the time domain [116].

3.2 *Principle of the Variable Projection Method (VARPRO)* [63, 111, 6, 53–55]

In this subsection we indicate how VARPRO eliminates the linear parameters of a model function from the nonlinear least squares procedure and then proceeds to estimate only the nonlinear parameters. This principle can of course be applied in any domain. Here we adhere to the time domain.

Model funtions comprising K sinusoids, e.g. those considered in Sect. 1.1.2, can be written in more abstract and condensed form as:

$$\hat{\bar{x}} = F\bar{c}, \tag{25}$$

where $\hat{\bar{x}}$ is a vector made up of the complex-valued model function $\hat{x}_n$, pertaining to an *arbitrary* set of N sampling times $t_n, n = 0, 1, 2, \ldots, N-1, F$ is an $N \times K$ matrix containing the K (arbitrarily damped) complex-valued sinusoids, with $n = 0, 1, \ldots, N-1, k = 1, 2, \ldots, K$, and the vector $\bar{c}$ comprises the K real-valued amplitudes $c_k, k = 1, 2, \ldots, K$.

Each sinusoid $f_k(t_n), k = 1, 2, \ldots, K$, depends *non*linearly on a number of parameters, namely a frequency v_k, and one or more damping factors $\alpha_k, \beta_k, \ldots$; see Sect. 1.1.2.2. For the purpose of algebraic manipulation, we suppose momentarily that the nonlinear parameters are known, which implies that F is also known. Given this supposition, it is easy to estimate the amplitude vector $\bar{c}$ using a standard linear least squares procedure. The well-known solution is:

$$\bar{c}_{\text{est}} = (F^{\dagger}F)^{-1}F^{\dagger}\bar{x}, \tag{26}$$

in which $\bar{x}$ represents the data, arranged in vector form. The right-hand side of Eq. (26) is now substituted for $\bar{c}$ in Eq. (25) and the nonlinear parameters regain their status of unknowns. The nonlinear least squares problem then transforms into the minimization of:

$$\sum_n |x_n - \hat{x}_n|^2 = |(E - F(F^{\dagger}F)^{-1}F^{\dagger})\bar{x}|^2, \tag{27}$$

where E is the $N \times N$ unit matrix. The solution of this problem was first given in Ref. [63]. Note that there are no restrictions to the form of the model function so long as differentiation with respect to the model parameters is possible. Once the nonlinear parameters have been estimated, the amplitudes can be obtained from Eq. (26). A computer program, based on the subroutine VARPRO (the latter written by J. Bolstad) of Ref. [63], is available from the authors on PC formatted floppy disk, on the same condition as mentioned at the end of Sect. 2.4.3.

See Sect. 3.4.1 and Refs. [53–55] for applications of VARPRO.

3.3 A Nonlinear Least Squares Gauss–Newton Implementation for Exponential Damping [11]

The fact that VARPRO is general and poses no restrictions to the model function can be very useful. However, when one wishes to entertain the exponentially damped sinusoid model function, then the generality of VARPRO entails that the favorable properties of this particular model function are not exploited, which in turn amounts to wasting computational time. One solution would be to convert the existing code into a version dedicated to exponentially damped sinusoids. Instead, we have chosen to devise a dedicated Gauss–Newton type [49, 50, 112, 113] nonlinear least squares fit program that uses analytical formulae pertaining to exponential damping [11]. Moreover, the updates of the model parameters in each iteration cycle can be judged and steered on spectroscopic rather than general mathematical principles. An example of the latter idea is that a spectroscopist knows that an update of a frequency should preferably be smaller than the width of the related spectral component.

In the Gauss–Newton method one expands the model function in a Taylor series about the current values of the nonlinear model parameters, and retains only the first order terms. In this way nonlinear dependence on parameters is approximated by linear dependence on small updates of these parameters. For example, for a model function satisfying Eq. (1) one may write for the Taylor expansion in iteration no. l (the phases $\varphi_k, k=1,\ldots,K$, are assumed known):

$$\hat{x}_n^l = \sum_{k=1}^{K} c_k \left(f_{kn}^{l-1} + \frac{\partial f_{kn}^{l-1}}{\partial \alpha_k} \Delta\alpha_k + \frac{\partial f_{kn}^{l-1}}{\partial \nu_k} \Delta\nu_k + \frac{\partial f_{kn}^{l-1}}{\partial t_0} \Delta t_0 + \frac{\partial f_{kn}^{l-1}}{\partial \varphi_0} \Delta\varphi_0 \right) \tag{28}$$

where the superscript l indicates the iteration to which an entity belongs, and f_{kn} stands for the sinusoid k at time t_n. Each iteration now amounts to a standard *linear* least squares fit of $3K+2$ model parameters, namely $c_k, \Delta\alpha_k, \Delta\nu_k, k=1,\ldots,K$, and $\Delta t_0, \Delta\varphi_0$. The solution is of the same type as Eq. (26), but now the parameter vector has $3K+2$ entries and the $N \times K$ matrix F is to be augmented to an $N \times (3K+2)$ matrix Φ. The columns $K+1$ through $3K+2$ of Φ contain the partial derivatives of f_{kn}^{l-1} seen in Eq. (28) (for t_0 and φ_0 the related partial derivatives are to be summed over k).

Note that the calculations involved in the least squares procedure can be worked out analytically to a considerable extent owing to the favourable properties of the exponentially damped sinusoids f_{kn}. First, provided the samples have been taken at regular intervals Δt, the calculation of the elements of the $(3K+2) \times (3K+2)$ product matrix $\Phi^\dagger \Phi$ merely involves the evaluation of sums of the simple type [11]:

$$\sum_{n=0}^{N-1} n^j (z_k^* z_{k'})^n, \quad j=0,1,2, \tag{29}$$

where * stands for complex conjugation, $k, k' = 1, 2, \ldots, K$, $z_k \stackrel{\text{def}}{=} \exp[(\alpha_k + i\omega_k)\Delta t]$, and $z_{k'} \stackrel{\text{def}}{=} \exp[(\alpha_{k'} + i\omega_{k'})\Delta t]$. Simple closed-form expressions for these sums have been given in [11], including the special case $z_k^* z_{k'} \to 1$. As a result, the computation of $\Phi^\dagger \Phi$ requires little time. Should the damping be nonexponential or the sampling times be nonuniform, then the computation of $\Phi^\dagger \Phi$ becomes a burden. Second, the computation of $\Phi^\dagger \bar{x}$ can be accelerated by using the recursive relation $f_{kn+1} = z_k f_{kn}$, which is valid for exponentially damped sinusoids and uniform sampling.

Finally, we indicate how to impose prior knowledge by way of a simple example. Suppose one knows from spectroscopic experience that $\alpha_1 = p\alpha_2 \stackrel{\text{def}}{=} \alpha_{1,2}$, p being a known constant. The number of independent parameters then reduces from $3K + 2$ to $3K + 1$. As a result, columns $\bar{\Phi}_{K+1}$ and $\bar{\Phi}_{K+2}$ of Φ, which are related to $\partial/\partial\alpha_1$ and $\partial/\partial\alpha_2$ respectively, are to be replaced by a single column related to $\partial/\partial\alpha_{1,2}$ and equal to the linear combination $\bar{\Phi}_{K+1} + p^{-1}\bar{\Phi}_{K+2}$. The consequences for $\Phi^\dagger \Phi$ and $\Phi^\dagger \bar{x}$ are easily worked out. Other linear relations between model parameters can be handled in the same vein and their implementation in the program is straightforward. As the number of relations increases, the number of degrees of freedom decreases accordingly, which in turn favors the stability of the solution and enables one to handle lower SNRs.

See Sect. 2.5.1 (Table 1) and Sect. 3.4.1 for applications to in-vivo NMR experiments. The method is also applicable to quantification of echo data and to restriction to only a limited frequency region (frequency-selective quantification, in the same vein as LP-ZOOM); see Ref. [118] for both. See Ref. [119] for another recent example of nonlinear least squares fitting of NMR time domain data.

3.4 Applications

3.4.1 Is the Voigt Function more Suitable than the Lorentz Function?

The matter of which lineshape to entertain is often problematic, if only because sufficient knowledge about the 'sample' under investigation is lacking. As a result one tends to use a simple model function, such as the exponentially damped sinusoid. Alternative shapes should be investigated, however. One such alternative is the Voigt function, mentioned in Sect. 1.1.1.2. In the present section we attempt to assess whether the Voigt function yields significant improvement over the Lorentz function. (Recall that the increased mathematical complexity of the Voigt function is minimized by adhering to the time domain). For this study use is made of VARPRO because its implementation can accommodate any function whose derivatives with respect to the independent model parameters are available. The case chosen is a typical in-vivo ^{31}P-NMR measurement on a human brain, at $B_0 = 1.5\,\text{T}$. An impression of the SNR can be obtained by perusing Fig. 11 which displays the FFT of the complete (i.e. *un*truncated) signal. Prior to the actual quantification, the first five data points were truncated

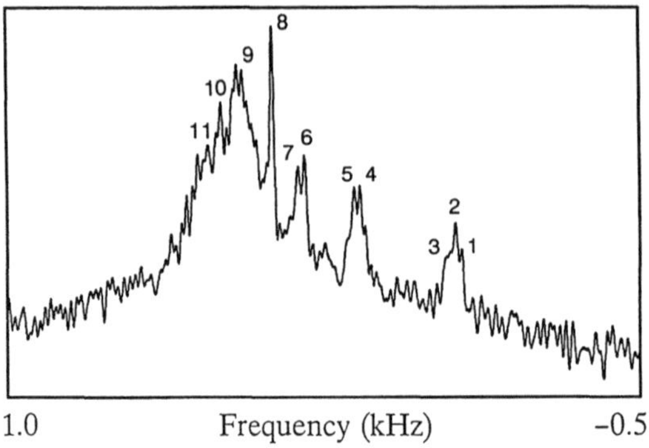

Fig. 11. FFT of a ^{31}P in-vivo NMR measurement ($B = 1.5\,T$) of a human brain used for investigating whether it is worthwhile to entertain the Voigt model rather than the Lorentz model. No initial data points were truncated for producing this figure. In the actual domain fit the first five data points were omitted. In this case use of the Voigt model yielded no significant improvement

with the aim to reduce the influence of the wide background structure. The zero and first-order phase correction were set at fixed values.

The results for the Lorentz and Voigt functions are compounded in Table 2. Note that constraints were imposed on the seven ATP components to augment the stability of the convergence. In particular, all seven damping factors β_k, $k = 1, \ldots, 7$, (see Fig. 11 and Table 2) were constrained to be equal. The remaining five β_k were not constrained. As expected, the root mean square of the residues

Table 2. Comparison of nonlinear least squares fitting of Lorentz shapes ($\beta_k = 0$) and Voigt shapes ($\beta_k \neq 0$) (in the time domain). The first five time data points were omitted so as to reduce the influence of the broad background peak. The FFT of the signal, including the five original data points, is displayed in Fig. 11; see text

k	ν_k (kHz)	α_k (kHz)	$\sqrt{(\lvert\beta_k\rvert)} \times \mathrm{sign}(\beta_k)$ (kHz)	c_k
1	−0.0814+0.0022[a]	−0.0508±0.0166	0.0	52± 7
	−0.0817±0.0021	−0.0672±0.0294	0.0172±0.0119	57±11
2	−0.0654±0.0022	−0.0508±0.0166	0.0	104± 13
	−0.0657±0.0021	−0.0672±0.0294	0.0172±0.0119	114± 22
3	−0.0494±0.0022	−0.0508±0.0166	0.0	52± 7
	−0.0497±0.0021	−0.0672±0.0294	0.0172±0.0119	57± 11
4	0.1555±0.0019	−0.0540±0.0146	0.0	104± 13
	0.1556±0.0017	−0.0666±0.0263	0.0172±0.0119	114± 22
5	0.1715±0.0019	−0.0540±0.0146	0.0	104± 13
	0.1716±0.0017	−0.0666±0.0263	0.0172±0.0119	114± 22
6	0.2865±0.0016	−0.0476±0.0122	0.	104± 13
	0.2863±0.0015	−0.0629±0.0262	0.0172±0.0119	114± 22
7	0.3025±0.0016	−0.0476±0.0122	0.0	104± 13
	0.3023±0.0015	−0.0629±0.0262	0.0172±0.0119	114± 22
8	0.3617±0.0006	−0.0250±0.0062	0.0	151± 27
	0.3617±0.0023	−0.0165±0.0221	0.0109±0.0147	135± 41
9	0.4382±0.0035	−0.3020±0.0457	0.0	1513±262
	0.4368±0.0021	−0.3419±0.0510	0.0591±0.0053	1645±288
10	0.4879±0.0020	−0.0180±0.0262	0.0	30± 30
	0.4882±0.0023	−0.0281±0.1253	0.0089±0.0865	36± 76
11	0.5309±0.0070	−0.2661±0.0788	0.0	629±223
	0.5307±0.0089	−0.2644±0.2176	−0.0324±0.3178	630±321

[a] The quoted errors equal 2 × the Cramér–Rao lower bounds pertaining to the estimated parameters

of the Voigt fit was lower than that of the Lorentz fit (51.37 versus 51.59), whereas the statistical errors in the parameters are generally larger than with the Lorentz fit. We infer two important points from Table 2. First, the amplitudes are not particularly sensitive to changing from Lorentz to Voigt. Second, except for β_{11}, all β_k are positive. Mathematically, the effect of a positive β is to counteract the exponential damping (governed by α) at large $t_n = n\Delta t$ (N = 256 $\Delta t = 0.333$ ms). This overall behaviour of the damping may be true, but in our opinion it is inappropriate (unphysical) to *model* it with a Voigt function with *positive* β. In view of these two points our verdict is that, at least in this particular case, there is no reason to replace the Lorentz function by the Voigt function. More investigations are desirable. Given the low value of the SNR attendant to in-vivo work, it may be advisable to carry out such studies with simulated annealing or the genetic algorithm.

Finally, we stress that the Voigt function should not be confused with a mere sum of a Lorentz function and a Gauss function (in either domain). The latter construction seems not very physical to us.

3.4.2 ^{1}H In-Vivo Spectroscopic Imaging [91]

The subject of ^{1}H spectroscopic imaging was already addressed in Sect. 2.5.3, in the context of removing the strong contribution of water. It was shown that HSVD is capable of the latter without affecting the contributions of interest, namely those of choline, creatine, *N*-acetyl aspartate, and lactate. In the present subsection we turn to quantification of the concentrations of the mentioned metabolites in each voxel, as treated in Ref. [91]. We recall that in Ref. [91] the time domain signal was an echo, only the second halv of which was used for processing.

Since in-vivo experimental conditions entail low SNR of the metabolite signals, it is recommendable to exploit prior knowledge. This implies recourse to interactive methods. In addition, the SNR does not warrant the use of model functions other than the simple exponentially damped sinusoid. In Ref. [91] these two considerations led to the choice of the fast, dedicated nonlinear least squares method described in Sect. 3.3. The prior knowledge imposed was the following: a) the relative frequencies as well as the damping factors of the four metabolites were fixed at predetermined values, b) the lactate signal was treated as a doublet with equally strong constituents at fixed distance to each other. As a consequence of these constraints, the only independent parameters to be fitted in each voxel were the zero order phase φ_0, the frequency offset, and four amplitudes (metabolite concentrations). This drastic reduction of the number of independent nonlinear parameters yielded increased stability of the least squares procedure and decreased sensitivity to the quality of the starting values of the parameters. As for the latter, only the procurement of the starting values of φ_0 of the first voxel of each row required some human involvement. The starting value of the offset frequency was derived from the parameters of the

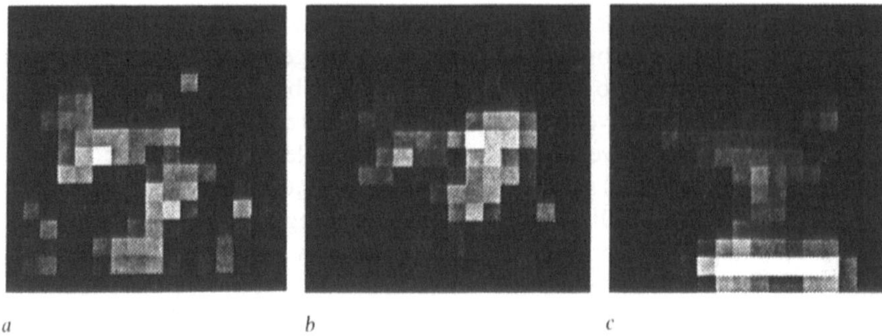

Fig. 12 a–c. ^1H in-vivo spectroscopic image of PCr in a human brain [91]. **a)** Water contribution subtracted by means of HSVD and remaining signal quantified with constrained nonlinear least squares time domain model fitting. **b)** Water contribution subtracted with HSVD and quantification of the remaining signal by FFT and integration. **c)** Water contribution removed by a shift-subtract procedure and quantification of the remaining signal by FFT and integration. Differences between the three images are apparent

water line. Finally, using the dedicated program the wanted metabolite concentrations could be quantified in each of the 256 voxels at little computational cost: The total time was about 15 minutes using a SUN-3 work station. See Fig. 12a for the resulting image of creatine.

We emphasize that judgement of Fig. 12a is difficult because the true image is unknown. All one can do is compare with results of alternative signal processing procedures that spend less effort on correcting for experimental imperfections. First we consider variations of the frequency offset which are caused by, among other things, eddy currents in metal walls surrounding the RF coil. The magnitude of these variations can be seen in Fig. 13, which shows the stacked modulus spectra of a row of voxels. Note that the peak positions are shifted to such an extent that frequency regions of choline and creatine of different voxels start to overlap. Consequently, quantification of concentrations

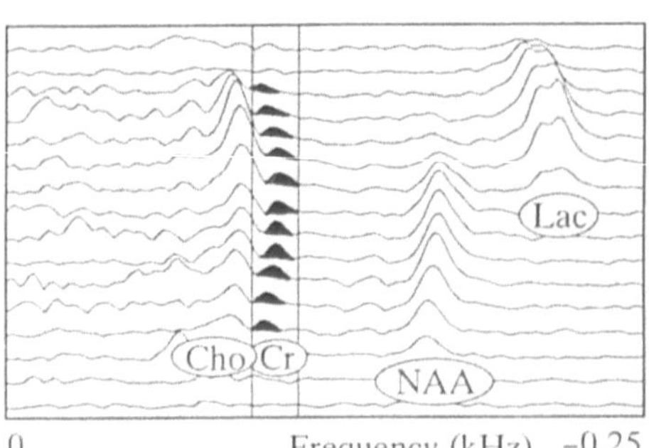

Fig. 13. Variation of the off-set frequency of a ^1H in-vivo NMR spectrum of a human brain in a number of consecutive voxels [91]. The two vertical lines illustrate that integration of the choline peak over a fixed frequency region would entail an error

by integration over fixed frequency regions in spectra not corrected for variation of the frequency offset, can entail significant errors. An image of creatine resulting from HSVD water removal and subsequent integration over a fixed frequency region is shown in Fig. 12b.

Next, we turn to the water signal subtraction technique. Rather than comparing to polynomial subtraction, we compare to a time domain shift-subtract procedure. An advantage of the latter technique is that it is very fast, but a disadvantage is that it affects the concentrations in the entire frequency region. An image of creatine obtained by water removal via shift-subtract and subsequent integration is shown in Fig. 12c.

Differences between the three images in Fig. 12 can easily be discerned. Ignorance about the true image obscures proper judgment, however. Nevertheless, Fig. 12a entailed fewest approximations.

3.4.3 ^{31}P In-Vivo Inversion Recovery T_1 Measurements. A 2D Approach [21]

The estimation of relaxation times T_1 from localized in vivo ^{31}P-NMR inversion recovery (IR) measurements is complicated by poor spectral resolution and poor SNR. In this subsection we treat how improvement can be achieved by using a 2D model function that encompasses the FIDs as well as the relaxation behaviour. *All* model parameters are fitted concurrently to the 2D data. In addition, prior knowledge about model parameters is imposed, and initial data points of each FID are skipped so as to reduce the contribution of immobile ^{31}P nuclei to the signal.

The basic setup of an inversion-recovery measurement is indicated in Fig. 14. The inversion-recovery time is T_I, the repetition delay time T_R, and the FID is observed during time t. Ideally, T_R should be made sufficiently long to allow for complete relaxation of all spins. However, for localized ^{31}P measurements of humans there is a practical limit to the total measurement time which prohibits waiting for establishment of thermal equilibrium. To incorporate this experimental condition in the quantification procedure one must set up a model function that simultaneously describes the relaxation behavior and the free

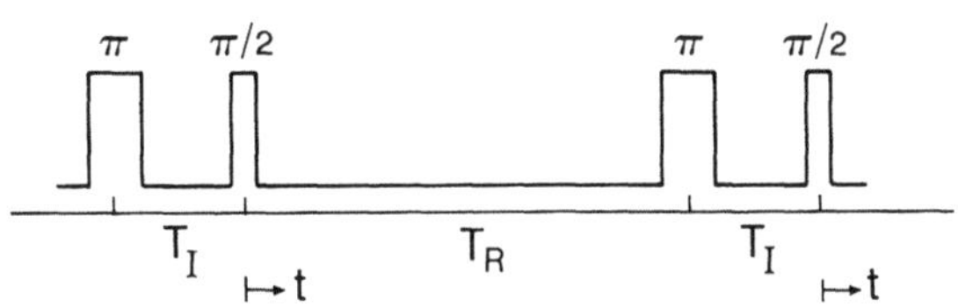

Fig. 14. Two successive π and $\pi/2$ pulse pairs of an inversion recovery measurement [21]. To limit the measurement time, the repetition delay time T_R is kept below the value needed to reach thermal equilibrium. The effect of non-equilibrium can be incorporated in the model function. As a result, the intensities resulting from the fit do pertain to thermal equilibrium

induction decay. This can be brought about as follows:

$$\hat{x}(t_n, T_{Im}) = \sum_{k=1}^{K} c_k f_k(t_n) g_k(T_{Im}), \tag{30}$$

where c_k, f_k, and g_k are respectively the (real-valued) amplitude pertaining to thermal equilibrium, the FID function, and the relaxation function, each of spectral component k. We assume the best approximation for the FID function to be the exponentially damped sinusoid, introduced in Sect. 1.1.2.1:

$$f_k(t_n) = \exp[(\alpha_k + i2\pi\nu_k)t_n + i(\varphi_0 + \varphi_k)], \tag{31}$$

The relaxation behavior is modelled as:

$$g_k(T_{Im}) = 1 - [2 - \exp(-T_R/T_{1k})]\exp(-T_{Im}/T_{1k}). \tag{32}$$

in which T_{1k} is the spin-lattice relaxation time of spectral component k, and T_I assumes the values $T_{Im}, m = 1, \ldots, M$. Equation (32) has been derived on the assumption that longitudinal magnetization is nulled by the $\pi/2$ pulse and perfectly inverted by the π pulse. Note that Eq. (32) accommodates that the relaxation is incomplete at the start of a new inversion pulse. For $T_R \gg T_{1k}$, one has $g_k(0) = -1$ and $g_k(\infty) = +1$, but with practical values of T_R this is not accomplished.

One way to determine the k relaxation times T_{1k} is to first fit M functions:

$$\hat{x}_m(t_n) = \sum_{k=1}^{K} c_{km} f_k(t_n), \quad m = 1, \ldots, M, \tag{33}$$

to FIDs no. $m = 1, \ldots, M$, respectively. In Eq. (33) c_{km} is an amplitude whose magnitude depends on T_I, according to $c_{km} = c_k g_k(T_{Im})$. Subsequently one fits each of the K relaxation functions of Eq. (33) to its associated T_I-dependent set of amplitudes $c_{km}, m = 1, \ldots, M$, which finally yields the wanted $T_{1k}, k = 1, \ldots, K$. A disadvantage of this *two-step* fitting process is that all spectral parameters have to be estimated over and again for each FID. Moreover, individual spectral components may require time consuming interactive processing at those T_I-values where their magnetizations are nulled. These unfavourable aspects are absent when a 2D model function of the type of Eq. (30) is fitted to all NM data points simultaneously, i.e. in a *single* step: Then all $4K$ parameters, $c_k, \alpha_k, \nu_k, T_{1k}, k = 1, \ldots, K$, are estimated concurrently, while at the same time the temporary weakness of individual spectral components at some T_I-values ceases to pose problems. Note that although φ_0 and t_0 were ignored in these considerations, they can easily be included in the fit [21], should such be desired.

In the study of Ref. [21] each FID comprised $N = 1024$ complex-valued data points, and the number of T_I-values equalled $M = 6$. The measurements were done with a Philips 1.5-T NMR scanner. T_R was *fixed to three seconds*, which is insufficient for reaching thermal equilibrium. In order to reduce the contribution of immobile nuclei, the first sixteen data points of each FID were not used for the quantification [1, 10, 119]. The last 512 of each FID were also discarded because their SNR was deemed too low. Excluding the time required

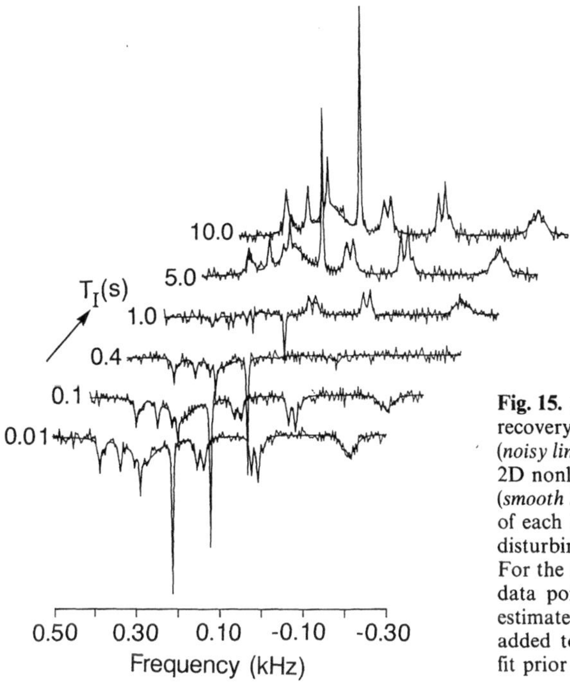

Fig. 15. FFT of a ^{31}P in-vivo inversion recovery measurement of a human brain (*noisy lines*) and FFT of a fast, constrained, 2D nonlinear least squares time dmain fit (*smooth lines*) [21]. The first 16 data points of each FID were removed to reduce the disturbing influence of immobile nuclei. For the sake of presentation these missing data points were reconstructed from the estimated model parameters and then added to both the measurement and the fit prior to FFT

to obtain adequate starting values, quantification of some fourteen spectral components with the dedicated nonlinear least squares method of Sect. 3.3 required less than 10 sec on a Convex C240. An example of a fit of localized in-vivo ^{31}P-NMR data from a healthy human brain is given in Fig. 15. As usual, the result is presented in the frequency domain. The effect on the spectrum of truncating the first sixteen data points is remedied by reconstruction of the missing part from the quantified parameters and subsequently appending this to both the fit and the data prior to FFT (see Fig. 1 in Sect. 1.1.3 for this aspect). Figure 15 shows that the combined model functions of Eqs. (30), (31), and (32) can yield a good fit. For tables of results for five cases and further details about the signal processing, the reader is referred to the original study [21].

4 Concluding Remarks

The main aim of this chapter was to point at advantages of quantifying (estimating) NMR model parameters *directly in the measurement* (time) *domain*, rather than to first transform the data to the frequency domain and carry out the quantification there. Notable advantages mentioned were:

a) The distribution of sampling times may be arbitrary. One simply substitutes the sampling times as they are into the time domain model function to be fitted to the data. Furthermore, by studying the Cramér–Rao lower bounds one can determine the sample distribution that yields the best overall result.

b) Point a) implies that truncation of data points does not entail errors. Among other things, this enables one to reduce the disturbing influence of a broad nondescript background line by simply discarding initial data points. In addition, a measurement may be ended before the signal has decayed into the noise. The latter is important for limiting the duration of multidimensional measurements.

c) If the distribution of sampling times is uniform, then black-box quantification methods can be used. Positive aspects of such methods are that human involvement can be minimized, and that signals of unknown shape can be parametrized. A price to pay for these properties is that the threshold SNR is not as low as that of other, interactive, quantification methods. As for interactive methods, if the *exponentially* damped sinusoid model function adequately describes the signal, then the computational cost can be reduced by a factor of several tens.

Various applications of both black-box and interactive quantification methods supporting the above points have been presented. Moreover, several examples have demonstrated that imposition of prior knowledge can yield significant improvement (over unconstrained quantification). We emphasize that the beneficial effect of prior knowledge pertains to all domains. See e.g. Ref. [120] for a very recent frequency domain example, pertaining to in-vivo NMR.

An appraisal of five methods mentioned in this chapter and used by the authors is summarized in Table 3. Note that various ratings in the table are based on an average performance, and that deviations in either direction are quite possible, depending on the particular application. Of the five methods, LP/VARPRO has so far been applied least. However, its potential for fast automatic processing of large 1D data files comprising relatively few sinusoids deserves more study. Note also the Hankel–Lanczos SVD (Sect. 2.2.2) which exploits the Hankel symmetry of the 1D data matrix, although promising, has not yet been sufficiently tested to be included in the table. Furthermore, simulated annealing, the genetic algorithm, and Bayesian estimation (like VARPRO each suitable for any domain and any model function) have been excluded from the table because the authors lack personal experience with these methods. General characteristics of these methods are high performance at low SNR, quality of starting values not critical, and long computation times.

Finally, we comment on an important problem, common to all domains, namely the (in-)adequacy of the entertained model function. First, recall that a black-box time domain method can automatically augment the number of exponentially damped sinusoids until the residue of the fit ceases to contain significant

Table 3. Appraisal of 1D capabilities of the main methods discussed in this chapter

Aspect	LPSVD	HSVD	LP/VARPRO	VARPRO	GN[a]
Maximum likelihood	no	no	yes	yes	yes
Iterative	no	no	yes	yes	yes
Black-box	+	++	□	–	–
Prior knowledge	□	□	□	++	++
Sample times	uniform	uniform	uniform	arbitrary	uniform
Model function	$e^{(\alpha+i\omega)t}$	$e^{(\alpha+i\omega)t}$	$e^{(\alpha+i\omega)t}$	arbitrary	$e^{(\alpha+i\omega)t}$
Unknown lineshape[b]	+	++	□[c]	–	–
Low SNR	□	□	□	++	++
Resolving power	+	+	□	++	++
Comput. speed[e]	–[d]	□[d]	+	□	++

[a] GN stands for dedicated (to exponential damping) Gauss-Newton nonlinear least squares
[b] Ability to parametrize an unknown shape
[c] Unsufficiently tested
[d] SVD by normal equations approach
[e] Large data sets
performance grades: – = low, □ = moderate, + = fairly good, ++ = good

spectral components. This amounts to expansion of each nonexponentially damped sinusoid into a sum of exponentially damped sinuoids. A useful application of this is automatic parametrization of the water contribution to an in-vivo NMR signal. Second, when applying interactive methods, entertainment of an adequate (if not the true) model function for each spectral feature is the best option. A practical criterion for adequacy is that the residue of the fit contain no significant spectral components (compared to the noise). When the SNR is high, this condition may be very demanding. In actual in-vivo NMR practice however, the SNR is often so low that inadequacy of the usual simple model functions is difficult to discern and study. Consequently, it seems not unreasonable to restrict oneself to these simple model functions. Note however that use of so-called 'natural lineshapes', derived from high-SNR reference data [121], may be a viable alternative in various circumstances. The latter approach should also be applicable in the time domain, after parametrization of the reference data by HSVD. More studies along these lines are desirable.

Acknowledgements: Part of the investigations described in this chapter were supported by the FOM/STW Research Foundations. The authors thank A. van den Bos, W.M.M.J. Bovée, P.M.T. Broersen, J.E. van Dijk, J.A. den Hollander, H.F.M. Lohman, P.R. Luyten, A.J.H. Mariën, A.F. Mehlkopf, and W.W.F. Pijnappel for stimulating discussions.

6 References

1. Lindon JC, Ferrige AG (1980) Prog NMR Spectrosc 14: 27
2. Haselgrove JC, Subramanian H, Christen R, Leigh JS (1987) Rev Magn Reson Med 2: 167
3. Barna JCJ, Laue ED, Mayger M, Skilling J, Worrall SJP (1987) J Magn Reson 73: 69
4. Manassen Y, Navon G (1988) J Magn Reson 79: 291
5. Stephenson DS (1988) Prog NMR Spectrosc 20: 515
6. Beer R de, Ormondt D van, Pijnappel WWF, Veen JWC van der (1988) Israel J Chem 28: 249
7. Bos A van den (1989) IEEE IM-38: 1005
8. Henrichs MP, Hewitt JM, Young RH (1986) J Magn Reson 69: 460
9. Heuer A, Haeberlen U (1989) J Magn Reson 85: 79
10. Starčuk Z, Starčuk Z, Halámek J (1990) J Magn Reson 86: 30
11. Beer R de, Ormondt D van, Pijnappel WWF (1988) in: Lacoume JL, Chehikian A, Martin N, Malbos J (eds) SIGNAL PROCESSING IV: Theories and applications. Elsevier, Amsterdam, p 1283
12. Abildgaard F, Gesmar H, Led JJ (1988) J Magn Reson 79: 78
13. Premoli A, Rastello ML (1989) Appl Spectrosc 43: 558
14. Slichter CP (1989) Principles of magnetic resonance, 3rd edn. Springer, Berlin Heidelberg New York
15. Beer R de, Ormondt D van, Pijnappel WWF (1989) Proc ICASSP, p 1504
16. Cope D, Khoury R, Lovett RJ (1988) J Quant Spectrosc Radiat Transfer 39: 163
17. Ouyang X, Varghese PL (1989) Appl Opt 28: 1538
18. Edzes HT (1990) J Magn Reson 86: 293
19. Twieg DB, Meyerhoff DJ, Hubesch B, Roth K, Sappey-Marinier D, Boska MD, Gober JR, Schaefer S, Weiner MW (1989) Magn Reson Med 12: 291. See however Maudsley AA, Twieg DB, Sappey-Marinier D, Hubesch B, Hugg JW, Matson GB, Weiner MW (1990) Magn Reson Med 14: 415
20. Segebarth CM, Balériaux DF, Beer R de, Ormondt D van, Mariën AJH, Luyten PR, Hollander JA den (1989) Magn Reson Med 11: 349
21. Ormondt D van, Beer R de, Mariën AJH, Luyten PR, Hollander JA den, Vermeulen JWAH (1990) J Magn Reson 88: 652
22. Beer R de, Ormondt D van, Pijnappel WWF, Veen JWC van der (1989) in: Skilling J (ed) Maximum entropy and Bayesian methods. Kluwer, Dordrecht, p 291
23. Kumaresan R, Tufts DW (1982) IEEE Trans. ASSP-30: 833
24. a) Barkhuysen H, Beer R de, Bovée WMMJ, Ormondt D van (1985) J Magn Reson 61: 465
    b) Barkhuysen H, Beer R de, Ormondt D van (1985) J Magn Reson 64: 343
    c) Barkhuysen H, Beer R de, Ormondt D van (1986) J Magn Reson 67: 371
25. Tang J, Lin CP, Bowman MK, Norris JR (1985) J Magn Reson 62: 167
26. Tang J, Norris JR (1988) J Magn Reson 78: 23
27. Vaals JJ van, Gerven PHJ van (1990) J Magn Reson 86: 127
28. Derby K, Hawryszko C, Tropp J (1989) Magn Reson Med 12: 235
29. Olejniczak ET, Eaton HL (1990) J Magn Reson 87: 628
30. Marple SL (1987) Digital spectral analysis with applications. Prentice-Hall, Englewood Cliffs
31. Kay SM (1988) Modern spectral estimation. Prentice-Hall, Englewood Cliffs
32. a) Barone P, Guidoni L, Massaro E, Viti V (1987) J Magn Reson 73: 23
32. b) Ormondt D van, Nederveen K (1981) Chem Phys Lett 82: 443
33. Beer R de, Ormondt D van (1989) in: Hoff AJ (ed) Advanced EPR. Applications in biology and biochemistry. Elsevier, Amsterdam, Chap. 4
34. Golub GH, Van Loan CF (1988) Matrix computations, 2nd edn. Johns Hopkins University Press, Baltimore
35. Huffel S van, Vandewalle J (1988) in: Deprettere EF (ed) SVD and signal processing (Proc. First International Workshop). Elsevier (North Holland), Amsterdam, p 189
36. Rahman MDA, Yu KB (1987) IEEE Trans ASSP-35: 1440
37. Tirendi CF, Martin JF (1989) J Magn Reson 85: 162
38. Laue ED, Skilling J, Staunton J, Sibisi S, Brereton RG (1985) J Magn Reson 62: 437
39. Wright KM, Belton PS (1986) Mol Phys 58: 485
40. Daniell GJ, Hore PJ (1989) J Magn Reson 84: 515
41. Hoch JC, Stern AS, Donoho DL, Johnstone IM (1990) J Magn Reson 86: 236

42. Högbom J (1974) Astron J Suppl 15: 517
43. Gladden LF, Carpenter TA, Klinowski J, Elliott RJ (1986) J Magn Reson 66: 93
44. Barna JCJ, Tan SM, Laue ED (1988) J Magn Reson 78: 327
45. Davies SJ, Bauer C, Hore JP, Freeman R (1988) J Magn Reson 76: 327, pertains to uniform sampling.
46. McKinnon GC, Burger C, Boesiger P (1990) Magn Reson Med 13: 145
47. Priestley MB (1981) Spectral analysis and time series, vols 1 and 2. Academic Press, London
48. Bos A. van den (1982) Parameter Estimation, Ch. 8 in: Sydenham PH (ed) *Handbook of measurement science, vol 1*. Wiley, London
49. Norton JP (1986) An introduction to identification. Academic Press, London
50. Tarantola A (1987) Inverse problem theory. Methods for data fitting and model parameter estimation. Elsevier, Amsterdam
51. Macovski A, Spielman D (1986) Magn Reson Med 3: 97
52. Börnert P, Dreher W (1990) J Magn Reson 87: 220
53. Veen JWC van der, Beer R de, Luyten PR, Ormondt D van (1988) Magn Reson Med 6: 92
54. Spielman D, Webb P, Macovski A (1988) J Magn Reson 79: 66
55. Spielman D, Webb P, Macovski A (1989) Magn Reson Med 12: 38
56. Horsfield MA, Hall C, Hall LD (1990) J Magn Reson 87: 319
57. Graaf AA de, JE van Dijk, Bovée WMMJ (1990) Magn Reson Med 13: 343
58. Evans AJ, Fischl R (1973) IEEE Trans AU-21: 61
59. Kumaresan R, Shaw AK (1985) Proc ICASSP, p 576
60. Kumaresan R, Scharf LL, Shaw AK (1986) IEEE ASSP-34: 637
61. Bresler Y, Macovski A (1986) IEEE ASSP-34: 1081
62. Park SW, Cordaro JT (1987) Proc ICASSP, p 1501
63. Golub GH, Pereyra V (1973) SIAM J Numer Anal 10: 413
64. Kung SY, Arun KS, Rao DVB (1983) J Amer Opt Soc 73: 1799
65. Barkhuysen H, Beer R de, Ormondt D van (1987) J Magn Reson 73: 553
66. Lawson CL, Hanson RJ (1974) Solving least squares problems. Prentice Hall, Englewood Cliffs
67. Klema VC, Laub AJ (1980) IEEE Trans AC-25: 164
68. Strang G (1988) Linear algebra and its applications, 3rd edn. Hartcourt Brace Jovanovitch, San Diego
69. Comon P, Kopp L (1989) Proc IEEE 77: 492 (See 'Author's Reply', p 493)
70. Cullum J, Willoughby RA (1985) Lanczos algorithms, vols. 1 and 2. Birkhäuser, Boston
71. Millhauser GL, Carter AA, Schneider DJ, Freed JH, Oswald RE (1989) J Magn Reson 82: 150
72. Hansen PC (1988) SVD of a Hankel Matrix. Internal Report, University of Copenhagen. See also Pijnappel WWF (1991) PhD Thesis, Univ. of Delft
73. Kats JM van, Vorst HA van der (1977) Automatic Monitoring of Lanczos schemes for symmetric or skew-symmetric eigenvalue problems. Acad. Comput. Centre, Report TR-7, Univ of Utrecht
74. Kailath T (1980) Linear systems. Prentice-Hall, Englewood Cliffs
75. Heinig G, Rost K (1984) Algebraic methods for Toeplitz-like matrices and operators. Birkhäuser, Boston
76. Schussheim AE, Cowburn D (1987) J Magn Reson 71: 371
77. Johnson BA, Malikayil JA, Armitage IM (1988) J Magn Reson 76: 352
78. Manassen Y, Navon G, Moonen CTW (1987) J Magn Reson 72: 551
79. Gorcester J, Freed JH (1988) J Magn Reson 78: 292
80. Luthon F, Blanpain R, Decorps M, Albrand P (1989) J Magn Reson 81: 538
81. Tirendi CF, Martin JF (1989) J Magn Reson 81: 577
82. Gesmar H, Led JJ (1989) J Magn Reson 83: 53
83. Zeng Y, Tang J, Bush CA, Norris JR (1989) J Magn Reson 83: 473
84. Madi ZL, Ernst RR (1988) J Magn Reson 79: 513
85. a) Rao BD, Hari KVS (1989) IEEE Trans ASSP-37: 1990
b) Rao BD (1990) IEEE Trans ASSP-38: 177
c) Li F, Vaccaro RJ, Tufts DW (1990) Proc ICASSP, p 2575
86. Tufts DW, Kot AC, Vaccaro RJ (1988) Proc ICASSP, p 2416
87. Kot AC, Parthasarathy S, Tufts DW, Vaccaro RJ (1987) Proc ICASSP, p 1549
88. Okhovat A, Cruz JR (1989) Proc ICASSP, p 2286
89. Pijnappel WWF, Beer de R, Ormondt D van (1991) in: Vaccaro RJ (ed) SVD and Signal Processing, II Rhode Island. Elsevier, Amsterdam, p 293

90. Barkhuysen H, Beer R de, Drogendijk AC, Ormondt D van, Veen JWC van der (1988) in: Maraviglia B (ed) Proc Int School of Physics "Enrico Fermi" on the *Physics of NMR Spectroscopy in Biology and Medicine*. Italian Physical Society, p 313
91. Beer R de, Ormondt D van, Mariën AJH, Luyten PR, Hollander JA den (1990) Proc SMRM, p 210
92. Beer R de, Bovée WMMJ, Graaf AA de, Ormondt D van, Chamuleau RAFM (1988) in: Deprettere EF (ed) SVD and signal processing, Elsevier (North Holland), Amsterdam, p 473
93. Lohman HFM (1989) MSc Thesis (in Dutch), Delft University of Technology
94. Barkhuysen H, Beer R de, Pronk BJ, Ormondt D van (1985) J Magn Reson 61: 284
95. Yan H, Gore JC (1988) J Magn Reson 80: 324
96. Henderson TL (1981) IEEE Trans ASSP-29: 982
97. Delsuc MA, Ni F, Levy GC (1987) J Magn Reson 73: 548
98. Yan H, Gore JC (1989) J Magn Reson 81: 400
99. Barkhuysen H, Beer R de, Bovée WMMJ, Brink AM van den, Drogendijk AC, Ormondt D van, Veen JWC van der (1986) in: Young IT et al. (eds) Signal processing III: Theory and applications. Elsevier, Amsterdam, p 1359
100. Bovée WMMJ, Beer R de, Graaf AA de, Montizaan J, Ormondt D van, Deutz NEP, Chamuleau (1987) Proc SMRM, p. 383
101. Galaszo JL, Bailey JE (1989) Biotechnol Tech 3: 13.
102. Gesmar H, Led JJ (1988) J Magn Reson 76: 183
103. Tang J, Norris JR (1988) J Magn Reson 79: 190
104. Dumelow T, Riedi PC, Abell JS, Prakash O (1988) J Phys F 18: 307
105. Hoffman RE, Levy GC (1989) J Magn Reson 83: 411
106. Sekihara K, Ohyama N (1990) Magn Reson Med 13: 332
107. Geen H, Wimperis S, Freeman R (1989) J Magn Reson 85: 620
108. Sharman KC, McClurkin GD (1989) Proc ICASSP, p 2716
109. Wu XL, Freeman R (1989) J Magn Reson 85: 414
110. Dennis JE (1984) Proc IEEE 72: 1765
111. Wolfe CM, Rust BW, Dunn JH, Brown IE (1987) Interactive nonlinear least squares program. National Bureau of Standards Report No NBS/TN-1238
112. Dennis Jr JE, Schnabel RB (1983) Numerical methods for unconstrained optimisation and nonlinear equations. Prentice Hall, Englewood Cliffs
113. Fraley C (1989) SIAM J Sci Stat Comput 10: 515
114. Box GEP, Tiao G (1973) Bayesian inference in statistical analysis. Addison-Wesley, Reading
115. Lee PM (1989) Bayesian statistics: an introduction. Oxford Univ Press, New York
116. a) Bretthorst GL (1990) J Magn Reson 88: 533
b) Bretthorst GL (1990) J Magn Reson 88: 552
c) Bretthorst GL (1990) J Magn Reson 88: 571
117. Reilly PJ, Wong KM, Reilly PM (1989) Proc ICASSP, p 2609
118. Dijk JE van, Beer R de, Ormondt D van, Bovée WMMJ (1990) Proc SMRM, p 1326
119. Montigny F, Brondeau J, Canet D (1990) Chem Phys Lett 170: 175
120. Graaf AA de, Bovée WMMJ (1990) Magn Reson Med 15: 305
121. Heineman FW, Berkowitz BA, Balaban RS (1990) Magn Reson Med 13: 490

Determination of Absolute Concentrations of Metabolites from NMR Spectra

Ernest B. Cady

Department of Medical Physics and Bio-Engineering, University College and Middlesex School of Medicine, London, United Kingdom

Table of Contents

In vivo NMR spectroscopy (MRS) has potential for the non-invasive estimation of absolute concentrations of metabolites in a variety of tissues. The additional information provided by absolute concentration estimates over and above that given by metabolite ratios and intracellular pH permits a deeper understanding of metabolic and pathological processes. Problems associated with the non-invasive, in vivo determination of absolute concentrations in tissue are discussed. The various methods which have been used to obtain this sort of information from biological material are dealt with in detail including techniques which have utilised spectroscopic localisation. Absolute concentration estimates obtained from various living tissues using a variety of MRS techniques are summarised. The results are compared with previously published values obtained using other techniques. The possibility that the choice of absolute concentration method can have an effect on the results obtained is discussed. It is concluded that techniques are now available that can both be applied non-invasively to a variety of living tissues in situ and provide reliable estimates of absolute metabolite concentrations. These developments should further enhance the role of in vivo MRS as a tool for clinical investigation and biological research.

List of Symbols

α	Greek lowercase alpha
β	Greek lowercase beta
γ	Greek lowercase gamma
δ	Greek lowercase delta
Δ	Greek uppercase delta
θ	Greek lowercase theta
ϕ	Greek lowercase phi
ρ	Greek lowercase rho
χ	Greek lowercase chi
π	Greek lowercase pi
τ	Greek lowercase tau
μ	Greek lowercase mu
ψ	Greek lowercase psi
$\sum$	Greek uppercase sigma (mathematical sum sign)
$\int$	mathematical integral sign
$\hbar$	Planck's constant/2π

1 Introduction

Although a great deal of useful information is obtained from relative concentrations, absolute concentrations can further enhance the usefulness of in vivo NMR spectroscopy (MRS). Because of the non-invasive nature of MRS, clinical measurements of absolute concentrations can be made (and repeated if required) on tissues such as neonatal brain that are normally inaccessible to biopsy. Furthermore, because observations can be made on tissue in situ, metabolite concentrations can be obtained without the risk of the freezing artefacts that can affect biopsy results. With the absence of reliable direct methods of absolute quantitation, early attempts to estimate metabolite concentrations in human tissue by MRS (mainly in muscle [1–5] and brain [6–10]) were indirect and depended on knowledge of either the total mobile phosphate (P_{tot}) concentration or the concentration of a particular, normally stable metabolite (e.g. adenosine triphosphate (ATP)). For instance, in studies of normal, resting, human muscle [1, 3] phosphocreatine (PCr) and inorganic orthophosphate (Pi) concentrations have been estimated from relative peak areas assuming either [ATP] = 5.5 mmol kg^{-1} wet wt or [P_{tot}] = 49.5 mmol kg^{-1} wet wt (see Table 1). In calculations based on [ATP], absolute concentrations were derived using the β-ATP resonance because, unlike the γ and α resonances, the β peak does not have contributions from other underlying resonances. (The γ and α peaks include contributions from adenosine diphosphate (ADP) and the α peak also includes contributions from nicotinamide dinucleotides (NAD + NADH).) In the calculations based on [P_{tot}] the following was assumed:

$$[P_{tot}] = [PCr] + [Pi] + 3 \times [ATP] + 2 \times [NAD + NADH] \qquad (1)$$

These two approaches gave very similar results and favourable comparison with biopsy data. In fact, due to the inadequate cessation of metabolic processes prior to and during the freezing of biopsy samples, it appears that MRS has

Table 1. The composition of resting human forearm muscle

Concentrations (mean ± SE; mmol kg^{-1} wet wt)	*Biopsy*	*^{31}P MRS* ([ATP] = 5.5 mmol kg^{-1} wet wt)	*^{31}P MRS* ([Mobile ^{31}P] = 49.5 mmol kg^{-1} wet wt)
	(n)	(n = 7)	(n = 7)
[ATP]	5.5 ± 0.1 (81)	5.5	5.1 ± 0.1
[PCr]	17.4 ± 0.2 (81)	29.0 ± 0.7	27.4 ± 0.2
[Pi]	10 (3)	4.4 ± 0.3	4.3 ± 0.3
[PCr + Pi]	31.6 ± 3.3 (11)	33.4 ± 0.8	31.7 ± 0.3
[PDE]		0.7 ± 0.1	0.8 ± 0.1

Reprinted with permission from Dawson MJ [1], Plenum Publishing Corp.

the potential for providing better measurements of [Pi] and [PCr] than biopsy. However, in many important situations these simple approaches are inadequate because of the invalidity of the assumption of stability for the concentration reference and therefore alternative methods are required.

In order to obtain estimates of absolute concentrations which reflect the actual concentrations of metabolites within living tissue, care and attention has to be paid to every stage of the MRS technique. In fact during the development of a useful absolute quantitation protocol the entire MRS process including patient handling, data acquisition and spectrum analysis has to be considered in detail. The accuracy obtained when applying absolute concentration data acquisition protocols will depend on factors such as the stability of spectrometer gain and coil tuning and matching. Correction factors for effects due to both incomplete spin–lattice relaxation and spin–spin relaxation (if spin echo or stimulated echo techniques are used) must be properly applied using the best T_1 and T_2 values available. System calibrations may be necessary and must be performed as diligently as the in vivo data collection itself. Effects which the method of data collection may have on resonance areas (e.g. due to nuclear relaxation during delays required for eddy current decay in localised spectroscopy) must be investigated and corrected for if possible. Furthermore, if the data analysis method whereby peak areas are measured is inadequate then this will affect the accuracy of the results.

A general review of absolute quantitation techniques has been presented by Tofts and Wray [11]. With the development and in vivo application of useful tools for spectroscopic localisation and the incorporation in these of absolute quantitation methods, the time has arrived for a further review. The following sections will be devoted mainly to describing the methods which have been shown to have potential for clinical investigations and in vivo research applications. However, in order to gain a full understanding of the problems associated with absolute quantitation in vivo, the basic methodologies utilised for performing the same function in vitro will be briefly described. Some absolute quantitation results obtained from various tissues in vivo will be described in Sect. 3.

1.1 Absolute Concentrations—Their Importance in the Interpretation of Spectra

Absolute concentrations are of particular importance in clinical or research situations in which gross tissue destruction or selective terminal cellular damage is to be expected due to various pathological causes. Qualitative descriptions of reduced NMR signal following hypoxic/ischaemic episodes in both ^{31}P studies of neonatal brain [12, 13] and ^{1}H studies of adult human brain [14] indicate the potential for increasing the information content of NMR spectra if absolute quantitation methods are used. Similarly reduced overall signal [15], and attempted quantitations [16, 17] have been reported also in NMR studies of Duchenne muscular dystrophy. Furthermore, absolute concentration estimates allow the calculation of thermodynamic parameters such as the phosphorylation potential [18, 19] and the free energy of ATP hydrolysis (ΔG) [5, 19].

Potential applications of direct absolute quantitation methods in the clinic include the sequential monitoring and estimation of damage to brain tissue after a hypoxic or ischaemic episode (e.g. birth-asphyxia, stroke, etc.) or following the progress of degenerative conditions such as Duchenne muscular dystrophy. One of the sequelae of perinatal hypoxia/ischaemia is the eventual appearance of porencephalic cysts on ultrasound scans and on images of the brain obtained using other modalities. At some stage, either during or following the initial insult, permanent tissue damage may be sustained which may affect the entire cell population or only a fraction thereof. This may affect a large volume of tissue or may be more focal in nature. At some stage in the proceedings, cellular contents are lost and damaged tissue in the affected region may be replaced by cysts containing cerebro-spinal fluid (CSF). CSF contains very low concentrations of phosphorylated metabolites. The net result of this in absolute quantitation terms is that the population of viable cells in a particular volume becomes depleted and bulk tissue metabolite concentrations (as measured in units of mmol kg^{-1} wet wt or mmol l^{-1} wet) are reduced. This may result in the detection of reduced overall NMR signal. The example of perinatal hypoxia/ischaemia is one of the situations in which the interpretation of absolute quantitation results is relatively easy. More complicated situations may arise if the development of the lesion includes the infiltration of remaining viable tissue by adipose, connective or other tissues which may have low, but not negligible, metabolite concentrations.

1.2 Problems with Absolute Concentration Estimation by NMR Spectroscopy

Although the title of this contribution refers to "the determination of metabolite concentrations", due to the poor resolution encountered in most biological spectra obtained in vivo, this is quite difficult in practice. Peak areas measured from spectra will nearly always include contributions from many metabolites although in certain situations a contribution from a particular compound may predominate. For instance, in a ^{31}P brain spectrum acquired using a 1.9 T system, the unresolved phosphomonoester (PME) peak (see Fig. 1) contains contributions from at least 10 metabolites (see Fig. 2 and Table 2) [20, 21]. The γ and α-nucleoside triphosphate (NTP) resonances (peaks 5 and 6 in Fig. 1) also contain contributions from other metabolites and it is for this reason that, when either NTP or ATP is utilised as an internal concentration reference (ICR), the β resonance is used. Furthermore, the γ, α and β-NTP resonances which are usually attributed to the three phosphate groups on ATP may also in part originate from other NTPs such as cytidine triphosphate (CTP), uridine triphosphate (UTP) or guanosine triphosphate (GTP) [22]. In brain these may

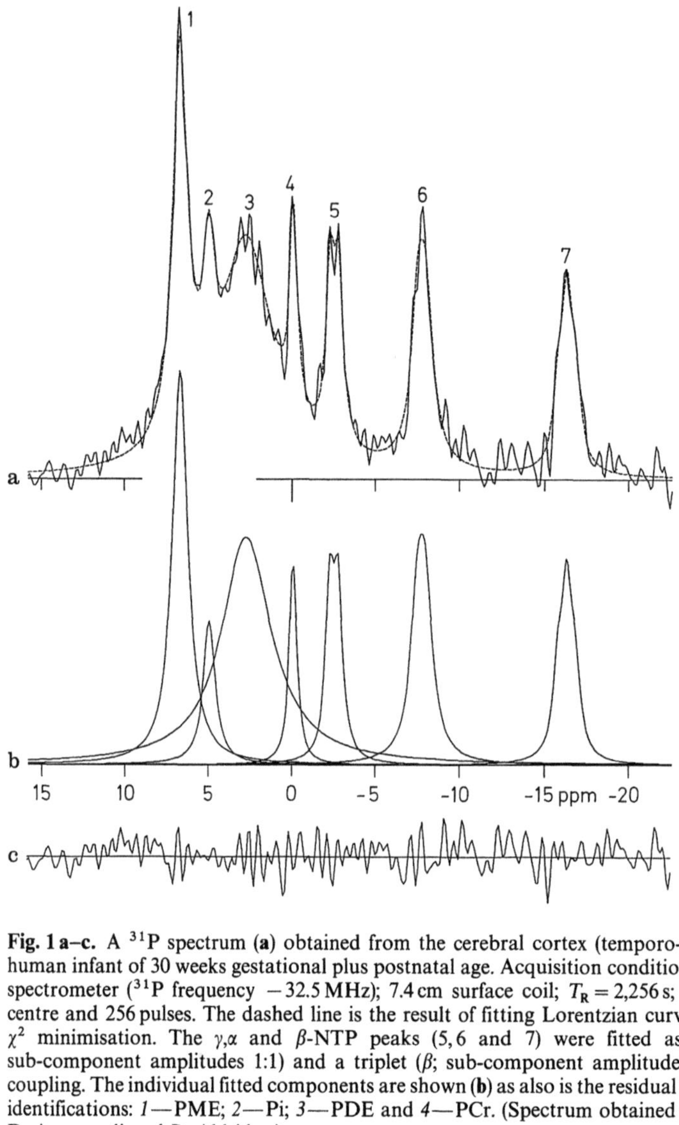

Fig. 1 a–c. A ^{31}P spectrum (**a**) obtained from the cerebral cortex (temporo-parietal) of a normal human infant of 30 weeks gestational plus postnatal age. Acquisition conditions: 1.9 T ORS-Bruker spectrometer (^{31}P frequency −32.5 MHz); 7.4 cm surface coil; $T_R = 2{,}256$ s; 90° flip angle at coil centre and 256 pulses. The dashed line is the result of fitting Lorentzian curves to the spectrum by χ^2 minimisation. The γ,α and β-NTP peaks (5, 6 and 7) were fitted as doublets (γ and α; sub-component amplitudes 1:1) and a triplet (β; sub-component amplitudes 1:2:1) with 16 Hz J coupling. The individual fitted components are shown (**b**) as also is the residual (**c**). Further resonance identifications: *1*—PME; *2*—Pi; *3*—PDE and *4*—PCr. (Spectrum obtained in collaboration with D. Azzopardi and R. Aldridge.)

contribute about 20% to the total NTP pool. The prominent peak seen at about 2.0 ppm in ^1H spin echo spectra of brain which is attributed to *N*-acetylaspartate (NAA) has been reported to contain between 10% and 25% *N*-acetylaspartylglutamate (NAAG) depending on the stage of development [23]. The presence of multiple components contributing to individual peaks in NMR spectra makes it difficult to estimate absolute concentrations from peak areas both from the point of view of the reliability of ICRs themselves (such as

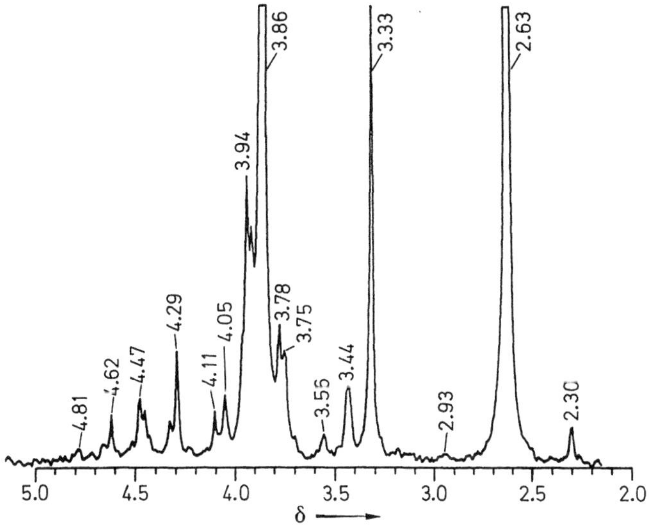

Fig. 2. The PME region of a proton-decoupled ^{31}P NMR spectrum acquired at 4.7 T (^{31}P frequency = 81 MHz) from a perchloric acid extract of guinea pig brain. The major signals are off-scale so that the minor sugar phosphate signals may be observed. Chemical shifts are referenced to glycerol-3-phosphorylcholine (GPC) at −0.13 ppm (PCr at −3.12 ppm). Resonance identifications can be obtained from Table 2 by converting from the GPC chemical shift reference to a PCr chemical shift reference. (Reprinted with permission from J. Neurochem. P-31 nuclear magnetic resonance analysis of brain: The perchloric acid extract spectrum. Glonek T, Kopp SJ, Kot E, Pettegrew JW, Harrison WH, Cohen MM 39: 1210 [20]. Raven Press.)

Table 2. Resonances contributing to the low-field in vivo "phosphomonoester" peak as observed by high-resolution ^{31}P MRS of guinea pig brain slices

Resonance	Chemical shift[a] (ppm)	Fraction of total extractable phosphorus (%)
U	7.74	0.97 ± 0.07
Hexose 6-phosphates	7.59	0.74 ± 0.09
U	7.45	0.48 ± 0.13
α-Glycerophosphate	7.41	0.89 ± 0.43
Fructose 1,6-diphosphate	7.23	2.54 ± 0.37
Fructose 1,6-diphosphate	7.17	
U[b]	7.06	2.17 ± 0.83
Phosphoethanolamine	6.96	8.39 ± 0.96
Inosine 5′-monophosphate	6.90	1.60 ± 0.14
Adenosine 5′-monophosphate	6.87	0.83 ± 0.24
NADP 2′-phosphate	6.68	0.62 ± 0.19
U	6.56	0.78 ± 0.04
Choline phosphate	6.45	1.66 ± 0.10

Mean ± SE; U—unidentified compound; [a] referenced to PCr; [b] complex resonance signal. Reprinted with permission from Cohen MM et al. [21], Raven Press.

NAA) and from the consideration that the peak of interest itself may not be due to a single metabolite. Hence the title of this chapter is somewhat erroneous in that the spectroscopist will almost always be estimating an absolute concentration pertaining to a cocktail of metabolites which together produce a single irresolvable peak in the spectrum. Further problems can arise when utilising certain ICRs. If the molarity of the ICR (e.g. ATP, P_{tot}, NAA or tissue water) changes due to physiological or pathological processes (for instance ATP depletion due to fatiguing exercise in muscle studies) then the internally referenced absolute quantitation system breaks down and alternative methods have to be sought.

2 Methods

The limitations of the use of ATP or P_{tot} as concentration references in ^{31}P spectroscopy have been discussed in Sect. 1. The main problem encountered with their use is that the concentration in the tissue under investigation may not be the same as the assumed value or it may alter during the course of the study. Hence two of the prime requirements for a useful concentration reference are an accurately known concentration and stability. It is also important to know how many nuclei per molecule contribute to the reference resonance and how many to the resonance for which an absolute concentration estimate is required. The resonance areas must be normalised according to the number of nuclei contributing per molecule. In many applications, the reference resonance should ideally be a single peak with a chemical shift well away from the spectrum of interest. This minimises contamination of the tissue spectrum by the wings of the reference peak and also facilitates measurement of the reference peak area. (If this area cannot be measured accurately, the error involved will compound with the errors associated with the measurements of areas for the peaks of interest in the tissue spectrum.) With respect to the reduction of systematic errors, it is advantageous for the reference substance to also have a short T_1 and thereby avoid the necessity of applying a large saturation correction to the area of the reference peak.

When surveyed in detail, there are several approaches to absolute quantitation in vivo. In general, these methods can be split up into two groups utilising either external concentration references (ECR) or ICRs. ECRs comprise of a phial, capillary or other convenient receptacle containing a standard solution and these are located outside the sample but within the sensitive volume of the detection coil. An ICR is either a stable metabolite occurring naturally in the tissue (endogenous) or a convenient compound which can be introduced into the organ being studied (exogenous). For reasons of safety only ECRs and endogenous ICRs have found application in clinical studies.

2.1 External Concentration References

External reference samples have been more commonly employed to determine chemical shifts but have been used also for estimating metabolite concentrations. Usually a capillary, filled with a standard solution of known volume, is placed close to the sample and, if possible, the sample and the reference should both be in a region of uniform coil sensitivity. This will not always be possible for in vivo studies and the standard has often to be placed away from the sample (e.g. immediately on the other side of a surface coil). There are many difficulties with the use of ECRs of this type including estimation of the volume of the tissue from which signal originates and the spatial variation of sensitivity and flip angle which is never exactly uniform (particularly when surface coils are used). The amplitude of the signal acquired from a sample (S_s) can be written as:

$$S_s = k\rho_s n V_s \phi_s \tag{2}$$

where k depends on the experimental conditions (e.g. the nucleus investigated, coil efficiency, receiver gain etc.), ρ_s is the sample concentration, n is the number of accumulated free induction decays (FIDs), V_s is the sample volume and ϕ_s is the factor by which the sample signal is reduced due to saturation of the nuclear spin system and flip angle (θ) given by [24]:

$$\phi_s = \frac{\sin(\theta)(1 - \exp(-T_R/T_1))}{1 - \cos(\theta)\exp(-T_R/T_1)} \tag{3}$$

where T_R is the pulse repetition time ($\gg T_2$). For an ECR, assuming that the coil sensitivity is the same for both sample and reference, the sample concentration is given by:

$$\rho_s = \frac{S_s V_{ref} \phi_{ref} \rho_{ref}}{S_{ref} V_s \phi_s} \tag{4}$$

The main problem with this approach is that V_s may not be known very well. In certain applications, the spectrometer response can be calibrated by using a standard solution in place of the tissue to be investigated. Problems of estimating tissue volume and accounting for the effects of coil loading produced by the sample conductivity are also encountered in this method. Changes in coil efficiency can be monitored by use of a further external reference common to both the in vivo and the phantom data collections. This will be dealt with in more detail in Sect. 2.4. Two interesting applications of ECRs have been to estimate [PCr] in a perfused skeletal muscle preparation [25] (see Sect. 2.3) and to estimate [ATP] in 3-dimensional spectroscopic imaging [26] (see Sect. 2.4). In both these applications, the reference substance resonated within the frequency band of the spectrum of interest. However, due to the data acquisition protocols employed, this was not a problem. Other applications of ECRs have employed substances which resonate outside the spectrum of interest. These have included the use of hexachlorocyclotriphosphazene (HCCTP; also known as phosphonitrilic chloride trimer; $\delta_{HCCTP} = 22.9$ ppm relative to PCr) [27] dissolved in C_6D_6 with added chromium acetylacetate (to reduce T_1) for the estimation of phosphorylated metabolites in perfused rabbit hearts [28, 29] and the same substance (dissolved in $CHCl_3$ with the same relaxation agent) in studies of human brain using slice-selective spectroscopy [30] (see Sect. 2.4 for further details). A further use of an ECR has been in localised studies using image selected in vivo spectroscopy (ISIS) [31]. The incorporation of ISIS into an absolute quantitation protocol required the use of an ECR for concentration calibration and a further external reference for monitoring changes in coil efficiency due to sample loading [32–34]. For the purpose of monitoring coil efficiency, hexamethylphosphorustriamide (HMPT; $\delta_{HMPT} = 125$ ppm relative to PCr) [32, 33] and methylphosphonic acid (MPA; $\delta_{MPA} = \sim 35$ ppm relative to α-ATP) [34, 35] were used in these studies. Absolute quantitation using ISIS will be dealt with in greater detail in Sect. 2.4.

2.2 Internal Concentration References

In many situations, and especially when localisation techniques are unavailable, methods using ICRs may be the best choice. The reference substance is within the tissue of interest (either naturally or introduced) and therefore can be located in the same region of the static magnetic field and the coil sensitive volume. This can avoid many of the quantitation difficulties associated with ECRs. A particular advantage with ICRs is that the volume occupied is often the same as that of the sample. Hence, for an ICR, Eq. (4) can be re-written as:

$$\rho_s = \frac{S_s \phi_{ref} \rho_{ref}}{S_{ref} \phi_s} \tag{5}$$

One problem however is that the absolute concentration of the ICR may not be known as precisely as that of an ECR. Further problems are that the [ICR] may change with pathology or physiological state and that the reference peak may contain contributions from other substances. A particular application of ICRs has been to estimate absolute concentrations in extracts. By adding a certain amount of a particular substance with a resonance in the spectrum to the extract and noting the increased area of its resonance, the original extract concentration can be determined. There are two types of ICR substances: homonuclear (in which the substance is observed as part of the spectrum of interest) and heteronuclear (for which a spectrum of another nuclear species must be collected and therefore the coil must have a multi-nuclear capability).

2.2.1 Homonuclear Internal Concentration References

These have the distinct advantage that they mostly resonate near or within the frequency band of the spectrum of interest. The process of quantitation is usually protected from errors due to system calibration inaccuracies, gain corrections or changes in spectrometer sensitivity because these factors affect the entire spectrum uniformly. However, a remaining source of potential error is non-uniformity of radiofrequency (RF) filter characteristics over the observed frequency band.

The uses of ATP or the total mobile phosphate as ICRs for ^{31}P spectroscopy and their limitations (dependences on the maintenance of normal physiological conditions and on the assumption that the tissue under investigation does not contain a necrotic fraction) have already been described. Because of these limitations, it appears that, except in physiologically stable circumstances, a heteronuclear approach may be best for absolute quantitation of ^{31}P spectra via internal referencing although one report on the invasive use of MPA in studies of rat kidney has been published [35] (see Sect. 2.3).

Homonuclear ICRs have been used to estimate absolute concentrations in ^{1}H NMRS studies. For this purpose mainly the combined PCr + creatine (Cr) signal [36–41] or NAA [38, 42] resonance have been utilised for in vivo brain studies. The water signal is usually reduced as much as is possible in ^{1}H NMRS studies. However there may be potential for the use of this resonance for absolute quantitation purposes [38]. A further use of the combined PCr + Cr resonances has been to quantitate lactate in ^{1}H studies of skeletal muscle [43]. In ^{1}H studies of brain extracts, absolute concentrations have been estimated by referencing to the methyl doublet of alanine after determining the concentration of this amino acid enzymatically [44] and by the addition of 1 mM 2,2,3,3-tetradeuteropropionate (TSP; normally used as a chemical shift reference) [23].

2.2.2 *Heteronuclear Internal Concentration References*

The estimation of metabolite concentrations by referencing to the signal from a different nucleus to that observed in the spectrum of interest has achieved some degree of success. The main application so far has been the use of tissue water for the determination of absolute concentrations of phosphorus metabolites [45–51] although, in principle, the method could be applied to other nuclei. Because concentrations are obtained by comparing signals obtained from different nuclei and hence from separate spectra, great care has to be taken to ensure that the spectrometer observe channels have stable gains. The application of this method depends on knowledge of the tissue water concentration, matching the spatial sensitivities of the coil for ^{31}P and ^{1}H, calibration of the ^{31}P and ^{1}H channels of the spectrometer using standard solutions and, if necessary, taking into account differential loading of the coil by the subject and the calibration solutions. If the spatial sensitivities of the coil are successfully matched for ^{31}P and ^{1}H, then this approach is applicable for situations with inhomogeneous B_1 (e.g. surface coils). This approach will be dealt with in more detail in Sect. 2.3.

2.3 Non-Localised NMR Spectroscopy

This section will deal with techniques which have been used for the acquisition of spectra without sophisticated localisation methods. In other words, situations in which localisation of the signal was due to the use of an isolated sample or/and the spatial sensitivity of the coil (e.g. that of a surface coil). Section 2.4 will describe applications in which gradient mediated localisation has been employed.

In 1977, Dawson, Gadian and Wilkie published a pioneering paper on the application of ^{31}P NMRS to the study of the contraction and recovery of perfused, living muscles [25]. In order to estimate [PCr] in the muscle fibres solely by NMR techniques, they perfused the muscles with Ringer's solution containing 10.0 mM phosphate buffer. Spectra were collected as shown in Fig. 3, i.e. muscle and perfusate (A) and perfusate only (B). In order to calculate [PCr], a resting muscle spectrum was also collected without phosphate buffer. This latter spectrum permitted the estimation of Pi/PCr in the resting muscle ($\mathrm{Pi/PCr_{rest}}$). From these spectra, [PCr] was calculated in the following manner. In spectrum (A), the Pi contributed from the muscle tissue is given by:

$$\mathrm{Pi_{muscle}} = x(\mathrm{Pi/PCr_{rest}}) \tag{6}$$

The peak area due to Pi in the perfusate is therefore:

$$\mathrm{Pi_{perf}} = y - x(\mathrm{Pi/PCr_{rest}}) \tag{7}$$

Hence $z - y + x(\mathrm{Pi/PCr_{rest}})$ is the difference in area of $\mathrm{Pi_{perf}}$ between spectra (a)

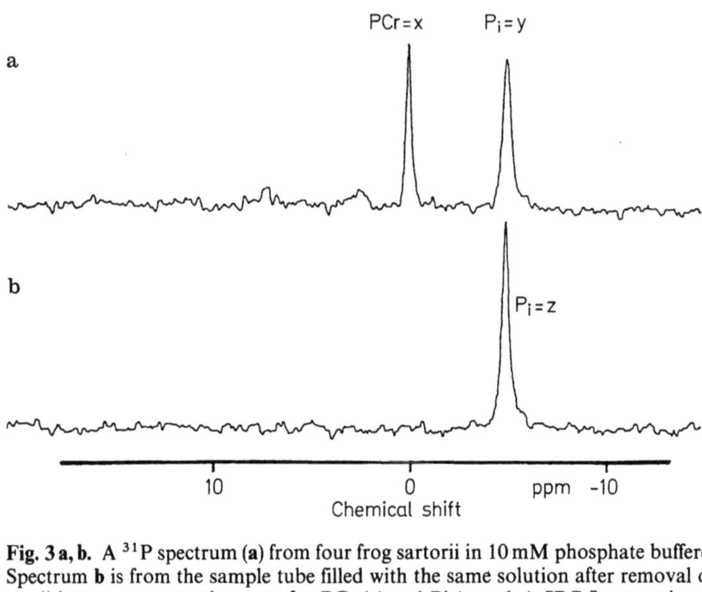

Fig. 3 a, b. A ^{31}P spectrum (**a**) from four frog sartorii in 10 mM phosphate buffered Ringer solution. Spectrum **b** is from the sample tube filled with the same solution after removal of the muscles. The small letters represent the areas for PCr (x) and Pi (y and z). [PCr] was estimated as described in the text. Acquisition conditions: 7.5 T (^{31}P frequency = 129.2 MHz); T_R = 16 s; 238 pulses. (Reprinted with permission from Dawson et al. [25], The Physiological Society (London).)

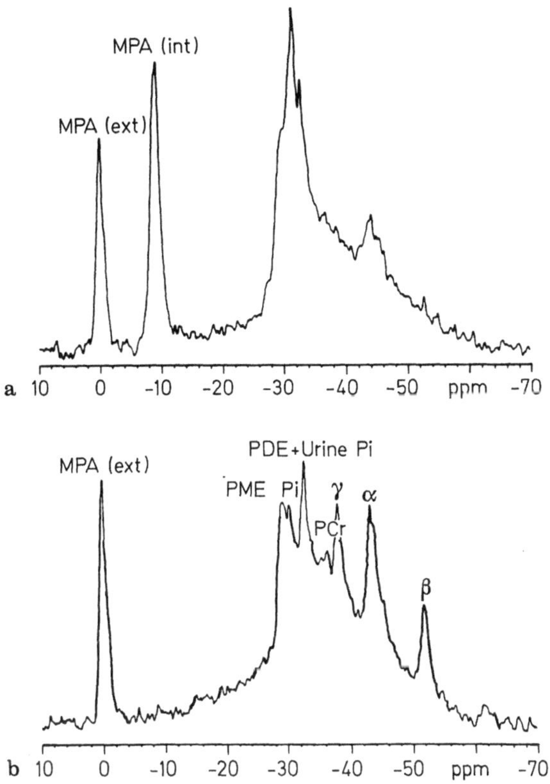

Fig. 4 a, b. ^{31}P NMR spectra of rat kidney in vivo before (**b**) and after MPA (4.0 ml of 150 mM i.v.) infusion and death (**a**). Spectra were obtained using a two turn implanted solenoidal coil with $T_R = 22$ s and 64 pulses. The peaks labelled γ, α and β are resonances originating from the three phosphate groups of ATP. The MPA(int) peak is due to MPA infused i.v. as an ICR. MPA(ext) represents MPA ($pH < 2$) in a capillary attached to the coil before implantation. (Reprinted with permission from Shine et al. [35], Academic Press, Inc.)

and (b). This part of the Pi peak area originates from the perfusate Pi which occupies the same volume as the sample of muscle tissue. The PCr signal comes from this same volume and hence:

$$[\mathrm{PCr}] = \frac{x}{z - y + x(\mathrm{Pi}/\mathrm{PCr}_{\mathrm{rest}})} \times 10\,\mathrm{mmol\,l^{-1}}\ \text{fibre volume} \qquad (8)$$

From their measurements, Dawson, Gadian and Wilkie estimated a [PCr] of 23.5 mmol l^{-1} fibre volume which agreed well with a value of 21.2 mmol kg^{-1} frozen muscle obtained by chemical analysis. In this study corrections for saturation were not applied because spectra (a) and (b) were acquired using a T_R of 16 s which was considered to be much greater than the T_1s of the metabolites concerned. In this example of the use of an ECR, the reference did not satisfy one of the criteria usually required i.e. the perfusate Pi had almost the same chemical shift as the intracellular Pi in the muscle fibres. However, because a substitution technique was used, this was not important.

In ^{31}P NMRS of in situ rat kidney, MPA has been utilised as a homonuclear ICR [35]. In these studies, fully relaxed spectra were obtained from the kidney using an implanted, two turn solenoidal coil (see Fig. 4a). This was followed by intravenous infusion of MPA for 30 min. A further spectrum (see Fig. 4b) was obtained after sacrificing the animal which ensured that a steady state renal MPA level was maintained. From these two spectra the ratio of the MPA and β-ATP peak areas was obtained. The absolute concentration of renal MPA was estimated using an NMR method in which the excised, MPA treated, kidney was placed in a phosphate-calibrated solenoidal coil (see Fig. 5). [ATP] was then estimated from [MPA(int)] and the ratio MPA(int)/β-ATP.

Homonuclear ICRs have been used also for in vivo ^{1}H NMRS investigations. Successful quantitation of histidine in the brain tissue of histidinemic mice has been achieved using combined surface coil spin echo [52] and binomial water suppression [53] techniques. Both PCr + Cr and NAA signals were utilised as concentration references [37]. Because spin echo techniques were used, in order to obtain quantitative values it was necessary to measure the T_2s of the resonances so that the areas of peaks in the spectrum could be related to the tissue concentrations. It was necessary also to apply corrections for incomplete spin lattice relaxation and an additional correction factor for the frequency dependence of the water suppression pulse sequence. Good correlation was found between the NMR results obtained both from in vivo tissue and extracts and with subsequent measurements on brain extracts using amino acid analysis (see Fig. 6).

A further interesting application of absolute quantitation in ^{1}H NMR has been the estimation of brain lactate concentration. The methyl proton resonance of lactate is often overlapped by other resonances, mainly from lipids. In spite of this problem, the application of various spectral editing techniques [54–58] has

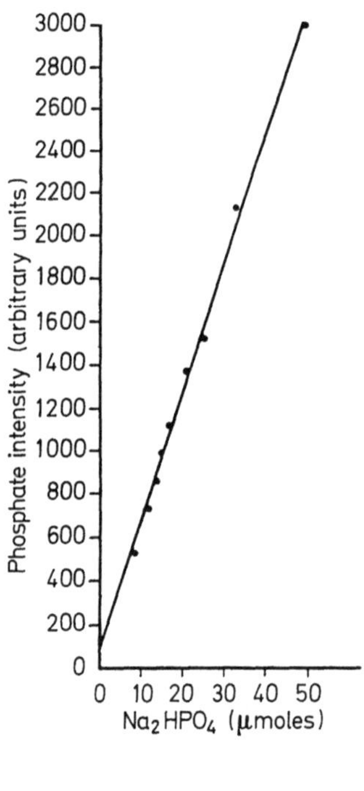

Fig. 5. After the spectra in Fig. 4 had been collected, the kidney was excised and placed in a six turn solenoidal coil (sensitive volume ~1.7 ml). This coil had previously been calibrated producing the above plot which shows ^{31}P NMR peak intensities versus μmol of phosphate for 1.7 ml solutions of Na_2HPO_4(Gd(DPTA)) doped to reduce the T_1 to ~2.2 s. (Reprinted with permission from Shine et al. [35], Academic Press, Inc.)

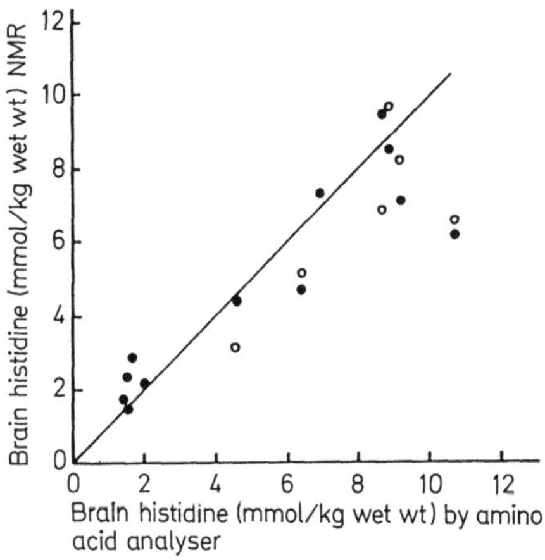

Fig. 6. A plot of concentration as measured in the brains of histidinemic mice by ^{1}H NMR at 8.5 T versus histidine concentration determined by amino acid analysis of extracts. ●—in vivo NMR data; ○—extract NMR data. The straight line represents the theoretical 1:1 equivalence. The 5 points at about 2 mmol kg^{-1} wet wt were obtained from *his/his* mice fed on a normal diet whereas the remaining points were obtained from *his/his* mice fed on a histidine enriched diet. (Reprinted with permission from Gadian et al. [37], Academic Press, Inc.)

enabled very convincing in vivo quantitation of brain lactate. Spectral editing reduces the signals from unwanted resonances whilst preserving the resonance of interest. Many methods depend on inhibiting the phase modulation [59] of the multiplet of interest on alternate data acquisitions. The phase modulated data is then subtracted from that with phase modulation inhibited to obtain an edited spectrum containing only the multiplet of interest. Phase modulation is inhibited either by using narrow band RF to saturate the coupled spins [54, 55, 57] or by employing binomial water suppression sequences [53, 56] so that the coupled spins remain unperturbed. The advantage of the use of spectral editing methods for absolute quantitation is that the resonance in the edited spectrum originates entirely from the metabolite of interest and its peak area contains no contributions from other metabolites. The uses of the resonances of Cr, NAA and brain tissue water as ICRs have been compared using binomial water suppression in the spectral editing sequence [38]. By comparison with lactate concentrations obtained by NMR measurements of extracts using added Na-2,2-dimethyl-2-silapentane-5-sulfonate as a concentration standard, NAA was found to be the best ICR of the three investigated. However, care must be taken with the use of NAA as a concentration reference in brain tissue because it has been shown that its concentration depends both on brain development [23, 60–62] and on anatomical location [60, 61]. Furthermore (as noted in Sect. 1.2), the NAA peak may include a contribution from NAAG [23].

As mentioned in Sect. 2.2.2, the use of tissue water as a heteronuclear ICR [45] has achieved a degree of success. Because this technique can be employed in applications with inhomogeneous B_1 (e.g. surface coils), it has found use for in vivo studies [46–51]. However, this approach cannot be applied in all situations. Two main requirements for its application are that the tissue water concentration does not change significantly as a result of pathological or physiological processes and that neither ^{1}H nor ^{31}P spectra are contaminated by signal from outside the tissue of interest. The nuclear magnetisation (M_0)

of a sample containing N spins at thermal equilibrium is given by [63]:

$$M_0 = N\gamma^2\hbar^2 I(I+1)B_0/3(k_B T_s) \tag{9}$$

where γ is the gyromagnetic ratio, $\hbar$ is Planck's constant divided by 2π, I is the nuclear angular momentum quantum number (spin), B_0 is the static magnetic field strength, k_B is Boltzmann's constant and T_s is the sample temperature. Hence the magnetisation is directly proportional to N (the number of spins of the nucleus under investigation in the sample) and it is therefore proportional to concentration also. If the detection coil is used both for transmitting and receiving and has a uniform RF field then the observed signal strength will be directly proportional to M_0 (and hence to N) provided the sample volume (V_s) is small compared with the sensitive volume of the coil (V_c). The signal amplitude (S) is given by [45]:

$$S = \mathrm{K}(B_1)_{xy}\sin(\theta)N \tag{10}$$

where K is a constant including the terms apart from N in Eq. (9), spectrometer gains and correction for saturation if necessary. $(B_1)_{xy}$ is the component of B_1 in the x, y plane produced by unit current flowing in the coil and θ is the flip angle. By substituting $\rho = N/(A_0 V_s)$ in Eq. (10), where ρ is the molar spin concentration and A_0 is Avogadro's number, and taking into account the fact that, in most practical situations (e.g. for a surface coil), neither B_1 nor θ will be uniform throughout the sample, one obtains on integrating over V_s:

$$S = \rho \int_{V_s} [\mathrm{K}'(B_1)_{xy}\sin(\theta)]\,\mathrm{d}V \tag{11}$$

where $\mathrm{K}' = A_0 V_s \mathrm{K}$. If the RF coil is tunable for two different nuclei (e.g. ^{31}P and ^{1}H) then $(B_1)_{xy}$ has an identical spatial distribution for both resonant frequencies [64–66]. Furthermore, if it is arranged that the flip angles at the coil centre are the same for both nuclei (by adjusting the amplitude and/or duration of the transmitter pulse) then θ will have the same spatial dependence at both frequencies. Hence the integral of $(B_1)_{xy}\sin(\theta)$ over V_s will also be the same for both nuclei. Therefore, the method can be used with coils having non-uniform B_1. Considering Eq. (11) applied separately to ^{31}P and ^{1}H, on division one obtains:

$$\frac{S_{31P}}{S_{1H}} = \frac{\mathrm{K}'_{31P}}{\mathrm{K}'_{1H}} \times \frac{\rho_{31P}}{\rho_{1H}} = D \times \frac{\rho_{31P}}{\rho_{1H}} \tag{12}$$

For this equation to hold true, identical steady state saturation conditions (ideally fully relaxed) must apply to both nuclei. It is an easy task to perform calibrations on standard solutions covering a range of [Pi] values (suitably loaded to imitate tissue i.e. with an ionic concentration equivalent to ~ 150 mM NaCl). For each standard solution S_{Pi}/S_{H_2O} is measured and plotted as a function of [Pi]. The relationship between S_{Pi}/S_{H_2O} and [Pi] should be linear. The slope is D/ρ_{H_2O} and hence the calibration constant can be determined. If tissue water is used as the ICR then Eq. (12) becomes:

$$\rho_{31P} = \frac{S_{31P}}{S_{H_2O}} \times \frac{111.1 \times f_{H_2O}}{D} \text{Mole}\,l^{-1} \tag{13}$$

where f_{H_2O} is the tissue water fraction (0.7 to 0.9 in most body tissues) and 111.1 is the molarity of pure water. This approach has recently been refined to take into account coil loading by conductive samples and the resultant changes in coil quality factor (Q) at the ^{31}P and ^{1}H frequencies [46, 67]. In these studies the coil was primarily tuned for ^{31}P; coupling to the sample was strong at this frequency and hence the effect of coil loading was significant. (As loading increases, coil sensitivity is degraded, $-Q$ is reduced, and a pulse of longer duration and/or larger amplitude is required to maintain a constant flip angle at the coil centre.) Tuning for ^{1}H was achieved by the incorporation in the coil circuitry of a selected length of co-axial cable and additional variable tune and match capacitors [68]. This gave only weak coupling with the sample at the ^{1}H frequency; the dependence on loading was minimal but signal strength was greatly reduced. A RF pulse of relatively large amplitude or duration was required to produce the same ^{1}H flip angle as that used for ^{31}P in order to match the spatial sensitivity. Coil loading at the ^{31}P frequency was accounted for by calibration of the match capacitance. However, if the coil is strongly coupled to the sample for both ^{31}P and ^{1}H, loading produces degradation of sensitivity and reduces θ for both nuclei. The ^{31}P signal at the pre-amplifier input is given by [69]:

$$S_{31P} = k_{31P}\rho_{31P}\frac{C_{31P}}{C_{m,31P}} \tag{14}$$

where k_{31P} is a constant dependent on the nucleus, coil and sample geometry and flip angle, $C_{m,31P}$ is the match capacitance and $C_{31P} = C_{m,31P} + C_{tune}$ where C_{tune} is the tuning capacitance. For a coil that is also strongly coupled for protons, a similar equation applies to the ^{1}H signal (S_{1H}) and the ratio of the ^{31}P and ^{1}H pre-amplifier input signals is given by:

$$\frac{S_{31P}}{S_{1H}} = k'\frac{C_{31P}}{C_{m,31P}} \times \frac{C_{m,1H}}{C_{1H}} \times \frac{\rho_{31P}}{\rho_{1H}} \tag{15}$$

where $k' = k_{31P}/k_{1H}$. Further RF coil theory [69] gives the relationship:

$$\frac{C_{31P}}{C_{m,31P}} = \frac{4\theta_{31P} a Z_1}{\gamma_{31P}\mu_0 v_{31P}\tau_{31P}} \tag{16}$$

where a is the coil radius, Z_1 is the characteristic impedance of the co-axial cable, μ_0 is the permeability of free space, v_{31P} is the peak transmitter voltage and τ_{31P} is the length of the RF pulse required to produce a flip angle θ_{31P} at the coil centre. A similar equation applies to $C_{1H}/C_{m,1H}$. Combination of these equations gives:

$$\frac{S_{31P}}{S_{1H}} = k'' \frac{\tau_{1H}}{\tau_{31P}} \times \frac{\theta_{31P}}{\theta_{1H}} \times \frac{\rho_{31P}}{\rho_{1H}} \tag{17}$$

where $k'' = k'(\gamma_{1H}/\gamma_{31P})(v_{1H}/v_{31P})$. Comparison with Eq. (12) indicates that $D = k''(\tau_{1H}\theta_{31P})/(\tau_{31P}\theta_{1H})$. Equation (17) implies that, if the ^{31}P and ^{1}H pulse lengths are held constant and the flip angles at both frequencies vary with loading in a similar fashion, then Eq. (17) can be applied directly without correction for loading. Alternatively the pulse lengths required to give matched flip angles in vivo can be measured (e.g. by the incorporation of a reference sample at the coil centre) [48]. Factors which can lead to erroneous results include subject movement and consequent coil detuning during the data collections, the possibility that the spatial distributions of water and phosphorus metabolites are not identical in the target tissue and contamination of spectra by signal from other tissues.

2.4 *Localised NMR Spectroscopy*

The incorporation of absolute quantitation techniques in various spectroscopic localisation protocols will be described in this section. Localisation here implies the application of various techniques to produce a well-defined volume of interest (VOI).

ICRs have been used in localised ^{1}H studies in order to estimate the concentration of lactate in human brain [14, 39–41]. In the earliest of these studies [39], 2-dimensional ISIS localisation [70] was combined with a DEPTH pulse [71] which was used to eliminate high-flux signals originating from tissue near to the surface coil. By incorporating a DANTE pulse [72] this combination of localisation techniques was used for lactate spectral editing. In order to estimate the [lactate], the edited methyl lactate resonance at ~1.33 ppm was compared with the PCr + Cr resonance at ~3.03 ppm. The [PCr + Cr] was assumed to be 10.5 mM and, after correcting for relaxation, this gave a normoxic [lactate] of about 0.5 mM. Similar ^{1}H studies using STEAM localisation gave a maximum [lactate] of 1.0 mM when assuming [PCr + Cr] = 10 mM [41] and a maximum [lactate] of 0.5 mM by assuming [NAA] = 6 mM [40]. However the assumption of [PCr + Cr] = 10 mM gave [NAA]s between 13 and 18 mM which were much higher than values obtained using analytical techniques. Using the same localisation technique, investigations of cerebral metabolism following acute stroke revealed a mean [lactate] of 16 mM in the affected region ([PCr + Cr] assumed to be 10.5 mM) [14]. This work also showed complete absence of NAA and greatly reduced signals from both PCr + Cr and choline containing compounds within the affected volume of tissue. These observations demonstrate that absolute concentrations obtained using ICRs may be erroneous in pathological tissue because the reference concentration may be altered. [PCr + Cr] has also been used in fully localised

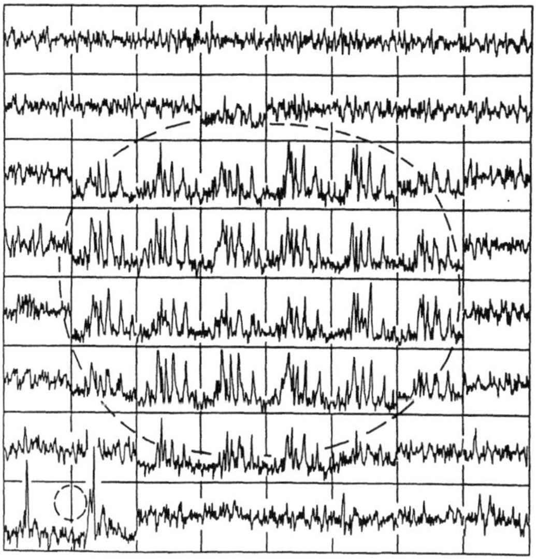

Fig. 7. A transaxial 8 × 8 × 256 point 3D FT ^{31}P spectroscopic image of adult human brain acquired in 34 min with 3 cm × 3 cm × 3 cm spatial resolution. Two voxels at the *bottom left* exhibit strong signals from a vial (*dashed circle*) containing 2 ml of 1 M H_3PO_4 which was used as an ECR. The *dashed line* locates the periphery of the head as determined by conventional ^1H imaging with the forehead at the left. (Reprinted with permission from Bottomley et al. [26], Academic Press, Inc.)

^1H homonuclear edited studies to estimate [lactate] in human skeletal muscle during recovery from exercise [43, 73]. [PCr + Cr] was assumed to be 42.5 mM and an average [lactate] of 14 mM was found after exhaustive exercise.

Because of the possibility that various pathological conditions may cause [ICR]s to change, in order to obtain more reliable clinical results ECRs must be utilised. One of the earliest attempts to obtain absolute concentrations from localised spectra using an ECR utilised 3-dimensional Fourier transform spectroscopic imaging (3D FTSI) [26]. In this technique, a selective RF pulse is used to define a slice from which data are to be acquired. Two-dimensional phase encoding is used to localise signals within the slice. A three-dimensional Fourier transform produces a "chequerboard" type of display with a spectrum within each square (see Fig. 7 which shows 3D FTSI results from an adult human brain obtained using a ^{31}P quadrature birdcage coil for increased sensitivity). To obtain absolute concentrations, an ECR (in this case 1M H_3PO_4) contained in a suitable vial is situated within the field of view of the spectroscopic image. (The signal from the ECR can be seen in the bottom left-hand voxels of Fig. 7.) The β-NTP peak area in voxel *ij* was corrected for saturation and $[NTP]_{ij}$ was estimated by comparison with the ECR signal in voxel *lm*:

$$[NTP]_{ij} = \frac{S_{\beta\text{-NTP},ij} V_{ECR,lm} \psi_{lm} \phi_{ECR} E_2(T^*_{2,ECR}) \rho_{ECR}}{S_{ECR,lm} V_{ECR} \psi_{ij} \phi_{\beta\text{-NTP}} E_2(T^*_{2,\beta\text{-NTP}})} \tag{18}$$

In the above equation, $S_{\beta\text{-NTP},ij}$ and $S_{ECR,lm}$ are the areas of the β-NTP and ECR resonances; V_{ECR} is the volume of ECR in the vial within voxel *lm* of volume $V_{ECR,lm}$; ψ_{lm}/ψ_{ij} is the relative coil sensitivity at voxels *ij* and *lm*; ϕ_{ECR} and $\phi_{\beta\text{-NTP}}$ are the saturation factors as defined in Eq. (3), and $E_2(T^*_{2,ECR})$ and $E_2(T^*_{2,\beta\text{-NTP}})$ are factors that account for signal loss during the magnetisation stabilisation delay between switching off gradients and commencing data acquisition. Because the signals from tissue and the ECR were acquired simultaneously, a correction for coil loading was not required. Saturation factors were determined using spectra obtained from the entire spectroscopic image slice without phase-encoding using $T_R = 10$ s and $T_R = 0.5$ s. Results obtained using this approach will be summarised in Sect. 3.2. An ECG gated variation of 3D FTSI has recently been used to obtain estimates for [PCr] and [ATP] in human heart muscle [74] (see Sect. 3.3). In this application, a 40 cm square surface coil was used as a ^{31}P transmitter while signal reception was achieved using a 10 cm diameter circular surface coil centred in the transmitter coil and coplanar with it. This provided some degree of B_1 homogeneity. Because a surface coil receiver was used, the metabolite and ECR signals had to be corrected for the spatial variation in coil sensitivity.

A further application of an ECR in which signal was localised only to a slice has been used to obtain rapid metabolite assays from human brain [30]. In this method, slice-selective NMRS was combined with ^{1}H NMR imaging to determine the cross-sectional area of the metabolite containing tissue (A_{met}). In the application of this slice-selective procedure to human brain, ^{31}P signal originated mostly from brain and cranial muscle. (It was assumed that bone and fat contributed negligibly to the total signal.) The areas of the ventricles were taken into account when estimating A_{met} and, from the ^{1}H images, it was estimated that muscle tissue contributed only 7% of the area from which metabolite signals were obtained. The ^{31}P ECR (1 M phosphonitrilic chloride trimer dissolved in $CHCl_3$ with chromium acetylacetonate added to reduce T_1) was contained in a vial of known cross-sectional area (A_{ECR}) which was situated within the field of view of the slice. Under conditions of spatially uniform excitation and detection sensitivity, a "hard" acquisition pulse and long T_R, the metabolite concentration (ρ_{met}) is given by:

$$\rho_{met} = \frac{A_{ECR} S_{met} \rho_{ECR}}{A_{met} S_{ECR}}. \tag{19}$$

Two methods for slice-selection were compared. One approach utilised a selective 180° pulse to invert the magnetisation within the desired slice. This preceded a 90° "hard" acquisition pulse on alternate scans and the resulting signals were subtracted as in one-dimensional ISIS [31]. The other method used slice-selective 90° pulses alone with a 2.2 ms delay between the centre of the pulse and the start of data acquisition. The former approach had the distinct advantage that there was no requirement for a magnet stabilisation delay between the 90° read-out pulse and the acquisition of data. Similarly to 3D FTSI, because signals from tissue and the ECR were acquired simultaneously, corrections were not required for coil loading. Results obtained using these pulse sequences will be discussed in Sect. 3.2.

Three-dimensional ISIS [31] has also been adapted to permit the estimation of absolute concentrations [32–34]. This involved applying the same ISIS localisation technique to a phantom containing a standard Pi solution following

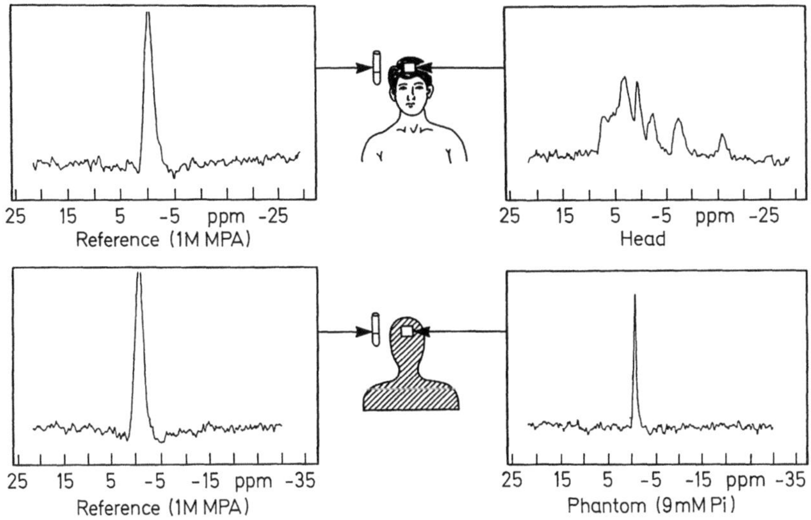

Fig. 8. The experimental protocol for obtaining absolute concentration measurements from human brain using ^{31}P ISIS localisation. The 1 M MPA sample is used to relate the differential effects of loading the coil by the head and the phantom. The standard solution in the phantom is used as an ECR. (Reprinted with permission from Luyten et al. [34] Academic Press, Inc.)

the in vivo data collection as shown in Fig. 8. External references (HPMT [32, 33] and MPA [34]) were used to calibrate the difference in coil efficiency, caused by sample loading, between the in vivo and the phantom studies. The external reference sample was present for both the in vivo and phantom data collections. Under these circumstances the metabolite concentration (ρ_{met}) was given by [32–34]:

$$\rho_{met} = \frac{\rho_{ph} S_{met} N_{ph} \phi_{ph} S_{ref(ph)}}{S_{ph} N_{met} \phi_{met} S_{ref}} \tag{20}$$

where S_{met}, $S_{ref(ph)}$, S_{ph} and S_{ref} are the peak areas of the metabolite to be quantitated, the external reference with the coil loaded by the phantom, the phantom standard, and the external reference with in vivo loading respectively. N_{met}, N_{ph}, ϕ_{met} and ϕ_{ph} are the numbers of scans and the saturation factors used for the in vivo and phantom data collections respectively; ρ_{ph} is the concentration of the phantom standard. This absolute quantitation method has been applied to normal human brain [32, 34] and various brain tumours [75], human skeletal muscle [32] and human liver [33] (see Sects. 3.1, 3.2 and 3.4). For the studies of brain, either a 16.5 cm diameter ^{31}P "headphone" coil [32] or an 18 cm diameter ^{31}P Helmholtz coil [34] were used in combined transmit/receive mode thereby producing a reasonably uniform B_1 field. For skeletal muscle, a 9 cm diameter surface coil was utilised (also in transmit/receive mode) [32]. The absolute concentration estimates for liver metabolites were obtained using either 9 cm or 14 cm diameter circular surface coils both as combined transmitter and receiver [33]. In the latter study, a computer simulation was used to correct for the chemical shift dependent VOI displacement. Because of the well-known decrease in sensitivity with distance from a surface coil this would otherwise give a dependence of apparent concentration on chemical shift.

2.5 *Spectrum Analysis*

The spectrum analysis method is an important factor in the estimation of absolute concentrations from spectra collected in vivo. Concentration estimates are directly proportional to the measured peak areas. In vivo spectra usually have less than optimum resolution and many peaks suffer from overlap. They are difficult to analyse rigorously because they contain signals from nuclei in a variety of chemical environments. Some peaks are relatively narrow and originate from mobile molecules whereas other peaks are broad and can underlie the entire mobile metabolite spectrum (e.g. the feature due to membrane phospholipids in ^{31}P spectra). Furthermore, due to factors including B_0 inhomogeneity, multiplicity caused by homo- and hetero-nuclear coupling and chemical shift anisotropy (e.g. the broad membrane phosphodiester (PDE) signal [76, 77]), peak profiles are not always pure Lorentzian functions. In order to obtain peak areas which are close representations of the metabolite signals involved, rigorous data analysis techniques that can take care of peak overlap must be used. Methods using peak heights and direct integration from the spectrum can lead to significant systematic errors when applied to in vivo spectra. Various techniques now exist which to some extent avoid the pitfalls of these more traditional methods. These include the fitting of exponentially decaying sinusoids to the FID (time domain analysis) in which four characteristics

Table 3. Absolute concentrations of phosphorylated metabolites in the cerebral cortex of normal newborn human infants estimated using integration, frequency domain analysis or time domain analysis (LPSVD)

	Concentrations (mmol litre^{-1} wet)			
	[PME]	[Pi]	[PDE]	[PCr]
Integration	14.3	4.0	8.6	4.1
	(2.6)	(0.8)	(1.5)	(0.7)
Frequency	12.9	2.5	18.5	2.4
	(1.5)	(0.4)	(5.7)	(0.6)
Time	8.4	3.2	15.3	3.2
	(0.9)	(0.9)	(2.7)	(1.2)

Mean (SD); $n = 5$; gestational plus postnatal ages at time of scan 35 to 37 weeks. Time domain data obtained in collaboration with M. Wylezinska. LPSVD software supplied courtesy of R de Beer have to be found for each resonance (i.e. amplitude, frequency, decay time constant and phase) and the fitting of Lorentzian or Gaussian (or a mixture of both) profiles to the spectrum (frequency domain analysis). Of the time domain methods, linear prediction single value decomposition (LPSVD) [78], and variable projection using prior knowledge (VARPRO) [79] have proven useful. (The prior knowledge consists of information about multiplets including multiplicity, *J* coupling and relative multiplet component amplitudes. This information can be used in frequency domain methods also.) Table 3 compares the absolute concentrations obtained from the brains of normal newborn infants by applying integration, frequency domain or time domain methods to the region of the spectrum suffering from severe overlap (i.e. between the PME and γ-NTP peaks). Large differences are apparent between the results obtained by these techniques. The frequency domain and time domain methods give results which agree much better with available values for mammalian brain [50, 51].

2.6 Coil Loading and Coil Design

Tissue samples are conductive and if positioned close to an NMR coil they introduce coil "loading" which results in a lower "Q" (efficiency) factor. Signal amplitude is reduced and RF pulses need to be of longer duration and/or greater amplitude in order to maintain the same flip angle as for an unloaded coil. For methods in which the ECR or ICR signal are collected simultaneously with the spectrum of interest (e.g. 3D FTSI [26, 74]), coil loading is not so important because it has the same effect on both the metabolite and reference signals. However loading is an important factor in absolute quantitation if the concentration estimates are made, for example, by comparing the ^{31}P in vivo signal with that obtained from a phantom in a further data collection [32–34] or if spectrometer calibration is required in a separate set of experiments (e.g. when tissue water is used as an ICR [45, 46, 48–51, 67, 69]. In these techniques corrections for coil loading have often to be applied. For example, when ISIS localisation is used and absolute concentrations are determined by comparing the in vivo signals with those obtained from a phantom study using an identical localisation protocol (see Sect. 2.4) [32–34], it is necessary to relate the in vivo and phantom coil loading conditions by incorporating a further reference sample which is used to measure the relative coil sensitivities. In this respect, coil designs in which the effects of sample loading are minimised can help to reduce errors. Such coils include the various balanced designs which have been published (e.g. balanced tuning [32, 80], and inductive coupling and series tuning [81]). Faraday shields between the coil and the subject have also been used to further reduce loading effects [32].

A further aspect of coil design which can improve absolute concentration estimates is the use of a transmitter coil which provides a homogeneous B_1 field and thereby a uniform flip angle and hence uniform saturation (for a particular resonance) throughout the tissue. This can be achieved by using a volume coil (e.g. a quadrature birdcage coil [26, 30], a headphone coil [32], a Helmholtz coil [34] or a slotted tube resonator [82, 83]). An almost homogeneous effective B_1 field can also be obtained by using a surface coil transmitter which is much larger than the receiver coil [74].

2.7 *Corrections for Spin–Lattice and Spin–Spin Relaxation*

Due to the limited time available for the conduct of many in vivo studies, it is not always possible to obtain fully relaxed spectra. If data from tissue or from a concentration reference are acquired with T_R less than five times the T_1 of any resonance, then the signal amplitude will be lower than that obtained under fully relaxed conditions. (This is because it takes at least $5T_1$ for the sample magnetisation to almost completely recover its unperturbed state. A further implication is that at least 5 pulses should be applied to the sample before commencing the accumulation of data in order that a steady state has been reached [84] otherwise the signal resulting from each of the first few pulses will be greater than that later on.) In order to obtain absolute concentration estimates which reflect those in tissue, corrections must be applied. This usually necessitates a second series of experiments in which either T_1s or saturation correction factors are measured for each of the resonances under investigation. The factor ϕ by which resonance areas are reduced when pulsing with $T_R < 5T_1$ has been defined in Eq. (3). In order to determine the resonance area which would be observed in a fully relaxed spectrum, the peak area must be divided by ϕ. This factor can be calculated using Eq. (3) provided that the flip angle (θ) and the T_1 of the resonance of interest are known. For the direct application of Eq. (3) it is important that B_1 is uniform throughout the sample otherwise θ will vary spatially. For localised studies using volume coils [32], B_1 is fairly uniform over the VOI and ϕ for a particular resonance can be calculated from the effective flip angle (i.e. that at the centre of the VOI) and the T_1. In 3D FTSI studies, ϕ must be calculated and applied separately to data from each voxel in the image [26] (see Eq. (18)). When surface coils are used, B_1 (and hence θ) can vary substantially throughout the sample. However, if the VOI is small enough an effective θ can be calculated at the VOI centre and used in Eq. (3) to determine ϕ [33, 74]. Improvements in localised absolute quantitation techniques can be obtained by using frequency modulated inversion and excitation pulses [34, 85] which give more uniform flip angles even if B_1 is inhomogeneous. If surface coil spectra are acquired without localisation (e.g. when tissue water is used as a heteronuclear ICR [45, 48–51]—see Sect. 2.3) saturation correction factors can be determined by comparing peak areas obtained at T_R and at $\geqq 5T_1$ in a separate set of measurements. It should be noted, however, that the spatial sensitivity of a surface coil depends on T_R [86]. (For $T_R \geqq 5T_1$ the sensitivity is greater near to the coil whereas for $T_R \sim T_1$ a more uniform response is obtained due to greater saturation in the vicinity of the coil.) Resonance areas from concentration references must also be corrected unless the T_1 is reduced by use of a relaxation agent [30]. If tissue water is used as an ICR, it may be possible to obtain fully relaxed ^{1}H spectra [49–51] because, due to the high water proton concentration, adequate signal to noise ratio can be obtained in a relatively short time.

In many of the currently employed localisation techniques, delays are required both during the pulse sequence itself and before data acquisition to allow for the decay of eddy currents induced by magnetic gradient switching. These delays lead to the possibility of signal losses due to T_1 or T_2 relaxations and this must be evaluated and corrected for if necessary. For instance, in the 3D FTSI measurements described in Sect. 2.4, the delay (T_d) between the selective RF pulse and the start of data acquisition was 2.7 ms in brain studies [26] or 1 ms in cardiac studies [74]. The factor E_2 in Eq. (18) gives the fraction of signal remaining after T_d and is approximately $\exp(-T_d/T_2^*)$ where T_2^* includes a component due to magnetic field inhomogeneity and can be estimated from line width measurements. (Note that, in equation [18], the metabolite and reference areas are divided by their respective E_2 values.)

If spectra have been acquired using spin echo techniques then corrections for T_2 relaxation during the pulse sequence have to be made in addition to the T_1 corrections that are necessary if $T_R < 5T_1$. In a "$\pi/2 - \tau - \pi - \tau -$ acquire data" experiment, the maximum amplitude of the echo at time 2τ is $\exp(-2\tau/T_2)$ times the amplitude at $\tau = 0$ [59]. Hence each peak area obtained in a spin echo study must be multiplied by $\exp(2\tau/T_2)$ in order to use the data for estimating absolute concentrations. An example of the use of this correction is the ^{1}H NMR quantitation of lactate by spectral editing in gerbil brain [38] mentioned in Sect. 2.3. However, it should be noted that three further corrections had also to be applied in this study. Firstly, due to the fact that the lactate resonance used for quantitation was the methyl doublet (J coupling = 7.3 Hz), a phase modulation [59] factor of $1 - \cos(2\pi J\tau)$ had to be applied. Secondly, because a surface coil was used, phase modulation was incomplete and the effect of this was estimated empirically using standard lactate solutions. Finally, the lactate and NAA signals had to be corrected for the chemical shift dependence of the sensitivity of the binomial water supression pulse [53].

2.8 Sources of Error

In order to estimate absolute concentrations by non-invasive MRS, numerous factors have to be taken into account. Hence the estimated concentrations are subject to significant systematic and statistical errors. For example, if one considers the 3D FTSI technique [26], Eq. (18) contains 11 terms each of which has its own error δ. The percentage absolute error in $[\mathrm{NTP}]_{ij}$ is:

$$\delta[\mathrm{NTP}]_{ij}\% = \delta[\mathrm{ECR}]\% + \sum_{\mathrm{NTP,ECR}} (\delta S\% + \delta V\% + \delta\psi\% + \delta\phi\% + \delta E_2\%) \tag{21}$$

Even a moderate error of 5% in each of the factors involved would lead to $\delta[\text{NTP}]_{ij}\% = 55\%$. In the $\pi_{\text{selective}} - \pi/2$ slice-selective method [30] metabolite concentrations were calculated using an equation similar to Eq. (18). Errors were reduced because, due to the spatially uniform excitation and detection, long T_R (compared with the metabolite T_1s) and the use of a hard $\pi/2$ read-out pulse, the ψ, ϕ and E_2 terms could be neglected. This reduced the number of terms involved as indicated in Eq. (19). The percentage absolute error for the slice-selective technique becomes:

$$\delta[\text{NTP}]\% = \delta[\text{ECR}]\% + \delta S_{\text{NTP}}\% + \delta A_{\text{ECR}}\% + \delta S_{\text{ECR}}\% + \delta A_{\text{NTP}}\% \qquad (22)$$

and the [NTP] error is approximately $\delta S_{\text{NTP}}\%$ since the errors on the other factors are relatively small. Hence, the general rule to be adopted when attempting to obtain absolute concentrations by in vivo MRS is to give a high priority to minimising errors.

2.9 Quality Assurance

There are two fundamental approaches to the assessment of the performance of an absolute quantitation method. Both of these approaches are complimentary. Under ideal conditions the method can be applied to a tissue (e.g. skeletal muscle) for which absolute concentrations are well documented in the literature. The estimated concentrations can then be compared with previously published values obtained using other reliable techniques. The other approach involves the design and construction of tissue-like phantoms which contain known amounts of metabolites. Concentrations estimated by MRS can then be compared with the known values. In order that the phantom is comparable with tissue, sophisticated design considerations, such as the incorporation of relaxation agents to give T_1s and T_2s similar to those found in living tissue, should be included. So that data acquisition and spectrum analysis methods are adequately tested, the relaxation times should cover the ranges encountered in vivo (e.g. an attempt should be made to include components which mimic the broad PDE and bone/membrane phospholipid features in the spectrum as well as the relatively narrow resonances from highly mobile molecular species). For use as a long term quality assurance facility, the design of such phantoms should include consideration of stability with respect to biological, physical and chemical processes.

3 Results Obtained by In Vivo NMR Spectroscopy

Table 4 gives some ^{31}P metabolite concentration estimates which have been obtained from a variety of mammalian (including human) tissues. These estimates have all been determined using non-invasive MRS techniques. Most of the

Table 4. Some estimated ^{31}P metabolite concentrations in normal and pathological, in situ, living tissues as determined by non-invasive absolute quantitation using MRS

Tissue	[PME]	[Pi]	[PDE]	[PCr]	[NTP]	Method	Ref.
Brain (human)	4.0 ± 0.8[a,e]	2.0 ± 0.5	11.5 ± 3.1	4.9 ± 1.0	2.5 ± 0.5	ISIS; ECR[h]	32
Brain tumours (human)[r]	1.1[f]	1.6	6.6	3.2	1.8	ISIS (modified); ECR[h]	75
Brain tumours (human)[s]	0.8[f]	0.9	2.3	1.3	1.2	ISIS (improved); ECR[h]	75
Brain (human)	—	2.6[b,d,e]	—	4.3	3.0	ISIS; ECR[i]	34
Brain (human)	3.7 ± 1.8[a,f]	1.5 ± 0.7	14.1 ± 3.5	5.2 ± 0.9	3.0 ± 0.5	$\pi_{selective} - \pi/2$; ECR[j]	30
Brain (human)	2.2 ± 0.9[a,f]	0.9 ± 0.4	6.4 ± 1.7	3.4 ± 0.7	2.9 ± 0.4	$\pi/2_{selective}$; ECR[j]	30
Brain (human)	—	—	—	3.6[b,f]	3.1	3D FTSI; ECR[k]	26
Brain (human, neonatal)[t]	12.8 ± 2.5[a,f]	2.5 ± 0.7	17.9 ± 4.8	2.4 ± 0.6	3.7 ± 0.6	ICR[l]	51
Brain, birth asphyxia (human, neonatal)[u]	2.8[f]	1.7	6.5	1.3	1.3	ICR[l]	50
Brain (rat)	—	—	—	—	2.8 ± 0.2[c,f]	ICR[m]; $n = 3$	46
Heart (human)	—	—	—	11.0 ± 2.7[a,g]	6.9 ± 1.6	ECG gated 3D FTSI; ECR[n]	74
Liver (human)	0.9 ± 0.4[a,g]	2.1 ± 0.3	5.4 ± 1.4	—	2.2 ± 0.8	ISIS; ECR[o]	33
Muscle (human, skeletal)	—	5.7 ± 0.4[a,e]	—	36.5 ± 3.5	7.3 ± 0.3	ISIS; ECR[h]	32
Muscle (human, skeletal)	2.2 ± 0.7[a,f]	4.7 ± 1.0	4.2 ± 0.8	32.0 ± 4.5	6.3 ± 0.8	ICR[p]	49, 50
Muscle (mouse, skeletal)	—	—	—	12.8 ± 1.7[a,e]	5.6 ± 1.0	ICR[q]	48
Muscle (rat, skeletal)	—	—	—	—	6.8 + 0.2[c,f]	ICR[m]; n = 9	46

[ECR—external concentration reference; ICR—internal concentration reference; [a] mean ± SD; [b] mean; [c] mean ± SE; [d] error ~ 15%; [e] mM; [f] $mmol\,l^{-1}$; [g] $mmol\,kg^{-1}$ wet wt; [h] HPMT and 200 mM Pi; [i] MPA and 100 mM Pi; [j] HCCTP ($[^{31}P] = 33\,M\,l^{-1}$); [k] 1 M H_3PO_4; [l] 100 M H_2O; [m] 88.9 M H_2O; [n] 0.2 M H_3PO_4; [o] HPMT and 200 mM KG_2PO_4; [p] 85.6 M H_2O; [q] 83 M H_2O; [r] 3astrocytomas, 1 gliosarcoma and 1 glioblastoma; [s] 5 astrocytomas, 2 meningiomas and 1 ganglioma; [t] gestational plus postnatal age 35 to 37 weeks; [u] age 6 days; data reanalysed using frequency domain fitting; HCCTP—phosphonitrilic chloride trimer; HPMT—hexamethylphosphoroustriamide; MPA—methylphosphonic acid.]

data are for physiologically normal tissue but some results from brain tumour and birth asphyxia studies are included. For further details of the methods used the reader should refer to the cited references although the techniques have been described to some extent in Sects. 2.3 and 2.4. The results obtained from skeletal muscle will be discussed first because human biopsy data are relatively plentiful. Furthermore, if an absolute quantitation method obtains reasonable results for skeletal muscle this provides some justification for its application to other tissues.

3.1 Skeletal Muscle

The [NTP] values for skeletal muscle in Table 4 compare very favourably with the biopsy [ATP] in Table 1 although the MRS absolute quantitation estimates tend to be slightly higher. (NB. NTP in skeletal muscle is almost entirely ATP.) This may not be significant because other published human biopsy data [87, 88] indicate an [ATP] range from 4.3 to 6.5 mmol kg^{-1} wet wt. For mammalian skeletal muscle, [ATP] values of 6.8 mmol kg^{-1} wet wt (rat) [89], 6.2 mmol kg^{-1} wet wt (rat) [90] and 5.0 mmol kg^{-1} wet wt (mouse) [90] have been reported.

Table 1 indicates that ^{31}P MRS gives consistently higher [PCr] and lower [Pi] when compared with data obtained by needle biopsy. It has been suggested that this is due to the hydrolysis of PCr, and the consequent production of Pi, during the interval between collection of the biopsy sample and freeze clamping [1]. For the purpose of comparison with human biopsy data, it may be better to use [PCr + Pi] in which case reasonable agreement is obtained although the values in Table 4 are slightly higher than the biopsy value given in Table 1. (The values for [PME], [Pi] and particularly [PDE] from references 49 and 50 may be subject to some over-estimation caused by the inclusion of signal from the underlying PCr wing due to the integration spectrum analysis method.)

3.2 Brain

The NTP concentrations for brain tissue in Table 4 agree well with values obtained from other studies of mammalian brain using conventional biochemical techniques in which ATP, GTP, UTP and CTP were all measured. The mean [NTP] values determined in these studies were 3.0 mmol kg^{-1} frozen wt (21 day old rat; whole brain) [91], 2.4 mmol kg^{-1} wet wt (adult rat; whole brain) [92] and 3.7 mmol kg^{-1} wet wt (adult rat; cerebral cortex) [93]. (A further value of 3.3 mmol kg^{-1} wet wt for the summed [ATP] and [GTP] in adult rat forebrain has been reported [94]). It should be noted that there is a possibility that results from brain may be dependent on the location of the tissue from which data are collected. There is strong evidence for a non-homogeneous anatomical distribution of ATP in the brain [95–97]. There is also a possibility that the ventricles may have occupied part of the VOI in some studies.

In addition to other metabolites (see Table 2 and Fig. 2), the brain PME peak is known to contain a major contribution from phosphoethanolamine (PEt) [98–100]. This has been reported to have a relatively high concentration in neonatal mammalian brain [61, 101–104]. In the brain of the newborn rabbit [PEt] is close to 8 mmol kg^{-1} wet wt but falls to about 1.5 mmol kg^{-1} wet wt by adulthood [102, 103]. The high [PEt] in neonatal brain probably explains the large [PME] for normal newborn infants in Table 4.

The brain [Pi] estimated by MRS is in reasonable agreement with a reported value of 2.2 mmol kg^{-1} wet wt from freeze blown adult rat brain [105].

The [PDE] results in Table 4 are difficult to interpret. This is because the PDE peak is complicated and appears to include contributions both from low molarity mobile metabolites (glycerolphosphorylcholine (GPC), glycerolphosphorylethanolamine (GPE) and an unidentified acid labile phosphodiester (ALP)) [20, 21, 106] and from a large broad component originating mainly from phospholipid bilayer but with some signal also from motionally averaged macromolecules [76, 77]. Extracts of adult guinea pig brain [20] give a [ALP + GPC + GPE] of about 2.5 mmol kg^{-1} wet wt while in canine brain a value of 7.9 mmol kg^{-1} dry wt has been reported [106]. The higher [PDE] values in Table 4 support the existence of significant contributory components in addition to ALP, GPC and GPE. In the results obtained using only slice selective localisation the value of 6.4 mmol l^{-1} obtained using the $\pi/2_{selective}$ method was less than half that obtained when the $\pi_{selective} - \pi/2$ sequence was employed [30]. In fact, all the estimated concentrations were lower when the $\pi/2_{selective}$ method was used. These reductions were due to dephasing of the sample magnetisation occuring in the 2.2 msec delay before commencement of data acquisition. This delay was necessary to allow for eddy current decay following magnetic gradient switching. Corrections for this effect were not applied to the $\pi/2_{selective}$ data. The [PDE] reported for neonatal brain is slightly higher than that for adult brain. This may be related to the use of a simple "single pulse—acquire data" sequence in which the delay before digitisation is minimal.

The MRS [PCr] estimates for adult human brain are slightly higher than reported values of 4.1 mmol kg^{-1} wet wt and 3.6 mmol kg^{-1} wet wt obtained from mature mammalian brain by freeze blowing [105, 107]. The [PCr] in neonatal brain is lower and compares well with reported values of 2.4 to 3.0 mmol kg^{-1} wet wt obtained from 1 and 5 day old rat brains and 2 day old canine brains frozen by immersion or blowing [91, 107, 108].

The metabolite concentration estimates obtained from brain tumours are all smaller than values for normal adult brain. [PME], [PDE] and [PCr] show the greatest reductions. It appears that tumour metabolite concentrations differ significantly from those in normal brain. More detailed inspection of the tumour data indicates that absolute concentrations may have potential for tumour characteristation [75]. The birth asphyxia data show significant reductions in all the metabolite concentrations. These results indicate that, within a few days of birth, absolute quantitation is capable of providing new information, additional to that provided by metabolite ratios and intracellular pH (pH_i), about the severity of brain tissue damage resulting from perinatal hypoxia/ischaemia.

3.3 Cardiac Muscle

The PCr and ATP concentrations for human myocardium given in Table 4 are both slightly higher than published values of 8 mmol kg^{-1} wet wt for PCr and 5.6 mmol kg^{-1} wet wt for ATP obtained from canine tissue using conventional, invasive biochemical techniques [109]. Results from high-energy phosphate assays for normal human heart do not appear to be available. Hence, any species variation between canines and humans is not currently resolvable.

3.4 Liver

The [ATP] estimated by MRS is about 20% less than the value of 2.5 mmol kg^{-1} wet wt reported from freeze-clamp studies of human liver [110]. However it has also been reported that 15% to 20% of rat liver cell ATP is mitochondrial and therefore this fraction may be invisible to MRS [111]. [Pi] estimated by MRS was less than half the value of 5.1 mmol kg^{-1} wet wt obtained by freeze-clamping of human liver tissue [112]. This may be due to the fact that MRS only detects free Pi whereas measurements using freeze-clamp extraction techniques include both free and bound Pi. It is also possible that [Pi] may be increased in liver biopsy due to pre-freezing hydrolysis.

4 Conclusions

There are several important conclusions that can be drawn from the previous sections of this chapter. One of the most important is that non-invasive, in vivo MRS techniques are now available that can provide realistic estimates of absolute metabolite concentrations. MRS methods can now be applied to obtain biochemical assays from tissues previously inaccessible to biopsy. Preliminary results from studies of human brain tumours and neonatal hypoxic/ischaemic brain injury indicate that absolute quantitation provides further useful information additional to that given by relative concentrations and pH. The estimation of absolute concentrations should greatly improve the prognostic and tissue characterisation capabilities of MRS in biomedical applications. Estimates obtained by MRS are not susceptible to the freezing artifacts associated with conventional human biopsy although random and systematic errors may be substantial. MRS concentration estimates (e.g. of high-energy metabolites such as PCr and ATP) may provide better indication of the true physiological state of the tissue. In order to use MRS to obtain metabolite concentrations that closely reflect those in intact tissue, both the data acquisition method and the data analysis method have to be considered in detail and as complimentary parts of the absolute quantitation protocol. Assurance of the quality of results requires the development of phantoms containing metabolites whose concentrations, chemical shift and relaxation conditions cover the ranges encountered in vivo or/and comparable studies on tissues (e.g. skeletal muscle) for which reliable literature values are available.

5 References

1. Dawson MJ (1982) Biosc Rep 2: 727
2. Taylor DJ, Bore PJ, Styles P, Gadian DG, Radda GK (1983) Mol Biol Med 1: 77
3. Wilkie DR, Dawson MJ, Edwards RHT, Gordon RE, Shaw D (1984) in: Pollack GH, Sugi H (eds) Contractile mechanisms in muscle. Plenum, New York, p 333
4. Taylor DJ, Crowe M, Bore PJ, Styles P, Arnold DL, Radda GK (1984) Gerontology 30: 2
5. Taylor DJ, Styles P, Matthews PM, Arnold DL, Gadian DG, Bore P, Radda GK (1986) Magn Reson Med 3: 44
6. Cady EB, Costello AM de L, Dawson MJ, Delpy DT, Hope PL, Reynolds EOR, Tofts PS, Wilkie DR (1983) Lancet 1: 1059
7. Prichard JW, Alger JR, Behar KL, Petroff OAC, Shulman RG (1983) Proc Natl Acad Sci USA 80: 2748
8. Delivoria-Papadopoulos M, Leonard JC, Subramanian VH, Eleff S, Leigh JS, Chance B (1984) Ann Neurol 16: 581
9. Tofts PS, Wray S (1985) J Physiol (Lond) 359: 417
10. Cady EB, Chu A, Costello AM de L, Delpy DT, Gardiner RM, Hope PL, Reynolds EOR (1987) J Physiol (Lond) 382: 1
11. Tofts PS, Wray S (1988) NMR Biomed 1: 1
12. Delivoria-Papadopoulos M, Chance B (1988) in: Guthrie RD (ed) Recent advances in neonatal intensive care. Churchill-Livingstone, Edinburgh, p 153
13. Azzopardi D, Wyatt JS, Cady EB, Delpy DT, Baudin J, Stewart AL, Hope PL, Hamilton PA, Reynolds EOR (1989) Ped Res 25: 445
14. Bruhn H, Frahm J, Gyngell ML, Merboldt KD, Hanicke W, Sauter R (1989) Magn Reson Med 9: 126
15. Newman RJ, Bore PJ, Chan L, Gadian DG, Styles P, Taylor D, Radda GK (1982) Brit Med J 284: 1072
16. Edwards RHT, Dawson MJ, Wilkie DR, Gordon RE (1982) Lancet 1: 725
17. Griffiths RD, Cady EB, Edwards RHT, Wilkie DR (1985) Muscle Nerve 8: 760
18. Erecinska M, Wilson DF, Nishiki K (1978) Am J Physiol 234: C82
19. Veech RL, Lawson JWR, Cornell NW, Krebs HA (1979) J Biol Chem 254: 6538
20. Glonek T, Kopp SJ, Kot E, Pettegrew JW, Harrison WH, Cohen MM (1982) J Neurochem 39: 1210
21. Cohen MM, Pettegrew JW, Kopp SJ, Minshew N, Glonek T (1984) Neurochem Res 9: 785
22. Shoubridge EA, Briggs RW, Radda GK (1982) FEBS Lett 140: 288
23. Bates TE, Williams SR, Gadian DG (1989) NMR Biomed 2: 225
24. Ernst RR, Morgan RE (1973) Molec Phys 26: 49
25. Dawson MJ, Gadian DG, Wilkie DR (1977) J Physiol (Lond) 267: 703
26. Bottomley PA, Charles HC, Roemer PB, Flamig D, Engeseth H, Edelstein WA, Mueller OM (1988) Magn Reson Med 7: 319
27. Gard JK, Ackerman JJH (1983) J Magn Reson 51: 124
28. Billadello JJ, Gard JK, Ackerman JJH, Gross RW (1985) Anal Biochem 144: 269
29. Gard JK, Kichura GM, Ackerman JJH, Eisenberg JD, Billadello JJ, Sobel BE, Gross RW (1985) Biophys J 48: 803

30. Bottomley PA, Hardy CJ (1989) Clin. Chem. 35: 392
31. Ordidge RJ, Connelly A, Lohman JAB (1986) J Magn Reson 66: 283
32. Roth K, Hubesch B, Meyerhoff DJ, Naruse S, Gober JR, Lawry TJ, Boska MD, Matson GB, Weiner MW (1989) J Magn Reson 81: 299
33. Meyerhoff DJ, Karczmar GS, Matson GB, Boska MD, Weiner MW (1990) NMR Biomed 3: 17
34. Luyten PR, Groen JP, Vermeulen JWAH, den Hollander JA (1989) Magn Reson Med 11: 1
35. Shine NR, Xuan JA, Koretsky AP, Weiner MW (1987) Magn Reson Med 4: 244
36. Behar KL, den Hollander JA, Stromski ME, Ogino T, Shulman RG, Petroff OAC, Prichard JW (1983) Proc Natl Acad Sci USA 80: 4945
37. Gadian DG, Proctor E, Williams SR, Cady EB, Gardiner RM (1986) Magn Reson Med 3: 150
38. Williams SR, Proctor E, Allen K, Gadian DG, Crockard HA (1988) Magn Reson Med 7: 425
39. Hanstock CC, Rothman DL, Prichard JW, Jue T, Shulman RG (1988) Proc Natl Acad Sci USA 85: 1821
40. Frahm J, Bruhn H, Gyngell ML, Merboldt KD, Hanicke W, Sauter R (1989) Magn Reson Med 9: 79
41. Frahm J, Bruhn H, Gyngell ML, Merboldt KD, Hänicke W, Sauter R (1989) Magn Reson Med 11: 47
42. Behar KL, Rothman DL, Shulman RG, Petroff OAC, Prichard JW (1984) Proc Natl Acad Sci USA 81: 2517
43. Pan JW, Hamm JR, Hetherington HP, Rothman, DL, Shulman RG (1988) Proc Soc Magn Reson Med Seventh Ann Meeting, 20–26 August 1988, San Francisco, USA, p 618
44. Cerdan S, Parrilla R, Santoro J, Rico M (1985) FEBS Lett 187: 167
45. Thulborn KR, Ackerman JJH (1983) J Magn Reson 55: 357
46. Wray S, Tofts PS (1986) Biochem Biophys Acta 886: 399
47. Shapiro JI, Chan L (1987) J Magn Reson 75: 125
48. Kuan WP, Scott KN, Briggs RW (1988) Proc Soc Magn Reson Med Seventh Ann Meeting, 20–26 August 1988, San Francisco, USA, p 938
49. Cady EB, Azzopardi D (1989) NMR Biomed 2: 305
50. Cady EB (1990) J Magn Reson 87: 433
51. Cady EB (1991) J Magn Reson 91: 637
52. Bodenhausen G, Freeman R, Turner DL (1977) J Magn Reson 27: 511
53. Hore PJ (1983) J Magn Reson 55: 283
54. Rothman DL, Behar KL, Hetherington HP, Shulman RG (1984) Proc Natl Acad Sci USA 81: 6330
55. Rothman DL, Arias-Mendoza F, Shulman GI, Shulman RG (1984) J Magn Reson 60: 430
56. Williams SR, Gadian DG, Proctor E (1986) J Magn Reson 66: 1530
57. McIntosh TK, Faden AI, Bendall MR, Vink R (1987) J Neurochem 49: 1530
58. von Kienlin M, Albrand JP, Authier B, Blondet P, Lotito S, Decorps M (1987) J Magn Reson 75: 371
59. Rabenstein DL, Nakashima TT (1979) Anal Chem 51: 1465A
60. Tallan HH (1957) J Biol Chem 223: 41
61. Okumura N, Otsuki S, Kameyama A (1960) J Biochem 47: 315
62. Nakada T, Kwee IL, Suzuki N, Houkin K (1989) Magn Reson Med 12: 172
63. Kittel C (1968) Introduction to solid state physics, 3rd edn. Wiley, New York
64. Hoult DI, Richards RE (1976) J Magn Reson 24: 71
65. Hoult DI (1978) Prog Magn Reson Spec 12: 41
66. Ackerman JJH, Grove TH, Wong GG, Gadian DG, Radda GK (1980) Nature 283: 167
67. Tofts PS, Wray S (1988) Magn Reson Med 6: 84
68. Gordon RE, Timms WE (1982) J Magn Reson 46: 322
69. Tofts PS (1988) J Magn Reson 80: 84
70. Hanstock CC, Rothman DL, Jue T, Shulman RG (1988) J Magn Reson 77: 583
71. Bendall MR, Pegg DT (1985) Magn Reson Med 2: 91
72. Morris GA, Freeman R (1978) J Magn Reson 29: 433
73. Hetherington HP, Hamm JR, Pan JW, Rothman DL, Shulman RG (1989) J Magn Reson 82: 86
74. Bottomley PA, Hardy CJ, Roemer PB (1990) Magn Reson Med 14: 425
75. Hubesch B, Sappey-Marinier D, Roth K, Meyerhoff DJ, Matson GB, Weiner MW (1990) Radiology 174: 401
76. Bates TE, Williams SR, Gadian DG (1989) Magn Reson Med 12: 145
77. Murphy EJ, Rajagopalan B, Brindle KM, Radda GK (1989) Magn Reson Med 12: 282
78. Barkhuijsen H, de Beer R, Bovée WMMJ, van Ormondt D (1985) J Magn Reson 61: 465

79. Van der Veen JWC, de Beer R, Luyten PR, van Ormondt D (1988) Magn Reson Med 6: 92
80. Murphy-Boesch J, Koretsky AP (1983) J Magn Reson 54: 526
81. Decorps M, Blondet P, Reutenauer H, Albrand JP (1985) J Magn Reson 65: 100
82. Alderman DW, Grant DM (1979) J Magn Reson 36: 447
83. Gruetter R, Boesch C, Muri M, Martin E, Wüthrich K (1990) Magn Reson Med 15: 128
84. Becker ED, Feretti JA, Gambhir PN (1984) Anal Chem 51: 1413
85. Silver MS, Joseph RI, Chen CN, Sank VJ, Hoult DI (1984) Nature 310: 681
86. Evelhoch JL, Crowley MG, Ackerman JJH (1984) J Magn Reson 56: 110
87. Edwards RHT, Young A, Wiles M (1980) N Engl J Med 302: 261
88. MacDougal JD, Ward GR, Sale DG, Sutton JR (1977) J Appl Physiol (Respirat Environ Exercise Physiol) 43: 700
89. Fitch CD, Jellinek M, Mueller EJ (1974) J Biol Chem 249: 1060
90. Beis I, Newsholme EA (1975) Biochem J 152: 23
91. Lolley RN, Balfour WM, Samson FE (1961) J Neurochem 7: 289
92. Mandel P, Edel-Harth S (1966) J Neurochem 13: 591
93. Chapman AG, Westerberg E, Siesjo BK (1981) J Neurochem 36: 179
94. Onodera H, Iijima K, Kogure K (1986) J Neurochem 46: 1704
95. Kogure K, Alonso OF (1978) Brain Res 154: 273
96. Hohn-Berlage M, Okada Y, Kloiber O, Hossman KA (1989) NMR Biomed 2: 240
97. Maudsley AA, Twieg DB, Sappey-Marinier D, Hubesch B, Hugg JW, Matson GB, Weiner MW (1990) Magn Reson Med 14: 415
98. Gyulai L, Bolinger L, Leigh JS, Barlow C, Chance B (1984) FEBS Lett 178: 137
99. Brenton DP, Garrod PJ, Krywawych S, Reynolds EOR, Bachelard HS, Cox DW, Morris PG (1985) Lancet 1: 115
100. Pettegrew JW, Kopp SJ, Dadok J, Minshew NJ, Feliksik JM, Golenk T, Cohen MM (1986) J Magn Reson 67: 443
101. Cohen MM, Lin S (1962) J Neurochem 9: 345
102. Agrawal HC, Davis JM, Himwich WA (1966) Brain Res 3: 374
103. Agrawal HC, Himwich WA (1970) in: Himwich WA (ed) Developmental neurobiology. CC Thomas, Springfield, Illinois, p 298
104. Burri R, Lazeyras F, Aue WP, Straehl P, Bigler P, Althaus U, Herschkowitz N (1988) Dev Neurosci 10: 213
105. Veech RL, Harris RL, Veloso D, Veech EH (1973) J Neurochem 20: 183
106. Cerdan S, Subramanian VH, Hilberman M, Cone J, Egan J, Chance B, Williamson JR (1986) Magn Reson Med 3: 432
107. Miller AL, Shamban A (1977) J Neurochem 28: 1327
108. Vannucci RC, Duffy TE (1977) Ann Neurol 1: 528
109. Jennings RB, Reimer KA (1981) Amer J Pathol 102: 241
110. Hultman E, Nilsson LH, Sahlin K (1975) Scand J Clin Lab Invest 35: 245
111. Murphy E, Gabel SA, Funk A, London RE (1988) Biochemistry 27: 526
112. Bode JC, Zelder O, Rumpelt HJ, Wittkamp U (1973) Eur J Clin Invest 3: 436

Author Index Volumes 21–25

MIX
Papier aus verantwortungsvollen Quellen
Paper from responsible sources
FSC® C105338

If you have any concerns about our products,
you can contact us on
ProductSafety@springernature.com

In case Publisher is established outside the EU,
the EU authorized representative is:
Springer Nature Customer Service Center GmbH
Europaplatz 3, 69115 Heidelberg, Germany

Printed by Libri Plureos GmbH
in Hamburg, Germany